Plenary and Main Lectures presented at the

INTERNATIONAL SYMPOSIUM ON
MACROMOLECULAR CHEMISTRY

Conférences plénières et principales présentées au

SYMPOSIUM INTERNATIONAL DE
CHIMIE MACROMOLÉCULAIRE

UNION INTERNATIONALE DE
CHIMIE PURE ET APPLIQUÉE
DIVISION DE CHIMIE PHYSIQUE
COMMISSION DE CHIMIE MACROMOLÉCULAIRE

et

FÉDÉRATION DES INDUSTRIES
CHIMIQUES DE BELGIQUE

LA CHIMIE
MACROMOLÉCULAIRE—4

Conférences plénières et principales présentées au

SYMPOSIUM INTERNATIONAL DE CHIMIE
MACROMOLÉCULAIRE

à Bruxelles–Louvain, Belgique
12–16 juin 1967

Springer Science+Business Media, LLC

INTERNATIONAL UNION OF
PURE AND APPLIED CHEMISTRY
DIVISION OF PHYSICAL CHEMISTRY
COMMISSION ON MACROMOLECULES

in conjunction with

BELGIAN ASSOCIATION OF
CHEMICAL MANUFACTURERS

MACROMOLECULAR CHEMISTRY—4

Plenary and Main Lectures presented at the

INTERNATIONAL SYMPOSIUM ON
MACROMOLECULAR CHEMISTRY

held in Brussels–Louvain, Belgium

12–16 June 1967

Springer Science+Business Media, LLC

First published by
Butterworth & Co. (Publishers) Ltd.

The contents of this book appear in

Pure and Applied Chemistry, Vol. 16. Nos. 2-3 (1968)

The book contains an additional chapter, pp. i–xxvi

Suggested U.D.C. *number:* 541·64

Suggested additional number: 542·952

Library of Congress Catalog Card Number 68-54464

ISBN 978-1-4899-6201-0   ISBN 978-1-4899-6407-6 (eBook)
DOI 10.1007/978-1-4899-6407-6

PLENARY LECTURES

MAIN LECTURES

SYNTHETIC POLYMERS IN THE MEDICAL SCIENCES

H. F. Mark

Polytechnic Institute of Brooklyn, New York, U.S.A.

1. INTRODUCTION

By far the most widespread and important use of synthetic organic polymers was stimulated by their attractive mechanical, thermal, electrical and optical properties. They invaded, and now dominate, the large and important industries which produce textiles, packaging materials, plastics, rubbers, electronic devices, coatings, adhesives and medical appliances and many thousand scientists are now engaged in stabilizing and upgrading the performance of synthetic polymers in connection with their application in these fields.

For instance in the domain of textiles the interest is concentrated on such mechanical properties as tensile strength, elongation to break, instantaneous and delayed recovery from small strains, loop and knot strength, and on such thermal characteristics as a high melting-point or softening-range and, from the electrical and optical point of view, on satisfactory antistatic behaviour and on soft and pleasant lustre. In the packaging industry tear strength, shock resistance, heat sealability, brilliant transparency and antistatic character are essential. Plastin materials, in the narrower sense, require rigidity, impact strength, mouldability, compatibility with fillers, electrical resistance and durability. Synthetic rubbers, again, must be endowed with long range, low modulus, reversible elasticity combined with ready curability, solvent resistance and superior ageing characteristics. Similar demands exist for coatings and adhesives.

In view of the preponderant importance of these properties most research and development activities have been focussed on the understanding of their relations to the fundamental structural properties of organic polymers. Their classification has, in fact, led to a high degree of predictability of the all-round behaviour of prospective new polymers which eliminates random and strictly empirical synthetic efforts and permits the successful designing of new useful materials.

During all this impressive progress the fact that in nature organic polymers not only provide for the firm and durable receptacles of life but also for the vehicles constructed of cellulose, lignin and resins but their growth and reproduction is controlled by proteins, hemicelluloses, starch and other polysaccharides. Animals are built of bones and of fibrous proteins but the life functions within their bodies are carried on by water-soluble or dispersible proteins, nucleic acids and polysaccharides. Since polymer science has been so successful in producing synthetic analogues for the group of natural life-protecting materials, why should it not be possible to produce the life-carrying organic polymers with similar success? In any event, it would be worthwhile to try. In doing so there would, of course, be other properties and property combinations which would be necessary and important for these new synthetics. Solubility in water, compatibility with other solutes in the liquids of a growing plant or of a living human body, capacity to form reversible or irreversible complexes with certain substrates, degradability and excretion under biological conditions. In this approach synthetic polymers are considered not as fibres, films, plastics, rubbers, coatings or adhesives but as chemical reagents, catalysts, activators or stabilizers, which will eventually fulfil all the applications of their natural originals.

Certain areas existed in which this concept was already rewarded with notable success, namely the synthesis and use of polymeric acids and bases in the form of ion-exchanging beads, membranes and fabrics and the corresponding activities in the field of chelate-forming polymers and of polymeric oxidation–reduction systems. In these applications it was necessary to combine certain physical properties—hardness, abrasion resistance, high softening-range—with the essential chemical character—diffusibility, ionic strength and ionic density or chelating capacity—and it has, indeed, been possible to produce reasonable successful combinations of these properties with a good prospect of achieving even better results in the not-too-distant future.

In this report consideration will be given to the efforts which are being made to use synthetic organic polymers to improve our understanding of the structure of natural proteins, to produce water-soluble analogues of them and to provide biochemical, biological and medical research and developments with new materials. The latter have already led to improved methods of diagnosis and cure and will, it is hoped, bring other even more important innovations in the not-too-distant future.

2. CONTRIBUTIONS TO THE PROBLEM OF PROTEIN STRUCTURE

Early studies of the supermolecular structure of proteins were concentrated on the fibrous types and were carried out by Brill and Polany[1] with silk and by Astbury and his coworkers[2] with wool and other keratin fibres using x-ray diffraction methods. The essential result was the discovery that the macromolecular chains of proteins exist essentially in two conformations: a folded 'α' form and an extended 'β' form. Similar information was gained by the x-ray study of myosin[3] which indicated that the reversible deformation of proteinic fibres was, at least partly, a consequence of conformational transitions from the 'α' to the 'β' form on extension and from the 'β' to the 'α' form on relaxation. No quantitative data existed on the relative stability of the two forms under given environmental conditions or on the free energy changes which accompany the conformational transition from one to the other state.

A new and significantly-refined level of our general understanding of protein texture came from the work of Pauling and his associates[4] with the proposal of an exact quantitative model for the α helix with 3·67 residues per turn and with an identity period of 1·5 Å per residue. This prediction soon found experimental confirmation from Perutz[5] and Bamford[6] whose x-ray tests established the existence of 1·5Å spacings in synthetic poly-peptides and native proteins. This was onw of the earliest and, at the same time, most significant contributions of synthetic polymers to the complete quantitative elucidation of protein structure because the synthetic poly-peptides could be made, at will, with different substituents, and spun and drawn into fibres readily and they furnished much clearer and better x-ray diagrams than the natural proteins thus permitting a more precise and reliable determination of the position and intensity of the crucial diffraction spots.

Another important experimental method for the study of natural proteins is infrared absorption analysis, particularly in its use as a differential tech-nique with two polarized beams; it permits not only far-reaching statements about conformational details but also provides information on intra- and intermolecular hydrogen bonding in the solid ordered and disordered states. The significance of many absorption bands in view of wavelength, intensity and state of polarization was established with the aid of synthetic polypeptides of exactly known chemical composition and configuration[7]. As a result of their studies of poly-γ-methyl-L-glutamate with polarized infrared radiation Ambrose and Hanby[8] were the first to demonstrate that the folded and helicoidal conformations of synthetic polypeptides—and logically those of proteins—are caused and maintained by intramolecular hydrogen bonding.

All this, with the help of synthetic polypeptides, to a large extent clarified the different preferred conformations of proteins in the solid state but did not provide any information on the nature of protein solution. When the answer to this question was sought with the aid of light-scattering, ultracentrifuge, ultraviolet absorption and electron microscopy, synthetic polypeptides again were of great value as model substances for the more complicated and more difficult to manage native proteins[9].

A particularly impressive and successful cooperation of exactly controlled synthetic chemical efforts and refined physical analysis was contributed by Goodman and his associated who prepared many oligomeric polypeptides of exactly-known degree of polymerization, chemical composition and con-figuration and used optical rotation and nuclear magnetic resonance in dilute solution[10]. Experimenting with a complete set of oligomers from the dimer ($n = 2$) to the nonamer ($n = 9$), they first found that in certain solvents—m-cresol, chloroform, dimethylformamide and dioxane—large positive specific rotations were observed, whereas in other solvents—dichloroacetic acid, trifluoroacetic acid and hexafluoroaceton—moderate negative rotations were found. This was interpreted by the obviously reasonable assumption that the first group of solvents supports the formation of helicoidal conformations through intramolecular hydrogen-bonding whereas the second group has a strong hydrogen-bridge-opening capacity; destroys the helices and reduces them to random coils. The capacity to form a helix is, however, limited to a certain minimum DP, namely $n = 5$, so that even in solvents of the first class the dimer, trimer and tetramer still show the earmarks of random conformation. This is shown in *Figure* 1 which is taken from the classical paper of Goodman and Schmitt[11] and shows a spectacular increase of the specific rotation of γ-L-glutamic acid esters in helix-supporting solvents above $n = 5$.

Another independent advance in the analysis of the conformational behaviour of dissolved proteins was carried out with the same oligomers by the same group through an extension of the systematic work of Moffit,

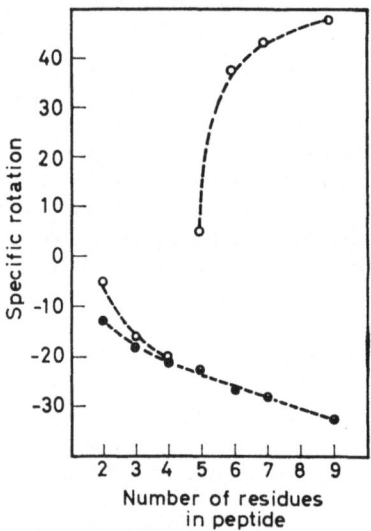

Figure 1. Optical activity of peptide derivatives as a function of solvent and number of residues. All rotations were measured in dioxane (open circles and dichloroacetic acid (half-filled circles) at 2% concentration except the hepta- and nonapeptide in dioxane solution. These rotations were measured on a 1·43% and 0·22% solution, respectively. Reproduced by permission of the *Journal of the American Chemical Society*[11]

Doty and Blount which employs optical rotatory dispersion[12]. This approach was initiated by Moffit, who added to the classical Drude equation a higher-ordered term with a coefficient (b_0) which was empirically correlated to the helix content of the proteinic solute in a tentative manner. Dispersion tests with the Goodman oligomers added convincing experimental evidence that the Moffit constant b_0 assumes values around $+50°$ for random coils and values around $-600°$ for helices. Interpolation between these two extremes for a given, actually-measured sample permits a *degree of helicity* to be established experimentally for the material under investigation and allows a study to be made of how the percentage of helix content can be affected by environmental changes such as temperature, nature of the solvent, pH or composition of a solvent mixture.

Although most facts nicely supported this procedure there was a flaw left in the argument because a certain assumption had to be made concerning the wavelength λ_0 of the absorption band in the Drude viz. Moffit equation which was usually taken to be at or near 200 Å. In order to clarify this dubious point, measurements of the specific rotation of several synthetic polypeptides were carried out in various solvents in the neighbourhood of the active absorption band (Cotton effect region) and it was established that the residue rotation at 2330 Å can be used as a good measure for the helical content of a dissolved protein or polypeptide by interpolating its value between $-1800°$ for the pure coil form and $-12\,700°$ for 100% helicity[13].

In a similar manner attempts are now being made—with notable success—to calibrate and rationalize the nuclear magnetic resonance signals of proteins in dilute solution using the corresponding signals obtained from synthetic polypeptides under the same environmental conditions[14].

Meanwhile it has been found that such important globular proteins as hemoglobin and myoglobin consist of α helical regions which are separated from each other by randomly-coiled chain segments. Both conformations in their cooperation and changing proportions are obviously essential for the life-supporting functions of the protein and it is, therefore, of considerable importance to understand their relative stability and convertibility. No wonder that Zimm, Gibbs and Lifson have devoted considerable work to the formulation of a theory of helix formation and stability[15]. It is very encouraging that their equations and conclusions can be tested on a large body of experimental data most of which have been collected with the aid of synthetic polypeptides.

3. SOLUBLE SYNTHETIC POLYMERS AS REAGENTS AND CATALYSTS

The synthetic use of organic synthetics in a number of industries resulted in the establishment of several *principles* which help in the design of specific polymers for specific uses. It became common knowledge that a good *fibre* former would have to be high melting (preferably above 230°C), substantially linear, readily orientable and crystallizable, it should possess a certain moisture regain, die absorptivity and antistatic character. A good synthetic rubber, on the other hand, should consist of chains having a low T_g, which means high internal flexibility, yet being stable to temperatures up to 200°C for short and to 150°C for long periods, offering the possibility of the formation of a firm three dimensional network and resisting the swelling action of water and organic solvents. Similar information was available for the molecular engineering of films, plastics, coatings and adhesives providing for reliable ground rules to eliminate many potential candidates and to focus attention on the most promising species.

As we turn to new applications for our synthetics, what are now the guiding principles for their design and construction? In view of the infancy of all efforts in this field, these principles still have to be more precisely formulated in the future, but at least a general pattern of them can be presented now.

All polymers of this new type consists of more than one component—they are co-, ter-, or quater-polymers of random or alternating character—and, eventually, block and graft copolymers with occasional moderate reticulation. All existing processes are used for their synthesis: initiation by free radicals, cations and anions, chain transfer and cross-linking, living polymeric segments, ring opening, cyclopolymerization and polycondensation steps of all kinds. Numerous (perhaps 200) new monomers have been synthesized already and many more will have to follow to incorporate into the final products all the desirable and necessary properties.

In terms of general design, these new polymers will consist of the following components:

(a) An inexpensive monomer which essentially serves for building the *backbone chain* and as a *spacer* which keeps the various reactive groups separated from each other. Monomers of this type are ethylene, pro-pylene, vinylchloride, butadiene, styrene and corresponding neutral units in the creation of long chains by polycondensation.

(b) A *solubilizing* monomer contributing a group which produces certain solubility characteristics, —OH, >NH, —COOH, —NH$_2$, $\rightarrow$C—O—C$\stackrel{\displaystyle <}{}$ etc. for hydrophilicity, phenyl-, alkyl- and *cyclo*alkyl groups for oleophilicity and >CO, $\rightarrow$CCl, —COOR, groups for compatibility with polar organic solvents.

(c) A monomer carrying the *reactive* group, which can be acidic (—COOH —SO$_3$H, —PO$_4$H), basic (—NH$_2$, —NHR, NR$_4^+$), reducing (—SH, —COH), oxidizing (—O—O—, —ClO$_2$), catalytically active (imi-dazole, benzimidazole, pyrimidine, phthalocyanine, hemocyanine) and eventually still others.

(d) A monomer introducing an *activating* or *promoting* group which is added with the idea that, as in natural proteins, the reactive group might be promoted by another group in its *vicinity* which cooperates with it by increasing its acidic or basic strength, raising its oxidation potential or improving its catalytic activity by environmental support. It was found that neighbouring groups can promote the original activity but, in certain cases can also reduce it and act as a *deactivator* or *modifier*.

(e) Under certain conditions it may also be desirable to add a monomer which acts as a site for *cross-linking* or *grafting* in order to produce certain compatibility characteristics and to influence the rheological properties of the solutions.

It is obvious that polymers of such structural complexity can serve many purposes and several cases will, therefore, now be enumerated in which such synthetic biopolymer models have been prepared and have already shown certain promising activities.

4. SYNTHETIC POLYMERS CARRYING THIOL GROUPS

A relatively early and notable successful effort in this direction was that pioneered in 1955 by C. G. Overberger and his coworkers; it led to the synthesis of many new monomers which promise interesting applications not only in biology and medicine but also in the domain of commercial coatings and adhesives[16].

It is known that the activity of many enzymes depends on the presence of sulphydryl groups, not all of which are of equal reactivity or importance for the proper functioning of the enzyme. Three types are now recognized depending on their reactivity with oxidizing, alkylating and mercaptide-forming agents. They are commonly classified as freely reactive, sluggishly reactive and masked. The masked groups become freely reactive when the protein is denatured, which is due to the unfolding of the peptide chains, making these hitherto inaccessible groups freely available. Resistance to oxidation of some sulfhydryl groups in enzymes is due to the distance between them in the native protein, which prevents the formation of a disulphide bond. The presence of electronegative groups in the chain also influences the ease of oxidation, because of a decrease in the degree of dissociation of the —SH bond resulting in a reduced oxidation rate.

There are reasons to believe that damage by irradiation is due to the oxidation of biologically-necessary cuprous ion to cupric. Protection can be obtained by shielding the cuprous ion against the oxidizing agents produced during irradiation. This protection is claimed to be due to chelation and, among the effective chelates are dithiols and β-mercaptoamines.

In order to understand more clearly the factors which govern the reactivity of enzymic sulphydryl groups and to point out useful pathways for improving chemical protection against the effects of radiation, Overberger thought that polymers containing sulfhydral groups would be useful models to study[17].

Consequently polymeric thiol compounds have been prepared via both addition and condensation polymerization. Direct synthesis of a free thiol polymer by vinyl polymerization is impractical since the sulfhydral group adds readily to an unsaturated linkage and, in addition, has a very high chain-transfer constant in radical initiated polymerization. Two special synthetic approaches had, therefore, to be used:

(a) the polymerization of a monomer containing the sulfhydryl group protected by a subsequently removable "blocking" group, and

(b) the introduction of the thiol group into a previously formed polymer by conversion of some convenient group already on the chain.

The preparation, polymerization and copolymerization of vinyl thiolacetate and its hydrolysis under air-free conditions, gave soluble polymers.

$$—(—CH_2—CH—)— \xrightarrow{H_2O} —(—CH_2—CH—)—$$
$$\qquad\qquad | \qquad\qquad\qquad\qquad | $$
$$\qquad\quad SCOCH_3 \qquad\qquad\qquad SH$$

Copolymerization with vinylene carbonate and methyl methacrylate and the hydrolysis of the resulting products gave the —SH group in the vicinity of —OH and —COOH groups:

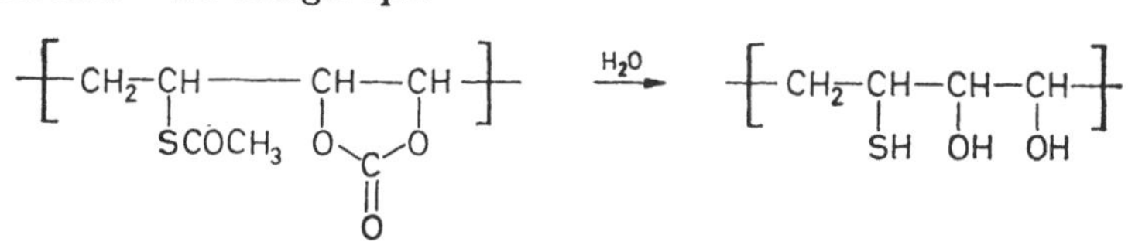

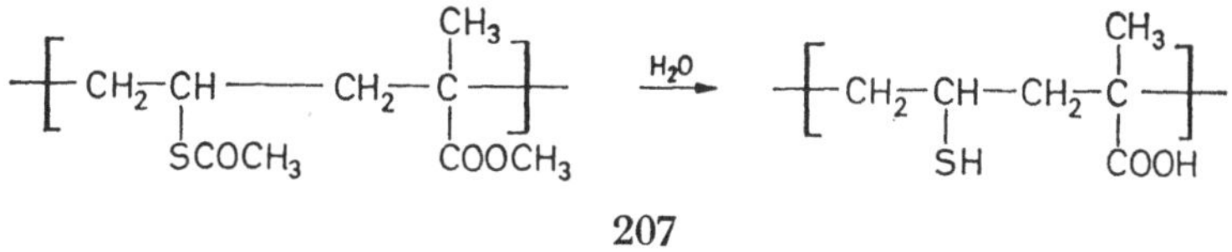

The cyclopolymerization of divinyldithiol carbonate and its hydrolysis afforded an alkali-soluble poly(vinyl) mercaptan.

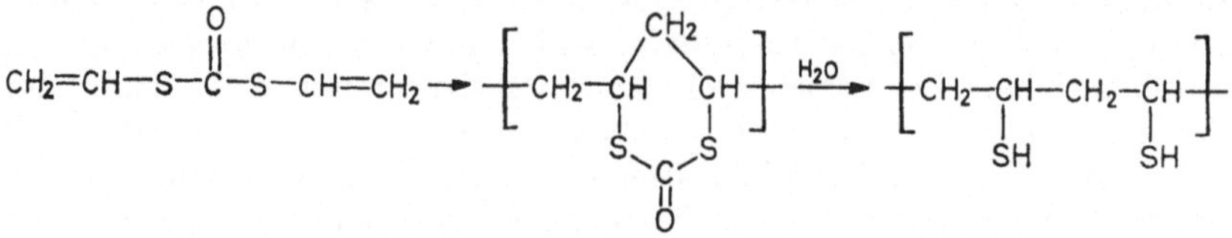

$$CH_2=CH-S-\overset{\overset{O}{\|}}{C}-S-CH=CH_2 \longrightarrow \left[CH_2-CH\underset{\overset{S}{\underset{C}{\diagdown}}\overset{}{\diagup}S}{\overset{CH_2}{\diagup\diagdown}}CH\right] \xrightarrow{H_2O} \left[CH_2-\underset{SH}{CH}-CH_2-\underset{SH}{CH}\right]$$

The hydrolysis of *p*-vinylphenyl thiolacetate yielded a pure homogeneous alkali-soluble homopolymer and copolymerization with methyl metha crylate gave, upon saponification, another synthetic sulfhydryl-containing material:

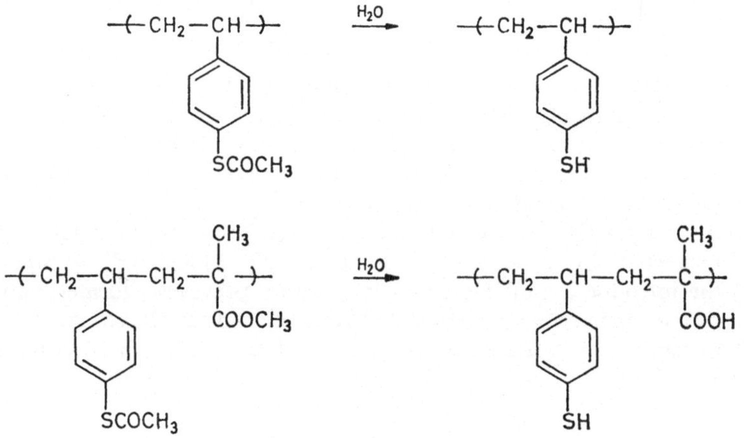

The preparation of blocked sulfhydral monomers and their polycondensation has led to the following products:

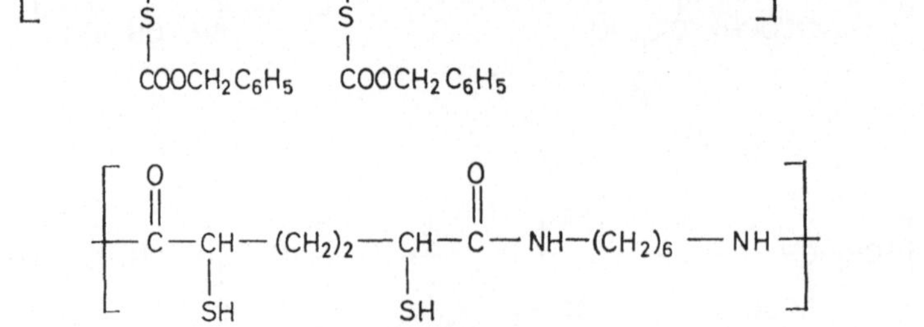

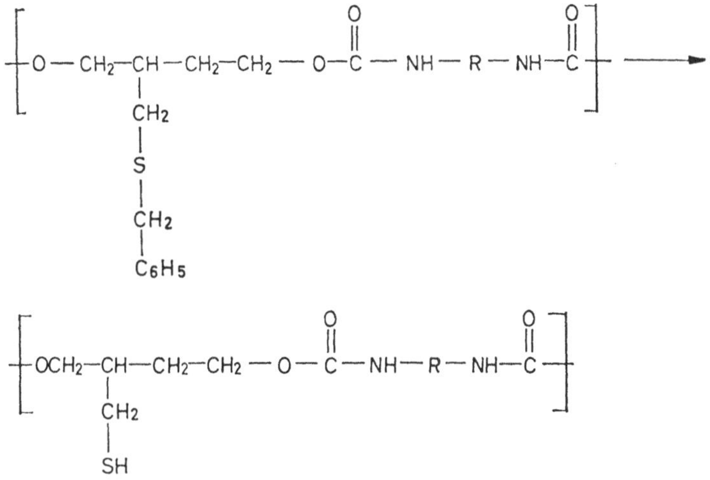

$$\left[O-CH_2-CH-CH_2-CH_2 - O-\overset{\overset{O}{\|}}{C} - NH-R-NH-\overset{\overset{O}{\|}}{C} \right] \longrightarrow$$

with the pendant chain:
$$CH_2-S-CH_2-C_6H_5$$

$$\left[OCH_2-CH-CH_2-CH_2 - O - \overset{\overset{O}{\|}}{C}-NH-R-NH-\overset{\overset{O}{\|}}{C} \right]$$

with the pendant chain:
$$CH_2-SH$$

Treatment of poly(vinyl alcohol) with hydrogen bromide and thiourea gives the isothiouronium salt which can be hydrolysed to polymeric, vinyl-mercaptan.

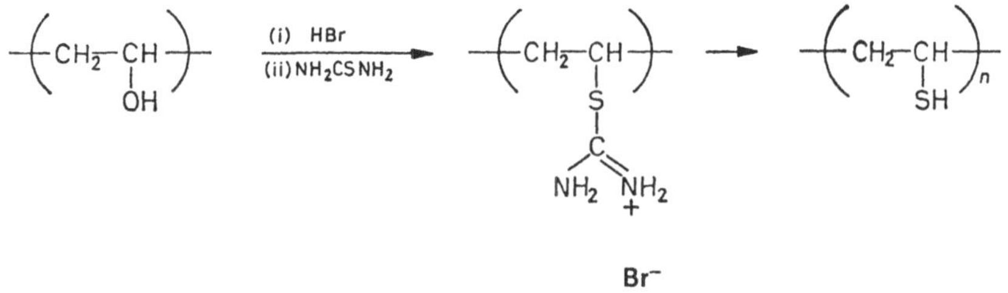

$$\left(CH_2-\underset{OH}{CH}\right) \xrightarrow[\text{(ii) NH}_2CSNH_2]{\text{(i) HBr}} \left(CH_2-CH\right) \longrightarrow \left(CH_2-\underset{SH}{CH}\right)_n$$

middle structure pendant:
$$CH_2-\underset{S}{\underset{|}{C}}\underset{NH_2\ \overset{+}{N}H_2}{\overset{\diagup\diagdown}{C}}$$

$$Br^-$$

Treatment of poly(vinyl alcohol) with chloroacetaldehyde yields a cyclic acetal with replaceable chloride. Reaction with thiourea followed by saponification leads to a polythiol.

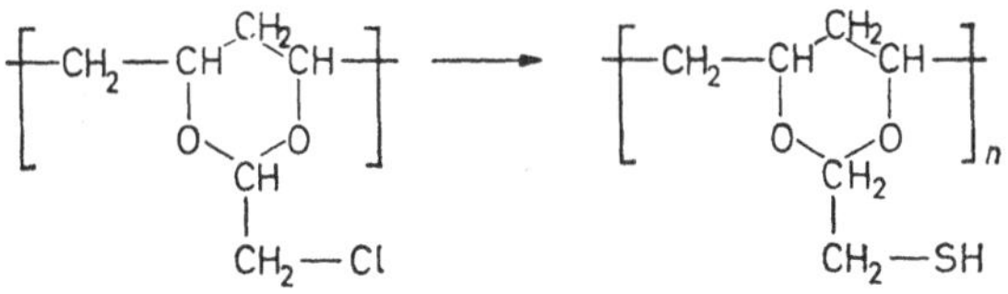

$$\left[CH_2-CH\overset{CH_2}{\diagup\diagdown}CH\right] \longrightarrow \left[CH_2-CH\overset{CH_2}{\diagup\diagdown}CH\right]_n$$

left ring: O—$\underset{CH}{}$—O, then CH_2—Cl

right ring: O—$\underset{CH_2}{}$—O, then CH_2—SH

Poly(hexamethylene adipamide) can be used as a backbone on which to graft the mercaptomethyl fragment by treatment with formaldehyde and ammonia. Subsequent treatment with thiourea and hydrogen chloride gives the isothiouronium salt from which a soluble polythiol is obtained.

209

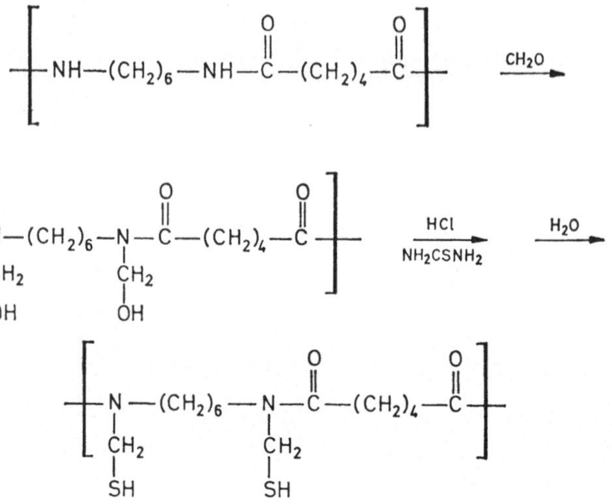

$$\left[\right]-NH-(CH_2)_6-NH-\overset{\overset{O}{\|}}{C}-(CH_2)_4-\overset{\overset{O}{\|}}{C}-\left[\right] \quad \xrightarrow{CH_2O}$$

$$\left[\right]-\underset{\underset{OH}{\overset{|}{CH_2}}}{\overset{|}{N}}-(CH_2)_6-\underset{\underset{OH}{\overset{|}{CH_2}}}{\overset{|}{N}}-\overset{\overset{O}{\|}}{C}-(CH_2)_4-\overset{\overset{O}{\|}}{C}-\left[\right] \quad \xrightarrow[NH_2CSNH_2]{HCl} \quad \xrightarrow{H_2O}$$

$$\left[\right]-\underset{\underset{SH}{\overset{|}{CH_2}}}{\overset{|}{N}}-(CH_2)_6-\underset{\underset{SH}{\overset{|}{CH_2}}}{\overset{|}{N}}-\overset{\overset{O}{\|}}{C}-(CH_2)_4-\overset{\overset{O}{\|}}{C}-\left[\right]$$

Soluble graft copolymers containing pendant mercaptan groups are obtained be treatment of poly(glycidyl methacrylate) with thioglycolic acid.

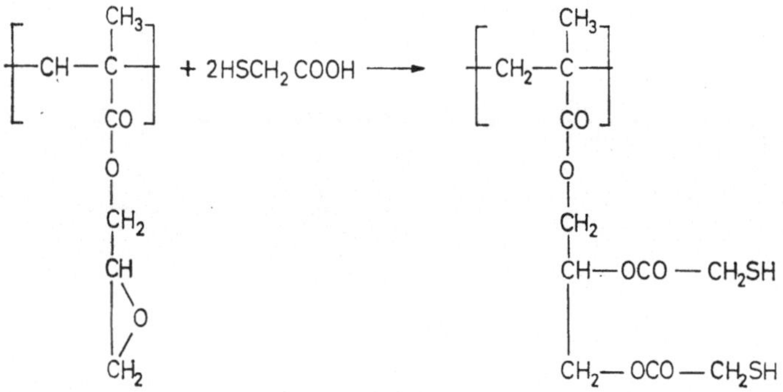

$$\left[-CH-\underset{\underset{\underset{CH_2}{\overset{|}{CH}}\diagdown\!\!\!\diagup O}{\underset{\overset{|}{CH_2}}{\overset{|}{O}}}}{\overset{\overset{CH_3}{\overset{|}{C}}}{\underset{}{\overset{|}{CO}}}}- \right] + 2HSCH_2COOH \longrightarrow \left[-CH_2-\underset{\underset{\underset{CH_2-OCO-CH_2SH}{\overset{|}{CH-OCO-CH_2SH}}}{\underset{\overset{|}{CH_2}}{\overset{|}{O}}}}{\overset{\overset{CH_3}{\overset{|}{C}}}{\underset{}{\overset{|}{CO}}}}- \right]$$

It is also possible to introduce thiol residues into proteins. Using thio-glycolides, highly-thiolated casein and ovalbumin was prepared:

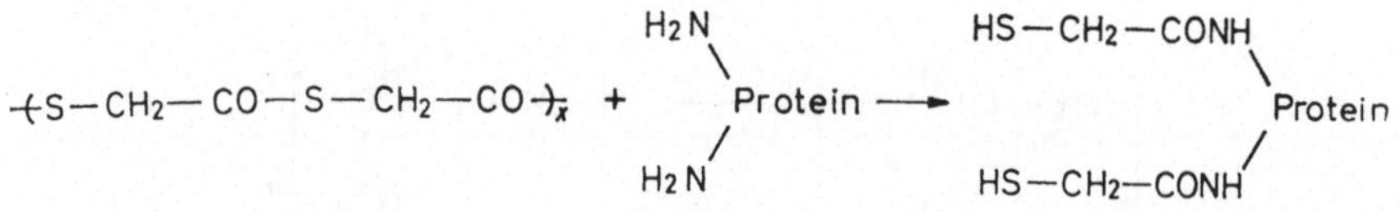

$$-(S-CH_2-CO-S-CH_2-CO)_x + \begin{array}{c} H_2N \diagdown \\ Protein \\ H_2N \diagup \end{array} \longrightarrow \begin{array}{c} HS-CH_2-CONH \diagdown \\ Protein \\ HS-CH_2-CONH \diagup \end{array}$$

Table 1 presents the specific oxidizability of various —SH-group-con-taining substances at pH 10; it permits several very interesting comparisons and conclusions. Compound VII—the repeating unit of the hydrolysed homopolymer VIII is almost 7 times as oxidizable as *p*-thiocresal (I).

Since the rates were determined in very dilute solutions it is reasonable to argue that a thiol group of VII will have a nearer neighbouring thiol group than will a thiol group in compound I. 2,4-Pentanedithiol (IX) oxidizes almost 20 times as fast as *p*-thiocresol and about 3 times as fast as VII which indicates that oxidation is increased as the ease of disulphide formation increases.

Table 1. Thiol oxidation rates at pH 10

Compound	Observed oxidation rate	Relative oxidizability
I *p*-Thiocresol†	1·17	1·00
II 2,2-Dimethyl-4-(*p*-mercaptophenyl) valeric acid	1·04	0·89
III Hydrolysed copolymer of vinyl thiolacetate and methyl methacrylate	1·21	1·04
IV 2,2-Dimethyl-4-mercaptovaleric acid	2·16	1·85
V Hydrolysed copolymer of *p*-vinylphenyl thiolacetate and methyl methacrylate	3·29	2·81
VI Thioglycolic acid	4·26	3·64
VII 2,4-Di(*p*-mercaptophenyl)-pentane†	7·51	6·43
VIII Hydrolysed homopolymer of *p*-vinylphenyl thiolacetate	8·88	7·58
IX 2,4-Pentanedithiol†	23·0	19·7

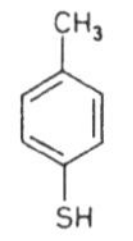

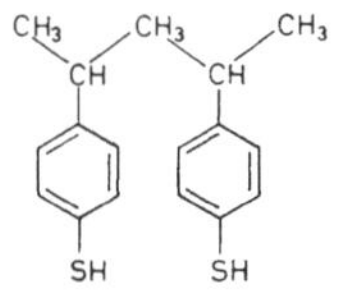

Another interesting fact is that the hydrolysed copolymer V oxidizes about three times as fast as the monomeric model II.

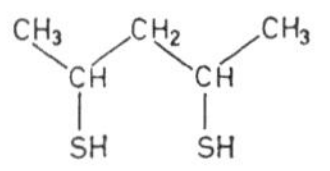

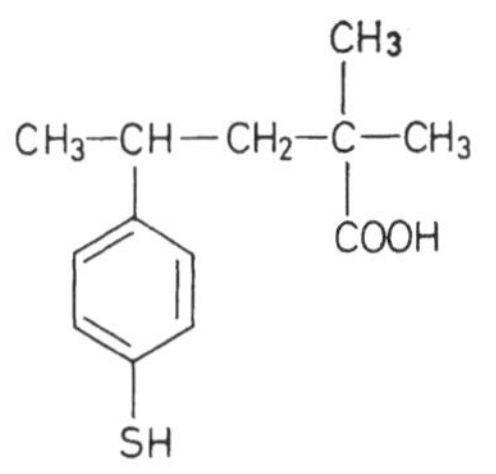

This can be understood by the observation that *p*-mercaptostyrene entered the copolymer faster than methyl methacrylate indicating that there exist blocks of *p*-mercaptostyrene. Since this sequence is known to enhance oxidizability (VII and VIII) it readily explains the increased rate.

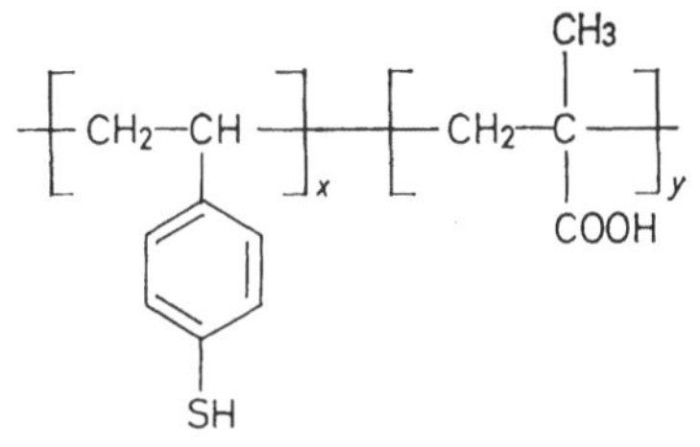

Table 2 shows thiol oxidizabilities in dimethyl formamide (DMF). The relative rates indicate that oxidizabilities increase as the distance between

211

Table 2. Thiol oxidation rates in DMF

Compound	Observed oxidation rate (μm ml^{-1} min^{-1}—SH^{-1}) $\times$ 10^5	Relative oxidizability
X β-Mercaptoethanol	1·7	1
XI 2,5-Hexanedithiol	3·9	2·3
XII 2,4-Pentanedithiol	9·7	5·7
XIII Polyvinylmercaptan	91·4	53·8

the thiols decreases. Compound XIII, polyvinylmercaptan is more than 50 times as fast as the monothiol, β-mercaptoethanol X, and almost 10 times as fast as its model XII.

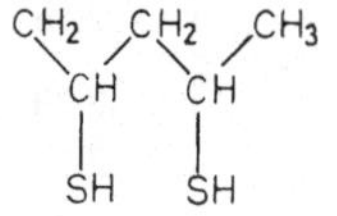

(XII) 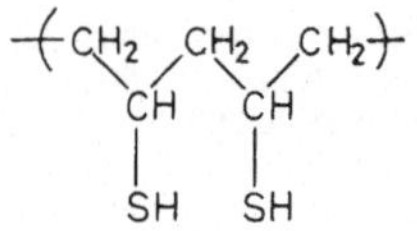   (XIII)

This is evidently a statistical effect. In the polymer, an individual thiol group has *two* nearest neighbours and many other —SH groups which approach it to a certain extent as the chain assumes various conformations while in compound XII, a single mercaptan has only one nearest neighbour with no other —SH groups in conformational vicinity.

5. SYNTHETIC POLYMERS CARRYING IMIDAZOLE GROUPS

Intensive and systematic research has centred about the study of the mechanism of enzyme catalysis; one significant result was that the proteolytic (hydrolytic) action of certain enzymes (e.g. papain) is retained even if a substantial degradation of the original material has occurred. This stimulated the idea of preparing and studying synthetic model polymers since, apparently, the catalytic activity of some enzymes depends on the co-operation of a limited number of functional groups and not on the entire, complicated protein molecule.

Overberger and his associates started (in 1961) to synthesize various co-polymers which contain imidazole and benzimidazole as substituents and to study their esterolytic action under various conditions[18]. Prior work on the catalytic activity of chymotrypsin indicated that the imidazole group of the histidine unit is the reactive centre of this enzyme and that its activity is enhanced by the vicinity of the hydroxyl group of a serine unit, which could form a tetrahedral addition complex with the imidazole group. With this background in the literature concerning the mechanism of enzyme action, polymeric materials were synthesized which had pendant imidazole groups, and other functional groups, such as hydroxyl or carboxyl groups, as required by the known facts concerning the mechanism. The imidazole group was provided by 4(or 5)-vinyl imidazole. Since these polymers would have a certain structure in solution, it was possible that by coiling of the chain an imidazole ring would be brought into the proper steric relationship with the other groups to provide greater reactivity than would be expected from a similar polymeric chain containing imidazole groups alone, or from simple imidazole compounds. The structure of a repeat unit of a hydrolysed copolymer of 4-vinyl imidazole and vinyl acetate (I), for instance, would approximate a peptide with neighbouring histidine and serine (II), such as prepared by Katchalski, and would be a model in accord with early proposals of the enzyme mechanism, involving acyl transfer from serine to imidazole.

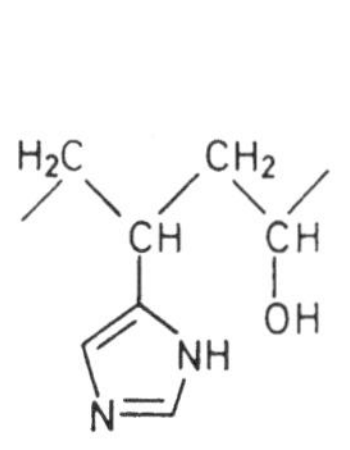

(I) 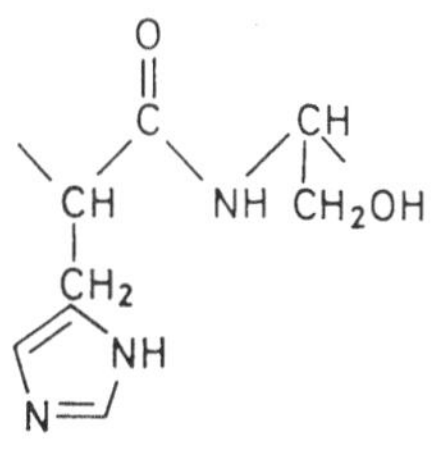   (II)

Since a copolymer with vinyl acetate would give a less-reactive secondary alcohol compared with the primary alcohol of serine, copolymerization with a monomer such as methyl acrylate (III) followed by reduction would give a primary alcohol in the polymer (IV). In addition, copolymerization with

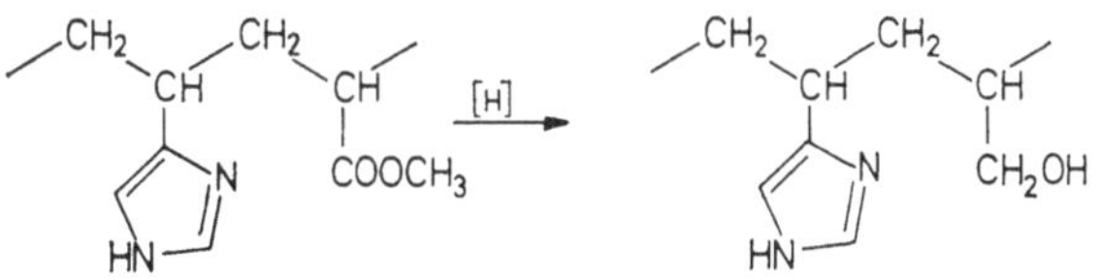

(III)   (IV)

vinylene carbonate would provide a hydroxyl group, after hydrolysis, on a carbon atom closer to the carbon bearing the imidazole ring. Since an acidic group is probably also necessary for the enzymic catalysis, this third site can be provided by terpolymerization, or by partial reduction of a copolymer with acrylic acid which had neighbouring carboxyl groups. Also, monomers other than 4-vinyl imidazole were considered, such as benzimidazole derivatives with the vinyl group on the benzene ring. These

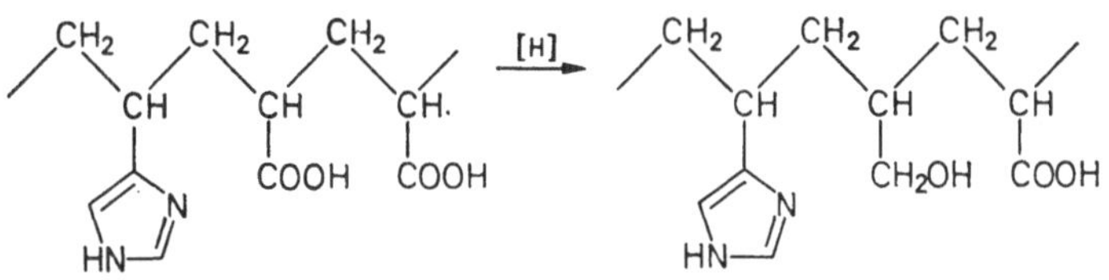

are only a few examples for many interesting possibilities by which different groups and different stereochemical situations may be attained on a polymeric chain with a hydrocarbon backbone by various combinations of copolymerization, hydrolysis and reduction.

213

Studying the hydrolytic action of such synthetic polymers, Overberger, in fact, found a pronounced catalytic superiority of the polymer over the monomer applied at equal concentration and pH. Using p-nitrophenylacetate as substrate they obtained solvolysis rate constants as shown in *Figure 2* which clearly indicate that in the pH range above 8·0 the polymer, poly-4(5)-vinylimidazole, is a much better catalyst than imidazole itself. Analogous

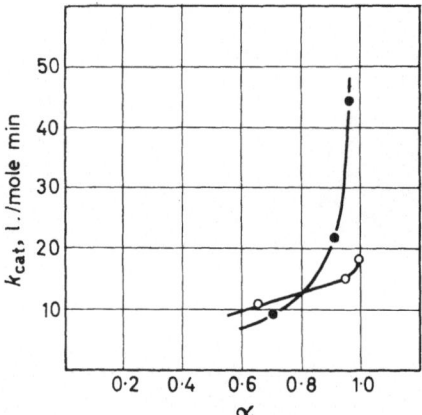

Figure 2. Solvolysis of p-nitrophenyl acetate catalysed by imidazole, ○, and by poly-4(5)-vinylimidazole, ●

results were found for the hydrolysis of 3-nitro-4-acetoxy-benzoic acid with the same polymer and both substrates were much more rapidly solvolysed by poly-5(6)-vinylbenzimidazole than by benzimidazole itself and at pH 10 the polymer was 50 times more active than the monomer.

These results prove convincingly that even in a homo-polymer the spatial arrangement of the pendant reactive groups (in this case imidazole groups) in the coiled chains produces a stronger catalytic effect than the same number of independent imidazole monomers.

Another significant level of clarification was achieved by Overberger and Morawetz[19] when they prepared a copolymer of 4(5)-vinylimidazole and acrylic acid, in which the chain can be negatively charged, and investigated its esterolytic activity on uncharged and charged substrates. They found that for the uncharged esters, such as p-nitrophenylacetate and 4-acetoxy-3-nitro-benzoic acid, the mono-4(5)-vinylimidazole was a better catalyst than the copolymer. But for a positively charged substrate, namely for 3-acetoxy-N-trimethylanilinium iodide the copolymer was four times more active than the monomer at pH values above 9. Apparently the ionized carboxyl groups of the acrylic acid which are distributed all along the length of the coiled chain keep the positively charged substrate for longer average periods in their own ionic atmosphere and hence in the vicinity of the catalytically-active imidazole groups which are also distributed all along the length of the same coiled chain.

In order to prove this point beyond any doubt Overberger and Morawetz also reversed the charges and prepared a copolymer which contained catalytically active imidazole groups together with methylimidazolium groups which—at low pH—assume a positive charge and, therefore, should be efficient promoters for anionic ester substrates. In fact they found that poly (1-vinyl-3-methylimidazolium iodide) catalysed the esterolysis of an anionic substrate, namely sodium 4-acetoxy-3-nitrobenzenesulphonate, noticeably more than uncharged esters.

Although these tests gave a reasonable proof for the binding of a charged substrate by oppositely-charged groups of a polymer chain with a resulting increase of the catalytic activity of a neighbouring reactive group Morawetz was not satisfied and performed an experiment which gave an even sharper and clearer confirmation of this idea[20]. He studied the effect of a highly-charged polyanion (poly-methacrylic acid and a 1:1 copolymer of maleic and acrylic acids at a degree of neutralization of 0·67) on the hydroxide-ion-catalysed hydrolysis of the double-charged cationic ester I. The hydrolysis proceeds in two steps and the time dependence of the ester concentration

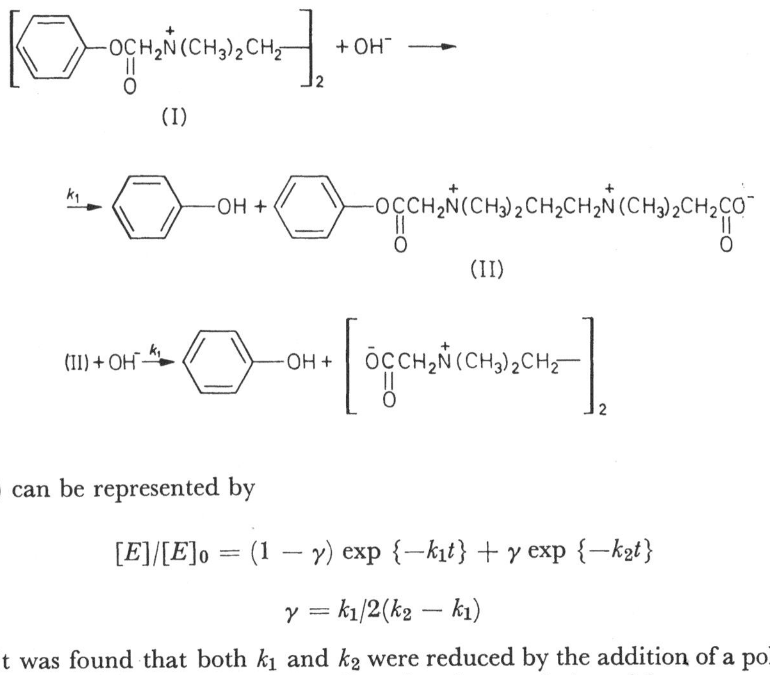

(E) can be represented by

$$[E]/[E]_0 = (1 - \gamma) \exp \{-k_1 t\} + \gamma \exp \{-k_2 t\}$$

$$\gamma = k_1/2(k_2 - k_1)$$

It was found that both k_1 and k_2 were reduced by the addition of a polymeric acid. This effect may be attributed to the association of the ester to the polyanion. Since the hydroxide ions are repelled by the high negative charge of the polymer, the ester groups are protected from their catalytic action. It was found by experiment that the polyanion reduces k_1 by a factor of 5 but k_2 by a factor of only 2. This result is obviously a consequence of the fact that species I, carrying two positive charges, is bound to polymethacrylate much more tightly than species II, which has only a single net positive charge. The model resembles in this respect an enzyme-substrate system, in which the reagent also has a much higher affinity for the enzyme than does the reaction product.

Another more detailed test was carried out with a 1:1 copolymer of maleic and acrylic acid with 2/3 of the carboxyls ionized

$$-\left[-CH-\!\!-\!\!-CH-\!\!-\!\!-CH-\!\!-\!\!-CH_2- \atop \quad COO^- \quad\ \ COOH \ \ \ COO^- \right]_n$$

as the inhibitor for the hydrolysis of the doubly-positive ester I. The pH was maintained at 8·96 the counterion concentration was also kept constant and it was found that the hydrolysis constant for the bound ester is about twenty times less than the value found at the same pH in conventional buffer solution.

This is another independent experimental proof that enzyme-like catalysis of synthetic polymers is strongly influenced by groups which hold a substrate at a charged chain and which repel the hydrolysing ionic species.

All these examples show that synthetic polymers can be designed in such a manner that they assume to a certain extent functions which are, in living systems, normally taken care of by natural proteins. These cases are, of course, only the first synthetic pioneering efforts in a new field, but there can be little doubt that they will soon be followed by many other more spectacular results.

6. GRAFTING OF SYNTHETICS ON PROTEINS

All the investigations described in the foregoing paragraphs attempted to simulate biological action through the preparation of entirely synthetic counterparts without the use of naturally-occurring components. There has however, existed for several years an approach which aims at new effects through the *combination* of synthetic or artificial species with active natural elements such as enzymes, antigens and antibodies and which culminated in the work of E. Katchalsky and his associates on chemical combination, particularly on the grafting of synthetic components on proteins. Such operations change the solubility characteristics of the naturally-occurring substances and produce water-insoluble derivatives of them which are of considerable theoretical interest and also have many important practical applications[21].

Water-insoluble enzymes have been used as heterogeneous specific catalysts in suspension or in column form; they can be readily removed from the reaction mixture and, if stable, can be employed repeatedly to induce specific chemical changes in relatively large amounts of substrate. Water-insoluble antigens and antibodies have been used specifically for adsorbing their corresponding antibody or antigen from a mixture of compounds. The adsorbed antibodies were subsequently recovered in pure form by elution. The theoretical interest in biologically-active proteins that are bound to well-characterized synthetic carriers comes from the fact that such protein derivatives provide simple models for the study of the effect of the micro-environment on the mode of action of enzymes, or other bioactive proteins, which act *in vivo* while embedded in membranes or other complex native surroundings.

216

One of the existing methods for the insolubilization of biologically-active proteins is their covalent binding to a suitable water-insoluble carrier and the covalent cross-linking of the protein by a suitable bifunctional reagent. The binding of a biologically-active protein to an insoluble carrier by covalent bonds must be carried out *via* functional groups on the protein which are nonessential for its biological activity and the binding reaction must be performed under conditions which do not cause denaturation. The functional groups of proteins suitable for covalent binding under mild conditions are the α- and ε-amino groups; the α-, β- and γ-carboxyl groups; the sulfhydryl and hydroxyl groups of cysteine and serine respectively; the imidazole group of histidine; and the phenol ring of tyrosine. Amino and hydroxyl groups readily react with acylating and alkylating agents, aldehydes, iso-cynates, and diazonium salts. Sulfhydryl groups react with organomercurial compounds and with alkylating agents. The imidazole and phenol groups will both couple with diazonium salts, and the imidazole group will, in addition, often react with various alkylating agents.

The most frequently used insoluble carriers are cellulose derivatives such as methyl-, benzyl-, acetyl- and benzoyl cellulose, polystyrene derivatives such as poly-*p*-aminostyrene and various copolymers such as that of ethylene and maleic anhydrides or the copolypeptide made of aminophenylalanine and leucine.

Table 3 contains a list of those enzymes which have already been chemically combined with synthetic polymers and gives an impressive picture of the research efforts which are exploring this method for the study of enzyme activity.

Figure 3 shows for the purpose of illustration how tyrosin chains can be attached to trypsin by the imitation of the polymerization of *N*-carboxy-tyrosine anhydride through the amino-groups of the trypsin and *Figure 4*

Table 3. Water-insoluble enzyme grafts

Enzyme	Carrier
Trypsin	*p*-Amino-DL-phenylalanine·L-leucine copolymer (APL)
	Carboxymethyl cellulose azide
	Maleic anhydride–ethylene copolymer
Chymotrypsin	Carboxymethyl cellulose
	p-Aminobenzyl cellulose
	APL
	Maleic anhydride–ethylene copolymer
Papain	*p*-Amino-DL-phenylalanine-L-leucine copolymer
	Collagen
Pepsin	Poly-*p*-aminostyrene
	APL
	Methacrylic acid–methacrylic acid-3-fluoro-4,6-dinitroanilide copolymer
Streptokinase	APL
Ribonuclease	Poly-*p*-aminostyrene
	p-Aminobenzyl cellulose
	APL
	Carboxymethyl cellulose
Urease	APL
Amylase	Poly-*p*-aminostyrene
	Methacrylic acid–methacrylic acid 3-fluoro-4,6-dinitroanilide copolymer

illustrates the preparation of a water-insoluble trypsin derivative by chemical reaction of the amino-groups of the enzyme with copolymers of ethylene
and maleic anhydrides (IMET).

The combination of the active compounds with synthetic components

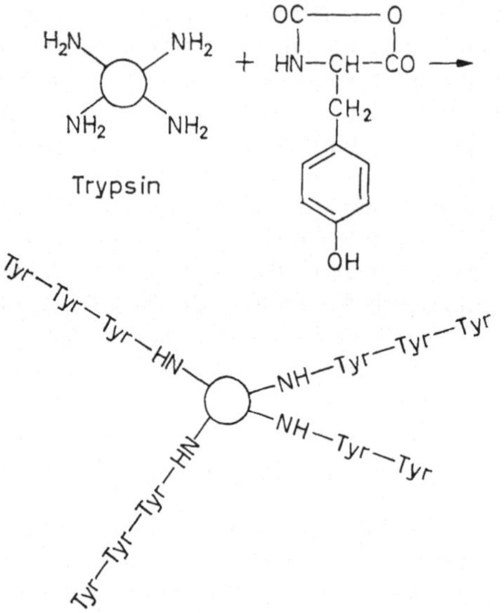

Figure 3. Preparation of polytyrosyl trypsin by the initiation of the polymerization of N-
carboxytyrosine anhydride with trypsin

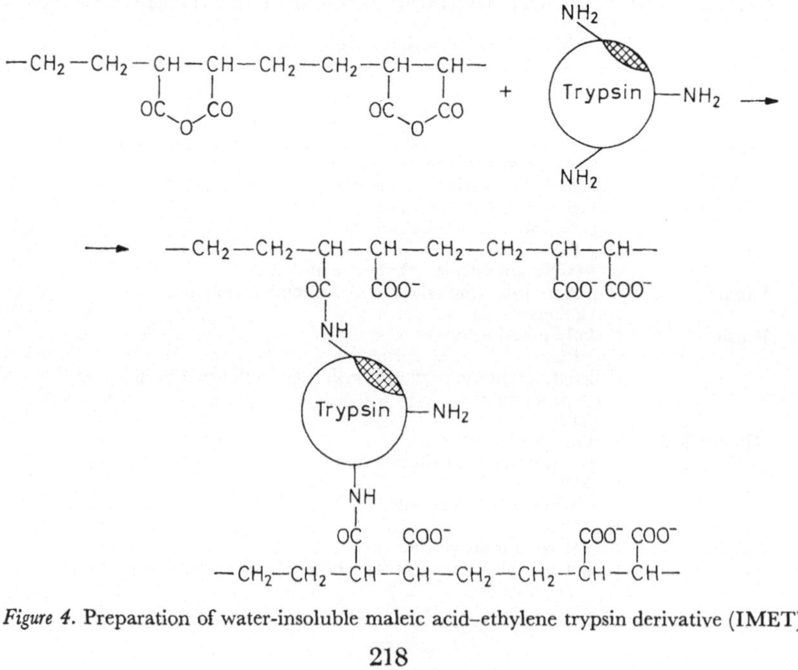

Figure 4. Preparation of water-insoluble maleic acid–ethylene trypsin derivative (IMET)

218

causes the optimum activity to move into another pH range as shown for IMET of various compositions and the availability of water-insoluble enzyme derivatives permits the preparation of columns and membranes with specific enzymatic activity. Such systems are extremely promising in the continuous preparation of larger quantities of pure products and in the exact regulation of the contact time between enzyme and substrate which permits exact degrees of conversion to be attained. At high flow rates one is able to establish very brief contact and to obtain and study the initial stages of enzyme–substrate interaction. It is also possible to expose a given substrate within a short time to the action of several enzymes in a continuous manner.

All the materials and methods previously described in this article are still in the stage of early development and this is also true for the combination of reactive proteins with synthetic species. However, the progress made during these pioneering efforts shows already that many new and important results can be expected in the not-too-distant future.

7. THE USE OF SYNTHETICS IN MEDICINE

A prominent American surgeon predicted in 1950 that "intramedullary fixation of bone fractures with thermostatic hemoplastic agents injected directly into the fracture site, cancellous in their form, setting to the strength and resilience of steel, stimulating callous formation and being themselves resorbed in that process will permit to establish all necessary weight bearing in nearly every fracture". In fact, hard implants to replace bones are already being widely used and their adequate specific gravity and their initial mechanical properties can be properly adjusted without great difficulties. There still exist, however, several problems which call for significant and necessary improvements in the chemical design of the surgical repair polymers. The research, which is obviously needed to find solutions of these problems is of a multidisciplinary character and calls for an intense, long-range cooperation between physicists, chemists, physiologists and surgeons over a long period. Accidental short-range contacts may lead to occasional favourable results but will not promote the neccessary fundamental understanding of all the contributing factors and will not produce progress on a broad front[22].

Several problems arise from the interaction between the body and the implant. Considering first the effects of the body and its metabolism on the repair polymer, one has, in general, to distinguish between two types *biostable* and *biodegradable*. For the former the effect of the host on the implant should be negligible and the repair piece should maintain its integrity for a lifetime. There exists, of course, a large body of information on the stability of many polymers under standard environmental outdoor conditions including the action of high and low temperatures, chemical reagents and radiation of all kinds. This does not, unfortunately, help very much to assess and create resistance against biodegradability. As a consequence, considerable systematic research is under way to establish the ground rules for the stability of the most important families of repair polymers in a living body. It has been revealed that the most important degradation processes

219

are *hydrolysis* (solvolysis) and *oxidation* under the influence of many types of enzymes. For example, polyamides can lose almost half of their tensile strength after only one year's implantation in certain locations, mainly by the scission of the amide bonds. Radioactive fragments from subcutaneous implantation in animals of tagged polyethylene are excreted after only 30 days, presumably as a result of an oxidative scission of the carbon–carbon bond in the chain. Corresponding tests were made with other polymers such as Teflon, polyesters (Dacron, Mylar), polyacrylics, polyurethanes and and silicones. The degree of degradation apparently does not only depend on the chemical nature of the bonds in the backbone chain but on many other factors, for instance the degree of orientation and crystallinity of the macro-molecules in the implant, on the fact whether the repair polymer—as a fibre, fabric, film or larger object—is under stress (constant or variable), on the flux of body liquids around and/or through the implant and, probably, on other as yet unknown factors. In general terms it would seem that fluoro-polymers, silicones and highly crystalline semiaromatic polyesters perform better than polyurethanes, polyacetals, polyamides and, even, polyolefins[23].

This selective resistivity of certain polymers stimulated research for improvements through the synthesis of specifically designed systems in which enhanced biostability is introduced by bond aromaticity (as in aromatic polyamides), by steric protection of the degradable bond (as in methylated polyesters, polyamides and polyolefins) and by the introduction of such stabilizing substituents as chlorine, fluorine and aromatics. Drawing on the vast amount of information coming from theoretical and experimental organic chemistry numerous new polymers are now synthesized and tested for reduced biodegradability.

As materials of this type become available they are not only used as such for the formation of repair parts but also as protective coatings on other— readily degradable—implants which are easier to obtain or offer superior weight-bearing characteristics.

While these studies permit the selection of the best biostable polymers with increasing success, they provide, at the same time, valuable information on the biosensitive species. These materials are supposed to be degraded at a predetermined rate by the influence of bioactive body reagents and to be eliminated through the normal excretory routes without being accumulated or stored in the vital tissues or organs. Elaborate studies have been made and are still in progress concerning the critical molecular weight (or DP) below which excretion occurs at desired rates and without ill effects on the kidneys and other organs. These results will permit a better selection to be made of material for sutures, blood plasma extenders and drug depositing carriers.

Much more complex and less clarified are the influences of the various implants on the host. They extend from short range effects such as blood-clot formation and thrombosis to such long-range consequences as the promotion and creation of sarcoma and cancer.

Some notable success has been achieved in the prevention of thrombosis by grafting heparin or other coagulation delaying ingredients on the surface of hard and soft repair polymers. This can be done, for example, by the formation of insoluble complexes between heparin and heparinoid materials with quaternary ammonium bases which are distributed on the surface of the implant by direct copolymerization or by an appropriately designed coating. *Table 4* contains figures which show how various polymers can be improved by heparinization in respect to the causing of thrombosis and

Table 4. Effect of surface heparinization

Surface	Clotting times		Hemoglobin (mg %)	
	Unheparinized (min)	Heparinized (h)	Unheparinized	Heparinized
Polystyrene	9	+24	16	50
Polyethylene	11	+24	25	200
Butadiene/pyridine rubber	12	1	37	75
Polyvinyl chloride	12	0·75	10	14
Cellophane	6	1	600	600
Natural rubber	10	1	13	—
Ethylene propylene rubber	5	1	15	20
Fluorosilicone rubber	18	1	15	20
Silicone rubber	15	1	5	40

proves that substantial delays in clotting time can be obtained. Considerable research efforts are now being made to follow this lead and to develop still more efficient systems of antithrombosicity[24].

The more remote effects of tissue injury, cell necrosis and general irritation depend largely on the chemical nature of the repair polymer. In preliminary research it was established that certain materials such as methyl methacrylate and methyl α-cyanoacrylate provoke inflammatory responses and cell necrosis, whereas others like polyesters, polyethylene, silicones and fluoropolymers elicit negligible response. In a first approximation these effects seem to go somewhat parallel with biostability, crystallinity and insolubility and it can be expected that new species which are prepared for superior biostability will also produce minimal irritation.

Using all previously described information some notable progress has been made in the design of porous plastic limb sockets. Before the advent of synthetics it was common to use leather for this purpose, which, however, degraded and deteriorated under the influence of perspiration. The plastic laminated sockets which were originally used instead of leather did successfully withstand degradation but were impervious to moisture and, as a consequence, sweat accumulated in and around the socket and became a source of serious discomfort and irritation for the amputee. Considerable progress was achieved when porous laminates were developed consisting of layers of fabric impregnated with and surrounded by a porous polymeric matrix. Such composites have a large number of fine capillaries which permit rapid evaporation of the sweat with minimal blocking of the sweat pores and which leave the fabric structure intact so that its strength and toughness are maintained. The idea for such a design is being developed at present in several ways, following essentially the well-known technology of fabric–foam and web–foam composites like Corfam, Ortex or Clarino. Notable and lasting success has already been achieved with knitted polyamide and polyester fabrics impregnated with epoxy-resins which are foamed with the aid of a volatile solvent or a blowing agent. Elaborate tests were performed in order to establish optimal conditions for strength, abrasion resistance and porosity as a function of resin content and cure time. Some representative results are given in *Tables 5* and *6*; it is obvious

Table 5. Strength and porosity of porous laminates as a function of resin content

Resin content (%)	Compression load (lb. avg.)	Per cent effective porosity
68	2660	67
67	2645	58
66	2197	66
64	1847	71
63	1847	74

Table 6. Compressive strength and porosity as a function of cure time

Cure time (h)	Compressive load (lb.)	Per cent effective porosity
$1\frac{1}{2}$	2450	61
$1\frac{3}{4}$	2440	58
2	2530	52
$2\frac{1}{4}$	2530	52
$2\frac{1}{2}$	2650	56

that they are of interest not only in the special use of limb sockets but in the much wider field of permeable membrane sheets and structural objects.

Very interesting and promising research is being done on a broad front in the design of denture bases, plastic teeth and restorative dental materials. Most denture base materials contain—at present—poly(methyl methacrylate) as the main ingedient with the occasional presence of small quantities of other monomers. A monomer–polymer paste is most commonly used for making dentures, but plastisol polymers that cure fully on heating are also available. The liquid component consists of methyl methacrylate (inhibited by traces of hydroquinone or butylated hydroxytoluene), small amounts of other acrylic monomers, plasticizers, and 5 to 15 per cent of a cross-linking agent, such as ethylene glycol dimethacrylate or divinylbenzene. The powdered polymer usually contains suspension-polymerized methyl methacrylate that may have been modified with small amounts of ethyl or butyl methacrylate or ethyl acrylate to produce a somewhat softer product. A catalyst, generally benzoyl peroxide in 0·5 to 1% concentration, is incorporated into the powder together with pigments, dyes and opacifiers. These slurries are cured between 75° and 100°C for a few hours. Preformed plastisols containing methyl methacrylate monomer and polymer are also commercially available. Instead of a heat-curing cycle, the slurry can be cured at room temperature if the monomer contains an accelerator, such as dimethylaniline or *p*-toluidine in 0·5 per cent concentration[25].

Table 7 gives the properties of typical acrylic, polyvinyl-acrylic, and polystyrene denture bases.

Approximately 30% of all artificial teeth sold in the United States are plastic; in the U.K. dentists use plastic teeth in 80 per cent of their work. Excessive wear, crazing, and discoloration of these teeth have been overcome by improved moulding, better cross-linking agents, and the complete removal of residual ingedients used in the suspension polymerization of the polymer for the fabrication of the teeth which are manufactured by injection or transfer moulding techniques.

Table 7. Properties of some denture base polymers

Property	Poly(methyl methacrylate)	Polyvinylacrylic	Polystyrene
Tensile strength (psi)	7000–9000	7500	6000
Compressive strength (psi)	11 000	10 000–11 000	15 000
Elastic modulus (psi)	5.5×10^5	3.3×10^5	5.3×10^5
Impact strength (ft.-lb./in. of notch)	60	180	50–60
Transverse strength	6000–8000	6000–8000	8000
Knoop hardness	16–22	14–20	14–20
Thermal coefficient of expansion (/°C)	80×10^{-6}	70×10^{-6}	$60–80 \times 10^{-6}$
Heat distortion temperature (°C)	70–90	55–77	70–100
Polymerization shrinkage (vol. %)	6	6	—
Dimensional stability	good	good	good
Water sorption (% in 24 h for $\frac{1}{8}$-in. specimen)	0·3–0·4	0·07–0·4	0·05–0·3
Water solubility (g/cm²)	0·02–0·06	0·01	0·01
Processing ease	good	good	good
Shelf life:			
Powder–liquid paste	good	fair	good
Plastisol	fair	fair	good

The desirable properties of artificial teeth are a more natural appearance, less breakage, reduction or elimination of clicking, better bond between tooth and resin base, and ease of grinding and polishing.

The materials at present used for these two important applications have enabled a vast amount of experience to be gained and attention has been focused on the need for certain significant improvements. There is little doubt that these improvements can be achieved if the basic knowledge of polymer science and the ever-increasing number of available materials and methods is fully evaluated.

Particularly intriguing problems are posed in the design and manufacture of satisfactory dental restorative materials. The present approach is based on the use of composite systems which offer a good compromise for hardness, toughness, thermal characteristics, adhesivity, appearance and long-range stability under oral environmental conditions.

Table 8 lists the thermal expansion coefficients of several polymers and of composite filler–polymer systems together with that of the dental crown (hydroxyapatite); it can be seen that these composites approach the tooth substance satisfactorily if about 70% (by weight) of the filler is used. Current research has established that 25–30 weight per cent of polymeric binder was sufficient to produce the necessary mechanical and physical characteristics if the problem of adhesivity is successfully solved.

For this purpose it is necessary to have a coupling agent between the inorganic filler and the main body of the polymer in the restorative material and another coupling agent between this polymer and the hydroxyapatite of the tooth material. The *inorganic filler* consists usually of fused silica in the

Table 8. Thermal expansion of organic polymers

Polymer	Coefficient of expansion per °C × 10⁶
Polyethylene	90
Polystyrene	80
Polytetrafluoroethylene	70
Polymethyl-methacrylate (PMMA)	80
PMMA plus 30% silica filler	60
PMMA plus 50% silica filler	40
PMMA plus 70% silica filler	20
Epoxydized PMMA plus 70% silica filler	13
Epoxydized PMMA plus 75% silica filler	11
Hydroxyapatite (dentin)	11

form of spherical particles (20–100 micron diameter) or of silicates which are hard and have low coefficients of thermal expansion†. In order to combine the main polymer with these particles through chemical bonds one applies trimethyl or triethyl silicate groups which ester-interchange with the hydroxyl groups of the filler and are attached to a chemically reactive group—epoxy-, amino-, acrylic, vinyl-, allyl-, isocyanate- through a short hydrocarbon chain. Ingredients of this type are represented in *Table 9* as *coupling agent A.*

The main *polymeric binder* must have high compressive strength, sufficient hardness, low shrinkage during rapid polymerization under oral conditions, low water-sorption and long-range biostability. The armoury of polymer science offers many candidates for the fulfilment of these conditions and work is under way with epoxydized acrylics and methacrylics which contain the reactive groups at the ends of relatively long segments in order to incorporate resilience and toughness whereas the hardness is controlled by the filler and by the density of the cross-links. The polymerization rate of such systems can be increased by small percentages of α-cyano-acrylates and similar kick-off ingredients. *Table 9* contains a few indications for the preferred composition of the structural binder on the basis of present practice and information.

In order to provide for this binder a firm attachment to the body of the tooth—*coupling agent B* (compare *Table 9*)—one synthesizes a monomer which combines chemically with the structural binder through such groups as epoxy, double bond, amino- or isocyanate and carries at the other end a group which can form a coupler with the calcium ions of the hydroxyapatite. One combination is the reaction product of *N*-phenyl-glycin-sodium with glycidyl methacrylate.

The successful design and preparation of such a sophisticated adhesive system represents an instructive example of what can be done if it is considered worthwhile. The efforts in the medical field can easily become pilot

† It may be interesting to note that there exist several compounds (for instance $LiAlSiO_4$, β-Eucryptite) that have negative coefficients of expansion between 0° and 50°C.

activities for other, wider applications particularly if and when the special monomers become more readily available and less expensive.

The notion that α-cyanoacrylates polymerize very rapidly under oral conditions and also promote rapid polymerization of the other monomers has stimulated another important application in surgery. If patients have

Table 9. Composition of dental filling

Inorganic filler :   Spherical particles of silica and silicates
Coupling agent A :   Compounds of the type

$$CH_3$$
$$|$$
$$CH_2\!=\!C\!-\!CO\!-\!O\!-\!CH_2\!-\!CH_2\!-\!CH_2\!-\!Si(OCH_3)_3$$

$$CH_2\!-\!CH\!-\!CH_2\!-\!O\!-\!CH_2\!-\!CH_2\!-\!CH_2\!-\!Si(OCH_3)_3$$
$$\diagdown\;\diagup$$
$$O$$

$$H_2N\!-\!CH_2\!-\!CH_2\!-\!CH_2\!-\!Si(OCH_3)_3$$

Polymeric binder :   Methyl-methacrylate, glycidyl-methacrylate, acrylic acid, tetra-
   ethylene glycol dimethacrylate together with α-cyanoacrylate-
   epoxides

$$CN$$
$$|$$
$$CH_2\!=\!C\!-\!CO\!-\!O\!-\!(CH_2\!-\!CH_2\!-\!O\!-\!)_n\!-\!CH\!-\!CH_2$$
$$\diagdown\;\diagup$$
$$O$$

and bis-alpha-cyano-acrylates

$$CN \qquad\qquad\qquad\qquad\qquad CN$$
$$| \qquad\qquad\qquad\qquad\qquad\qquad |$$
$$CH_2\!=\!C\!-\!CO\!-\!(-CH_2\!-\!CH_2\!-)_n\!-\!O\!-\!CO\!-\!C\!=\!CH_2$$

Coupling agent B :

$$OC\!-\!CH_2\!-\!N\!-\!CH_2\!-\!CH$$
$$O \qquad\qquad\qquad\qquad O\!-\!H$$
$$Ca$$
$$/|\backslash$$
$$Dentin$$

N-phenyl-glycin-Na plus glycidyl methacrylate

---

clotting deficiencies or have been receiving anticoagulants many surgical operations create difficult problems and there has, therefore, been a systematic investigation of how α-cyanoacrylates can be successfully used as tissue adhesives during and after surgery[26].

Table 10 presents the setting times of several α-cyanoacrylates and shows that the setting times decrease with increasing length of the alkyl group and that the surface-active character of the monomer is increased. If a bleeding wound is treated with these liquid monomers, or mixtures thereof, they can be administered in drops or in the form of an aerosol. *Table 11* contains information on the relative setting rates of the individual monomers in the dry and wet state according to the two presently available techniques and *Table 12* lists a number of surgical operations during which the hemostatic action of α-cyanoacrylate preparations was already successfully applied.

Table 10. Setting times of α-cyanoacrylates

Alkyl group in the monomer	Setting time (sec)	
	Dry	Wet
Methyl	40–50	40–50
Ethyl	30–40	30–40
n-Propyl	20–30	20–30
n-Butyl	5–10	5–10
n-Amyl	rapid	rapid
n-Hexyl	rapid	rapid
n-Heptyl	rapid	rapid

Table 11. Setting times according to different methods of applications

Alkyl group in the monomer	Setting time (sec)			
	Drop technique		Spray technique	
	Dry	Bloody	Dry	Bloody
Methyl	55	50	30	30
Ethyl	40	40	25	25
Propyl	30	25	20	15
Butyl	10	8	4	3
Amyl	7	5	Instantly	Instantly
Hexyl	5	5	Instantly	Instantly
Heptyl	6	4	Instantly	Instantly

Table 12. Surgical operations during which acrylic tissue adhesives were successfully applied

Surgery	Monomer	Method	Setting time (sec)
Gastrointestinal			
Enteroenterostomy	Butyl	Spray	4
By approximation	Butyl	Spray	4
By serosa patch	Butyl	Spray	4
Renal			
Hemostasis of renal wound	Butyl	Spray	3
	Heptyl	Spray	Instantly
Pulmonary			
Wedge resection and sealing of lung	Butyl	Spray	3
Vascular			
End-to-end asastomosis, jugular artery	Isobutyl	Drop	15
Patch graft, jugular artery	Isobutyl	Drop	14

A group of dogs was noted bleeding five minutes after heparinization which continued, while no bleeding was noted where the acrylic tissue adhesive was used. Prompt hemostatis was also noted following the application of the monomer in the bleeding suture line. The arteriogram and histologic examination revealed no difference in the patency rate of the anastomosis between the cases in which conventional anastomosis was done or the invagination adhesive was used.

In another group dramatic and prompt hemostasis was seen in all cases when the monomer was applied while no hemostasis was obtained in five cases with the use of conventional ligatures and sutures.

In a third group because of suture-line bleeding none of the procedures could be performed without the use of monomer. The prothrombin time averaged 40 seconds in these cases. When the shock experiments were terminated, no blood was seen in the thoracic or abdominal cavities in any of the animals.

This proves that the cyanoacrylate tissue adhesives can be used advantageously for hemostasis in heparinized subjects. No difference in the efficacy of the monomer was observed whether heparinization was started before or after surgery. The same prompt and successful results were obtained when the monomer was used as a primary method or as a supplementary measure to seal the suture line. Therefore, it appears to be preferable to use the monomer technique as a primary tool for various surgical procedures to accomplish the anastomosis, closure, hemostasis, or coaptation of various organs. The operative time was less when the monomer was used compared to the conventional suture method. With conventional suture technique, it was absolutely necessary to use the monomer in addition to control bleeding on the suture lines in the anticoagulated subjects. The successful, prompt hemostasis of liver wound in the anti-coagulated canine seems to indicate that hemostasis in other solid organs, such as the kidney, may be achieved by the same method. It therefore appears that anticoagulated patients may be operated on without increasing the risk of postoperative hemorrhage if the cyanoacrylate tissue adhesive is used.

Since the monomer was used successfully for the hemostasis of the heparinized and/or fibrinolized subject, these compounds may also be used successfully for the patient who has a clotting deficiency.

References

[1] R. Brill. *Liebigs Ann.* **434**, 204 (1923);
   cf. also K. H. Meyer and H. Mark. *Aufbau* Akad. Verlag, Leipzig (1930).
[2] W. T. Astbury *et al. Proc. Roy. Soc.* **A150**, 533 (1935).
[3] W. T. Astbury *et al. Proc. Roy. Soc.* **B129**, 307 (1940).
[4] cf. L. Pauling and R. B. Corey. *J. Amer. Chem. Soc.* **72**, 5349 (1950); *Proc. Roy. Soc.* **B141**, 21 (1953).
[5] M. F. Perutz. *Nature, Lond.* **167**, 1053 (1967).
[6] cf. C. H. Bamford, H. Elliott, and W. E. Hanby. *Synthetic Polypeptides*, Academic Press, New York (1960).
[7] J. D. Bath and J. W. Ellis. *J. Phys. Chem.* **45**, 204 (1941).
[8] E. J. Ambrose and W. E. Hanby. *Nature, Lond.* **165**, 921 (1949).
[9] E. R. Blout *et al. J. Amer. Chem. Soc.* **76**, 4492 (1954);
   P. M. Doty *et al. J. Amer. Chem. Soc.* **76**, 4493 (1954).
[10] cf. M. Goodman and J. S. Schulman. *J. Polymer Sci.* **C12**, 23 (1966).
[11] M. Goodman and E. E. Schmitt. *J. Amer. Chem. Soc.* **81**, 5507 (1959).
[12] W. Moffit. *J. Chem. Phys.* **25**, 467 (1956).
[13] E. R. Blout *et al. J. Amer. Chem. Soc.* **83**, 4766 (1961); **84**, 3193 (1962).
[14] cf., e.g., M. Goodman. *J. Polymer Sci.*, **C12**, 23 (1966).
[15] B. K. Zinner and J. K. Bragg. *J. Chem. Phys.* **28**, 1246 (1958); **31**, 526 (1959).
   J. H. Gibbs and E. A. DiMarzio. *J. Chem. Phys.* **28**, 1247 (1958); **30**, 271 (1959).
   S. Lifson and A. Roig. *J. Chem. Phys.* **34**, 1963 (1961).
[16] cf. C. G. Overberger *et al.* I.U.P.A.C. lecture, Tokyo, Sept., 1966.

227

[17] C. G. Overberger *et al. J. Polymer Sci.* **27**, 381 (1958); *J. Amer. Chem. Soc.* **80**, 5431 (1958).

[18] C. G. Overberger *et al. J. Amer. Chem. Soc.* **85**, 3513 (1963); **87**, 296, 4310 (1965).

[19] H. Morawetz *et al. Polymer Letters* **4**, 409 (1966).

[20] H. Morawetz and J. A. Shafer. *Biopolymers* **1**, 71 (1963).

[21] cf., e.g., E. Katchalsky *et al. Nature, Lond.* **169**, 1095 (1952); **176**, 118 (1955); *Science* **123**, 1129 (1956).

[22] A large group of prominent experts has been active in this field for many years under the auspices of the U.S. Government at the N.I.H. in Bethesda, Md. and at the N.B.S. in Gaithersburg. cf., e.g., F. Leonard. *Army Res. Dev. Mag.* March, 1966, p. 21. Also F. Leonard. *ASTM Monograph* **386**, Introduction (1964).

[23] R. I. Leininger. *ASTM Monograph* **386**, 71 (1964);
S. M. Atlas and H. Mark. *ASTM Monograph* **386**, 63 (1964).

[24] C. S. Woodward. *ASTM Monograph* **386**, 77 (1964).
cf. also the interesting work of O. Wichterle on hydrophilic gels.

[25] cf., e.g., R. L. Bowen. *J. Amer. Dent. Assoc.* **64**, 378 (1962); **66**, 57 (1963); **69**, 481 (1964) and *J. Dent. Res.* **44**, 690, 895, 903, 906 (1965).
J. M. Brauer. *J. Amer. Dent. Assoc.* **72**, 1151 (1966).

[26] F. Leonard *et al. Arch. Surg.* **94**, 187 (1967).

THERMODYNAMIC TREATMENT OF MEMBRANE TRANSPORT

A. Katchalsky

*Polymer Department, Weizmann Institute of Science,
Rehovoth, Israel*

1. INTRODUCTION

The advances which have been made in the synthesis of polymeric films have led to the development of several industrial processes based on transport through selective membranes. Desalination techniques based on electrodialysis and on reverse osmosis, and methods for the isolation of valuable substances from gaseous or liquid mixtures by filtration across suitable films, are well-known examples. The versatility of the new membranes also permits extensive experimental and theoretical studies of the physicochemical characteristics of transport across simple and complex layers.

Major interest in membrane behaviour is still concentrated, however, on biological processes. The earliest investigators in physiology were already aware that cell and tissue covers regulate selective accumulation and excretion of necessary and waste materials. Studies indicated that the evolution of osmoregulatory membranes governed the transition of plants and animals from the primeval sea to sweet water and land; and it is now recognized that living membranes are true organs, which have made possible some of the remarkable adaptations to extreme conditions of life. Thus, the ability to survive in hot and salty environments is related to the development of powerful desalination mechanisms, such as the kidneys of the mammalia, the tear glands of reptiles and birds, and the secretion devices of plants.

Even more intriguing is the recent finding of the cell biophysicists that a high percentage of cellular material is organized in the form of intracellular membranes. These two-dimensional patterns of cellular organization seem to contribute the transition step between the unidimensional biopolymers and the three-dimensional structures of the cell as a whole.

Further progress in biological and technological membrane research depends to a large extent on mastery of the laws governing membrane transport. A convenient method of deriving these laws is through the study of synthetical model systems the structure of which is known and may be regulated at will. The thermodynamic discussion presented below, while primarily concerned with the behaviour of polymeric films, has an evident relevance to the interpretation of several biological phenomena.

2. THERMODYNAMIC BACKGROUND[1]

2.1. The fundamental equation which served as starting point for numerous further developments is that of Nernst and Planck. It is based on the assumption that the steady velocity of transport in a viscous medium $\vec{v}$ is directly proportional to the driving force $\vec{X}$, the proportionality factor being the constant mobility ω. Thus,

$$\vec{v} = \omega \vec{X} \tag{1}$$

For diffusional flows

$$X_i = -\operatorname{grad} \tilde{\mu}_i \tag{2}$$

where X_i is the force acting on the ith component and $\tilde{\mu}_i$ its electrochemical potential. The flux of the ith component (J_i) is the product of velocity and concentration, or,

$$\vec{J_i} = c_i \vec{v_i} = c_i \omega_i \vec{X_i} \tag{3}$$

In the case of unidirectional diffusion of a non-electrolyte, equation (3) yields readily the conventional equation of Fick. Assuming that the system is dilute in the ith component, and that ideality may be attributed to the chemical potential, $\mu_i = \mu_i^0 + RT \ln c_i$

$$X_i = -\frac{d\mu_i}{dx} = -\frac{RT}{c_i}\frac{dc_i}{dx} \tag{4}$$

Inserting (4) into (3) we obtain

$$J_i = RT\omega_i \left(-\frac{dc_i}{dx}\right) = D_i \left(-\frac{dc_i}{dx}\right) \tag{5}$$

where

$$D_i = RT\omega_i \tag{6}$$

is the coefficient of diffusion according to the well-known equation of Einstein.

For the flow of an electrolyte, the electrochemical potential, $\tilde{\mu}_i$, may be written as

$$\tilde{\mu}_i = \mu_i^0 + RT \ln c_i + z_i F\psi \tag{7}$$

where ψ is the local electrical potential and z_i the valency of the ith component. Inserting equation (7) into (2) we obtain

$$X_i = -\frac{d\mu_i}{dx} = -\left(\frac{RT}{c_i}\frac{dc_i}{dx} + z_i F\frac{d\psi}{dx}\right) \tag{8}$$

which upon insertion into (3) gives the renowned Planck expression

$$J_i = RT\omega_i \left[\left(-\frac{dc_i}{dx}\right) + \frac{c_i z_i F}{RT}\left(-\frac{d\psi}{dx}\right)\right] \tag{9}$$

Equation (9) has been tested extensively for diffusion in liquid and solid media and was applied successfully to membrane transport by Meyer and Sievers[2] and by Teorell[3].

2.2. A serious limitation of equation (3), and the derived expressions (8) and (9), is that they deal with independent flows, i.e. with cases in which the transport of each component is not influenced by the concomittant flows passing through the system. It is known, however, since the pioneering work of Reuss in 1809[4], that we should in general assume a *coupling* between flows, which may profoundly change the transport pattern. In his ingenious experiments on the electrokinetic behaviour of porous media, Reuss demonstrated that the flow of electricity is accompanied by a volume flow, while a flow of water, induced by a mechanical pressure head, results in a flow of electrical current. A series of other coupling phenomena were discovered during the XIXth century, such as thermoelectricity, thermo-osmosis and thermodiffusion. It is therefore evident that a comprehensive theory of membrane transport should treat explicitly the coupling between flows and should provide a measure for the transport interaction.

In the study of biological systems, there is a special interest in the interpretation of coupling between diffusional transport and metabolic processes taking place within the membrane or in its vicinity. Such coupling, known in physiology as *active transport*, which plays an important role in the regulatory function of cells and tissue, will be considered later.

2.3. A suitable theoretical introduction to the analysis of coupling phenomena is provided by the thermodynamics of irreversible processes. Although kinetic and statistical mechanical treatments are better suited for the visualization of the processes under consideration, the statistical mechanics of irreversible processes is still an inadequate tool for the description of condensed systems.

Despite its inherent limitation as a formal and 'empty' conceptual system, nonequilibrium thermodynamics has the advantages of simplicity and consistency in making its statements a useful guide in membrane study. It is indeed the apparent simplicity of the equation which misleads the inexperienced who may believe that the thermodynamic statements are trivial and 'self evident' . . .

A convenient starting point for the development of the phenomenology of nonequilibrium thermodynamics is the treatment of the total change in entropy, written in the form

$$dS = d_e S + d_i S \tag{10}$$

where $d_e S$ is the entropy exchanged with the surrounding and $d_i S$ is the entropy created within the system by all irreversible processes. In the terms of equation (10) the second law may be written as

$$d_i S \geqslant 0 \tag{11}$$

or, *the irreversible entropy change is positive definite*. $d_i S$ equals zero for equilibria and is positive for all irreversible processes.

The thermodynamic description of the rate processes is based on the increase in inner entropy per unit time or on the entropy formation $d_i S/dt$. For isothermal processes, it is convenient to use the dissipation function introduced by Lord Rayleigh, $\Phi = T(d_i S/dt)$, which measures the degradation of free energy per unit time due to irreversibility.

The work of Onsager, Prigogine, deGroot, Meixner and their coworkers[5] led to the conclusion that for cases in which the equation of Gibbs, $dU = TdS - pdV + \Sigma\mu_i d\mu_i +$ etc., may be applied, i.e. for phenomena in which *local* equilibria may be assumed and the thermodynamic parameters of state maintain their validity, the dissipation function may be written as

$$\Phi = \Sigma J_i X_i \geqslant 0 \tag{12}$$

Here the J_i's are the irreversible flows taking place in the system (such as diffusional, electrical, thermal and chemical flows) while the X_i's are the conjugated thermodynamic forces. Some of the forces are well known, e.g. the electrical field intensity and the negative gradients of electrochemical potentials, while other forces are more sophisticated, such as the affinity of a chemical reaction

$$A = - \underset{k}{\Sigma}\nu_k\mu_k \tag{13}$$

which drives the flow of a chemical process. Equation (12) may be directly used by stipulating that the choice of flows and conjugate forces must be such that their product should have the dimension of dissipation per unit time. Only such flows and forces are applicable in the phenomenological relations discussed below.

Superficial consideration of equation (12) may give the impression that what it states is self-evident. If it is assumed that the direction of the flows is always the same as that of the conjugate forces, their product will evidently be always positive, whether the flows and forces are positive or negative. Equation (12), however, does not imply the positivity of all binary terms. It is only the sum total which has to be positive definite. Indeed, the more interesting cases are those in which part of the terms are negative, the overall positive dissipation being provided by other terms. A negative term means that the flow proceeds in a direction opposite to its own force and is a 'contragradient' flow. Such processes may be regarded as *driven* by those forces which provide the dissipation. Numerous cases of this type are well known and are naturally formulated through the thermodynamics of irreversible processes.

Thus, for two concomittant diffusional flows,

$$\Phi = J_d{}^1 X_d{}^1 + J_d{}^2 X_d{}^2 \geqslant 0 \tag{14}$$

If flow (1) proceeds in a direction opposite to that of the negative gradient of the concentration of component (1), we have a case of incongruent diffusion. The driving process is here the diffusion of component (2) which provides the required dissipation.

Similarly, for two chemical coupled processes $J_r{}^1$ and $J_r{}^2$, the dissipation function is

$$\Phi = J_r{}^1 A_1 + J_r{}^2 A_2 \geqslant 0 \tag{15}$$

where A_1 and A_2 are the affinities. Here again, equation (15) permits reaction (1) to proceed against its own affinity, if reaction (2) provides the dissipation. This is clearly the type of biochemical coupling in which the entropy reducing synthetic processes are based on coupling with dissipation providing metabolic processes. Finally, the coupling of diffusional with chemical processes makes possible a contragradient transport which can be related to chemical-metabolic dissipation

$$\Phi = J_d X_d + J_r A \geqslant 0 \qquad (16)$$

Equation (16) is thus a thermodynamic formulation of active transport.

2.4. In the further development of a thermodynamic treatment of flow processes, explicit relations have to be established between flows and forces. The simplest relation is given in equation (3) which may be rewritten as

$$\vec{J_i} = L_i \vec{X_i},$$

where

$$L_i = c_i \omega_i \qquad (17)$$

is the phenomenological coefficient relating J_i and X_i. It should be noted that L_i is not a constant, but a function of the parameters of the state. It is, however, independent of the flows and forces.

To extend the relation for all coupling possibilities, Onsager[6] suggested the general set of linear equations

$$J_1 = L_{11}X_1 + L_{12}X_2 + \ldots + L_{1n}$$

$$L_2 = L_{21}X_1 + L_{22}X_2 + \ldots + L_{2n}X_n \qquad (18)$$

$$J_n = L_{n1}X_1 + L_{n2}X_2 + \ldots + L_{nn}X_n$$

or

$$J_i = \Sigma L_{ik}X_k$$

where the coefficients L_{ii} are straight coefficients and L_{ik} are the coupling coefficients relating the ith flow J_i to the kth force X_k.

The linear set of equations (18) holds rigorously only for slow flows, close to equilibrium. Its range of validity is, however, sufficiently wide as to make it useful in the treatment of numerous natural phenomena.

As shown by Onsager, the matrix of the coefficients L_{ik} is symmetrical, so that

$$L_{ik} = L_{ki} \qquad (19)$$

This important theorem was checked experimentally in many fields of physics and physical chemistry and found to hold under a wide range of experimental conditions[7]. The main utility of equation (19) is in the possibility of deriving a large number of cross relations such as

$$\left(\frac{J_i}{X_k}\right)_{X_i=0} = \left(\frac{J_k}{X_i}\right)_{X_k=0} \qquad (20)$$

$$\text{(for all } i \neq k) \qquad \text{(for all } k \neq i)$$

which play the same prominent role in thermodynamics of transport phenomena as the Maxwell relations in classical thermodynamics.

3. TREATMENT OF SIMPLE MEMBRANES

3.1. Let us now consider thermodynamically the simplest case of the transport of a binary solution, say that of a nonelectrolyte (s) in water (w), across a homogeneous membrane.

For the common case of stationary flows—i.e. when the parameters of state, such as temperature, pressure and concentration, do not change with time, although they may vary with position—it is found that the dissipation per unit area is given by

$$\Phi = J_s \Delta\mu_s + J_w \Delta\mu_w \tag{21}$$

It will be noted that, in this case, the flows J_s and J_w are constant throughout the membrane and equal to the flows of solute and water, as measured in the adjacent compartments. Other novel features of equation (21) are the overall forces, $\Delta\mu_s = \mu_s{}^0 - \mu_s{}^i$ and $\Delta\mu_w = \mu_w{}^0 - \mu_w{}^i$, which represent the difference in chemical potentials across the membrane, instead of the local gradients given in equations (4) and (8).

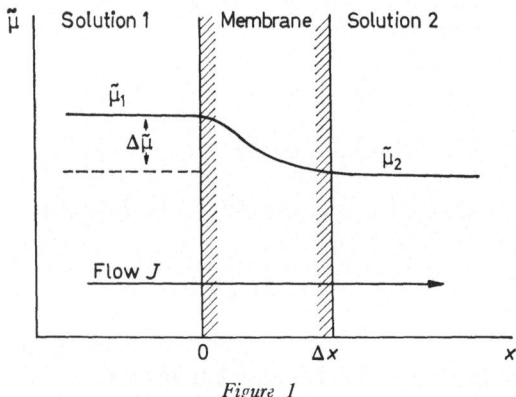

Figure 1

Equation (20) may be 'transformed' to another combination of forces and flows, provided that the dissipation function remains invariant. A convenient choice is that of a volume flow J_v driven by a hydrostatic pressure head Δp, and a diffusional flow J_D driven by the difference in osmotic pressure $\Delta\pi$†. With this choice, we obtain for the dissipation function

$$\Phi = J_v \Delta p + J_D \Delta\pi \tag{22}$$

† *Note:* The relation between volume flow J_v and the flows of the components J_s and J_w is: $J_v = J_s \bar{V}_s + J_w \bar{V}_w$, where $\bar{V}_s$ and $\bar{V}_w$ are the partial molal volumes of s and w.
The diffusional flow is given by

$$J_D = \frac{J_s}{\bar{c}_s} - \frac{J_w}{\bar{c}_w}$$

and on the basis of equation (3) it may be written as $J_D = \vec{V}_s - \vec{V}_w$ i.e., the diffusional flow is the relative velocity of solute to solvent.

and the corresponding phenomenological equations[8]

$$J_v = L_p \, \Delta p + L_{pD} \Delta \pi$$
$$J_D = L_{Dp} \Delta p + L_D \Delta \pi \tag{23}$$

The significance of the coefficients L_p, L_{pD}, L_{Dp} and L_D is readily under-stood. The application of a pressure head across a membrane separating two solutions of equal concentration ($\Delta \pi = 0$) causes a volume flow, linearly proportional to the pressure difference Δp; L_p is therefore the filtration coefficient, given by

$$L_p = \left(\frac{J_v}{\Delta p} \right)_{\Delta \pi = 0} \tag{24}$$

Similarly, the coefficient L_D is a kind of diffusion coefficient relating the diffusional flow J_D to the difference in osmotic pressure, at zero pressure head

$$L_D = \left(\frac{J_D}{\Delta \pi} \right)_{\Delta p = 0}$$

The coupling coefficients are especially interesting, and have the following significance: the coefficient L_{pD} relates to the phenomenon that a volume flow across a membrane can be induced also by a difference in osmotic pressure, when $\Delta p = 0$, i.e.,

$$(J_v)_{\Delta p = 0} = L_{pD} \Delta \pi \tag{25}$$

This volume flow is the well-known[9] osmotic flow, and L_{pD} is therefore *the coefficient of osmotic flow*. On the other hand, L_{Dp} relates the pressure head to the diffusional flow, which occurs through a membrane even if the adjacent solutions have equal concentrations, i.e.,

$$(J_D)_{\Delta \pi = 0} = L_{Dp} \Delta p \tag{26}$$

A fundamental observation in colloid chemistry is that of diffusional flow of solute and solvent with separation of components during filtration. The process is known as ultra-filtration. Thus, L_{Dp} is the *coefficient of ultra-filtration*, and Onsager's relation

$$L_{pD} = L_{Dp} \tag{27}$$

proves to be a non-trivial *physical* statement, namely, that the coefficient of osmotic flow equals that of ultra-filtration.

Inspection of equation (23) permits a better understanding of an ordinary osmotic experiment. The osmometric determination of molecular weights is based on the determination of molar concentrations through the measure-ment of the hydrostatic pressure (Δp) when volume flow stops (i.e., $J_v = 0$). From equation (23) we obtain:

$$(\Delta p)_{J_v=0} = -\frac{L_{pD}}{L_p}\cdot\Delta\pi \tag{28}$$

instead of the expected van't Hoff relation

$$\Delta p = \Delta\pi = RT\Delta c \tag{29}$$

which is generally used for the determination of Δc from known Δp. The difference between equations (29) and (28) is that equation (29) holds only for ideal, semipermeable membranes, through which no solute flow is allowed and a true osmotic equilibrium is established at $J_v = 0$. Equation (28), on the other hand, is valid for any membrane which allows both solute and solvent transport. The coefficient

$$-\frac{L_{pD}}{L_p} = \sigma \tag{30}$$

which determines the deviation of the membrane from semipermcability was named by Staverman[10] the *reflection coefficient*—when $\sigma = 1$ the solute molecules are fully reflected from the membrane. It is an important parameter which may be regarded as an indicator for the selectivity of the membrane; it represents the ability of the membrane to distinguish between solute and solvent molecules. For $\sigma = 0$, the membrane does not distinguish between the components; for negative σ, as found for instance in electrochemical systems, the solute permeates more readily than the solvent. The introduction of equation (30) into (23) gives a useful equation for volume flow as a function of both osmotic and hydrostatic pressures

$$J_v = L_p(\Delta p - \sigma\Delta\pi) \tag{31}$$

This expression has been applied extensively to describe the behaviour of both biological and synthetic membranes used in reverse osmosis. A selection of values of L_p and σ for various systems is given in *Table 1*.

It is often advantageous to have an explicit expression for solute flows J_s, instead of the diffusional flows J_D. A straightforward calculation gives

$$J_s = \omega\Delta\pi + \bar{c}_s(1-\sigma)J_v \tag{32}$$

where ω is a solute permeability coefficient, based on a combination of L_p, L_{pD} and L_D, and $\bar{c}_s$ is an average solute concentration. The over-simple approach to solute permeability leads to the conventional expression $J_s = \omega\Delta\pi$, as found in many textbooks of physiology and physical chemistry. Equation (32) shows, however, that a direct proportionality between solute flow and osmotic difference holds only when $J_v = 0$. If volume flow accompanies the solute flow, solvent drag effects have to be taken into consideration and a suitable correction introduced. Column 4 of *Table 1* presents some values of ω, indicating its variation range. For the sake of completeness, it is worth mentioning that in the physiological literature the permeability coefficient is usually given as

$$P = RT\omega$$

236

Table 1. Properties of membranes

Membrane	Solute	Solute permeability, ω $\left[10^{-15} \dfrac{\text{mole}}{\text{dyne sec}}\right]$	Reflection coefficient, σ	Filtration coefficient, L_p $\left[10^{-11} \dfrac{\text{cm}^3}{\text{dyne sec}}\right]$
Toad skin[a]	Acetamide	0·0041	0·89	0·4
	Thiourea	0·00057	0·98	
Nitella translucens[b]	Methanol	11	0·50	1·1
	Ethanol	11	0·44	
	Isopropanol	7	0·40	
	Urea	0·008	1	
Human red blood cell[c]	Urea	17	0·62	0·92
	Ethylene glycol	8	0·63	
	Melonamide	0·04	0·83	
	Methanol	122	—	
Visking dialysis tubing[d]	Urea	20·8	0·013	3·2
	Glucose	7·2	0·123	
	Sucrose	3·9	0·163	
Dupont 'wet gel'[d]	Urea	31·6	0·0016	9·7
	Glucose	12·2	0·024	
	Sucrose	7·7	0·036	

[a] B. Andersen and H. H. Ussing, Acta Physiol. Scand. **39**, 228 (1957).
[b] J. Dainty and B. Z. Ginzburg. Biochim. Biophys. Acta **79**, 102, 112, 122, 129 (1964).
[c] Values of ω from unpublished data of D. Savitz and A. K. Solomon: σ from D. G. Goldstein and A. K. Solomon. J. Gen. Physiol. **44**, 1 (1960): L_x from V. W. Sidel and A. K. Solomon, J. Gen. Physiol. **41**, 243 (1957).
[d] B. Z. Ginsburg and A. Katchalsky, J. Gen. Physiol. **47**, 403 (1963).

so that

$$(J_s)_{J_v = 0} = \omega \Delta \pi = P \Delta c_s$$

3.2. For systems involving electrolyte transport across a charged membrane we must *a priori* consider three flows driven by three conjugated forces. The flows may be chosen to be: the flow of salt, J_s, the flow of water, J_w, and the flow of electrical current I. The corresponding forces are $\Delta \mu_s$, $\Delta \mu_w$ and the electromotive force E, which is measured with reversible electrodes. Following the formalism developed heretofore, the dissipation function is

$$\Phi = J_s \Delta \mu_s + J_w \Delta \mu_w + I \cdot E \tag{33}$$

which may be transformed into other convenient forms. Whatever the dissipation function, the phenomenological equations are rather bulky, since the full matrix of coefficients relating flows to forces comprises nine terms, six of which are independent by Onsager's symmetry theorem. The determination of six coefficients requires six independent methods of measurement, generally an ungratifying task. We shall consider therefore only the simple case in which membranes permit no water transport and the dissipation function is reduced to[11]:

$$\Phi = J_s \Delta \mu_s + I \cdot E \tag{34}$$

237

In this case, the phenomenological relations may be reduced to the following practical form

$$J_s = \omega \Delta \pi + \frac{t_1}{F} \cdot I$$

$$I = \kappa \left(E + \frac{t_1}{F} \Delta \mu_s \right)$$

(35)

Here, the flow of salt means the flow of the ion which does not participate in the electrode reaction. Thus, if the electrode is an Ag–AgCl electrode, the flow of the cations would be identified with J_s. The coefficient ω is again the salt-permeability coefficient, while the transport number t_1 is the coupling coefficient measured at $\Delta \pi = 0$

$$t_1 = \left(\frac{J_s F}{I} \right)_{\Delta \pi = 0}$$

(36)

The straight coefficient κ is the conductance of the membrane, and may be determined at equal salt concentrations in the adjacent solutions, i.e. when $\Delta \mu_1$ for the cation, and $\Delta \mu_2$ for the anion equal zero.

It is often useful to substitute the reversible electromotive force E by the potential difference $\Delta \psi$ measured with two calomel electrodes. Inserting $\Delta \psi$ into equation (35) we obtain $I = \kappa [\Delta \psi + (1/F)(t_1 \Delta \mu_1 - t_2 \Delta \mu_2)]$ where 1 and 2 denote the cation and anion respectively.

We shall find in the following paragraphs that equations (35) and (36) are useful in the evaluation of the properties of complex membranes.

4. FACILITATED TRANSPORT

4.1. In the discussion on transport across simple membranes, it was assumed that the permeant passes the membrane matrix without interaction. In certain ion-exchange films, it can be assumed that the membrane may be approximated by a system of water-filled capillaries, the behaviour of which is adequately described by the methods of classical colloid chemistry. The permeability of most biological membranes and many polymeric films does not however fit the Helmholtz–Quincke and Gouy model[12] even to a rough approximation. It was found that biological covers may be highly permeable to substances which are of low solubility in the membranes and which should behave essentially as non-permeants. Furthermore, it is well established that the flow does not increase linearly with the concentration difference, but at sufficiently high concentrations it may reach a limiting value, or in general, exhibit flow saturation phenomena. The prevailing explanation of this behaviour is based on the assumption that a biological membrane contains a specific 'carrier' which combines readily with the permeant and facilitates its transport across media in which it dissolves with difficulty. Moreover, the carrier has a finite number of adsorption sites the saturation of which with the permeating substance puts a limit to the facilitation of transport. Although the mechanism of carrier

238

transport is unknown, and saturation phenomena may be observed also in two dimensional lattice models[13], numerous attempts have been made to reproduce facilitated transport with synthetic systems. Several interesting models were described recently by Eisenman *et al.*[14] and by Shean and Sollner[15]. J. Gabbay, in this laboratory[16] obtained facilitated transport of amino-acids across ion-exchange membranes, the carrier being the hydrogen ion. He found that when a zwitterionic amino-acid enters through one surface of a sulphonated resin membrane in the hydrogen form, it reacts according to the scheme:

$$A^{\pm} + RSO_3^- H^+ \longrightarrow AH^+ + RSO_3^- \tag{37}$$

The form which passes readily the membrane is the carrier-substrate complex AH^+. On the other surface, the complex dissociates and liberates the free amino-acid to the outer solution.

The last example serves as a model for the simple carrier mechanism which has been adequately studied by cell physiologists[17].

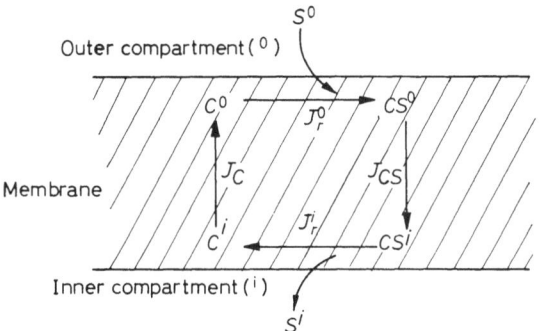

Figure 2. Schematic representation of carrier (C) mediated transport of the solute S. J_r^0 is the rate of adsorption of S^0 to the free carrier C^0 in the outer compartment to give CS^0. J_r^i is the rate of solute desorption in the inner compartment. J_C is the flow of free carrier from the inner to outer compartment. J_{CS} the flow of the solute-loaded carrier from the outer to the inner compartment

We shall consider here the general case in which carrier and substrate combine and dissociate throughout the membrane phase[18].

4.2. The continuity equation for the local change of concentration in every volume element is given by

$$\frac{\partial c_i}{\partial t} = -\frac{\partial J_i}{\partial x} + \nu_i J_r \tag{38}$$

where the diffusional flow J_i proceeds only in the x-direction, J_r is the local rate of chemical reaction (per unit volume) and ν_i the stoichiometric coefficient of the ith component. The reaction considered here is between permeant and carrier

$$C + S \rightarrow CS \tag{39}$$

where J_r is positive for association and negative for dissociation. Thus

$$\left(\frac{\partial S}{\partial t}\right)_x = -\left(\frac{\partial J_s}{\partial x}\right)_t - J_r; \left(\frac{\partial C}{\partial t}\right)_x = -\left(\frac{\partial J_c}{\partial x}\right)_t - J_r$$

and

$$\left(\frac{\partial CS}{\partial t}\right)_t = -\left(\frac{\partial J_{CS}}{\partial x}\right) + J_r$$

(40)

In equation (40), S, C and CS denote the concentrations of substrate, carrier and carrier–substrate complex, respectively. A stationary state is characterized by time independence of all parameters of state, so that every $(\partial c_i/\partial t)_x = 0$. Hence,

$$-\frac{\partial J_s}{\partial t} = J_r; -\frac{\partial J_c}{\partial x} = J_r \text{ and } \frac{\partial J_{CS}}{\partial x} = J_r$$

(41)

Equation (41) leads immediately to some interesting observations on facilitated transport. Adding the second and third statements in equation (41) we obtain

$$\frac{\partial(J_c + J_{CS})}{\partial x} = 0 \text{ or, } J_c + J_{CS} = \text{const.}$$

(42)

Similarly, the addition of the first and third expressions in equation (41) gives

$$\frac{\partial(J_s + J_{CS})}{\partial x} = 0 \text{ or, } J_s + J_{CS} = \text{const.}$$

(43)

Now, facilitated transport is characterized by the requirement that the *total* turnover within the membrane must cancel out—the same substance has to enter and exit from the membrane. That is the association on one side has to be precisely compensated for by the dissociation on the other, or,

$$\int_0^{\Delta x} J_r dx = 0$$

(44)

where 0 and Δx are the boundary values of the x coordinate at the membrane surfaces. Inserting into (44) the values of J_r from (41), we obtain

$$\int_0^{\Delta x} (\partial J_i/\partial x) \, dx = 0 \text{ or } J_i^0 = J_i^{\Delta x}$$

(45)

Since no external flow of carrier or carrier-substrate complex is permissible

$$J_c^0 = J_c^{\Delta x} = 0 \text{ and } J_{CS}^0 = J_{CS}^{\Delta x} = 0$$

(46)

Inserting these expressions into equation (45) we obtain the important conclusion that

$$Jc + Jcs = 0 \tag{47}$$

Equation (47) shows that at every point within the membrane, the flow of carrier is compensated by a counter flow of carrier–substrate complex. The total carrier may, therefore, be regarded as a circulating vehicle transporting substrate from one coast to another. It is further apparent that

$$Js + Jcs = Js^0 = Js^{\Delta x}$$

but since the permeant flow is continuous, $Js^0 = Js = Js^{\text{ext}}$ where Js^{ext} is the total solute flow measurable in the external solution. Thus,

$$Js + Jcs = Js^{\text{ext}} \tag{48}$$

indicating that although the permeant may move as a free component (Js) or as a complex (Jcs), the total amount transported per unit time and unit area is constant throughout the membrane.

4.3. From a thermodynamic point of view, the dissipation of free energy due to the transport of a single component is given by

$$\Phi = Js^{\text{ext}} \Delta \mu_s \tag{49}$$

whatever the mechanism of transport. It is, however, interesting to demonstrate that the detailed treatment of local carrier reaction introduces no changes into equation (48). The local dissipation for a volume element is given by

$$\phi = Js \left(-\frac{d\mu_s}{dx} \right) + Jc \left(-\frac{d\mu_c}{dx} \right) + Jcs \left(-\frac{d\mu_{cs}}{dx} \right) + J_r A \tag{50}$$

where A is the affinity of reaction (39)

$$A = \mu_s + \mu_c - \mu_{cs} \tag{51}$$

Using equations (47), (48), (41) and (50) we obtain

$$\Phi = Js^{\text{ext}} \left(-\frac{d\mu_s}{dx} \right) + Jcs \frac{d}{dx} (\mu_s + \mu_c - \mu_{cs}) + J_r A$$

$$= Js^{\text{ext}} \left(-\frac{d\mu_s}{dx} \right) + Jcs \frac{dA}{dx} + \frac{dJcs}{dx} A \tag{52}$$

$$= Js^{\text{ext}} \left(-\frac{d\mu_s}{dx} \right) + \frac{d}{dx} (JcsA)$$

Equation (52) is readily integrated over the thickness of the membrane to give the total dissipation per unit area

241

$$\Phi = \int \Phi dx = J_S{}^{\text{ext}} \int_0^{\Delta x} \left(-\frac{d\mu_S}{dx}\right) dx + \int_0^{\Delta x} \frac{d}{dx} (J_{CS}A) \, dx$$

$$\tag{53}$$

$$= J_S{}^{\text{ext}}\Delta\mu_S + (J_{CS}{}^{\Delta x} A^{\Delta x} - J_{CS}{}^0 A^0)$$

Taking into account equation (46), equation (53) reduces however to equation (49), i.e., $\Phi = J_S{}^{\text{ext}}\Delta\mu_S$.

$$Q.E.D.$$

4.4. The advantage of a 'microscopic', detailed, treatment lies therefore not in the possibility of overcoming the severe limitations imposed by the 'macroscopic', external view, but in providing means to interpret the relation between flows and forces which the thermodynamic treatment leaves undetermined. Thus, the phenomenological equation for single component facilitated transport is

$$J_S{}^{\text{ext}} = L\Delta\mu_S \tag{54}$$

which does not provide any information on facilitation or saturation flow with increasing concentration. The aim of our 'microscopic' analysis is therefore to make the coefficient L meaningful and to give it an explicit formulation. A fuller treatment is given elsewhere (Blumenthal and Katchalsky)[18]. We shall consider here only a simple case which suffices to illustrate the essential features.

It is assumed that S, C and CS are moving in a matrix or solution in such a manner that the màin frictional resistance to flow is due to interaction with the medium. In this case, hydrodynamic coupling between the components may be neglected and the independent flows can be described by the classical equation of Fick:

$$J_S = D_S \left(-\frac{dS}{dx}\right); J_C = D_C \left(-\frac{dC}{dx}\right) \text{ and } J_{CS} = D_{CS} \left(-\frac{dCS}{dx}\right) \tag{55}$$

where the D_is are constant. From the second and third equations of (55) and equation (47) we obtain

$$-\frac{d}{dx} (D_C C + D_{CS} CS) = 0 \quad \text{or} \quad D_C C + D_{CS} CS = C^{\text{tot}}, \text{ a const.} \tag{56}$$

The first and third equations of (55) inserted into (48) give:

$$J_S{}^{\text{ext}} = D_S \left(-\frac{dS}{dx}\right) + D_{CS} \left(-\frac{dCS}{dx}\right)$$

and, upon integration across the membrane,

$$J_S^{ext} = D_S \frac{\Delta S}{\Delta x} + D_{CS} \frac{\Delta CS}{\Delta x} \tag{57}$$

An appreciable simplification is obtained if, following the procedure of Wilbrandt and Rosenberg, it is assumed that surface equilibria exist between permeant and carrier. Then

$$\frac{C_0 S_0}{CS_0} = \frac{C_{\Delta x} S_{\Delta x}}{CS_{\Delta x}} = K_s \tag{58}$$

where K_s is an equilibrium constant. Using equations (56) and (58), we obtain

$$\Delta CS = CS_0 - CS^{\Delta x} = C^{tot} K_S D_C \frac{(S_0 - S_{\Delta x})}{(K_S D_C + S_0 D_{CS})(K_S D_C + S_{\Delta x} D_{CS})}$$

$$= C^{tot} K_S D_C \frac{\Delta S}{(K_S D_C + S_0 D_{CS})(K_S D_C + S_{\Delta x} D_{CS})} \tag{59}$$

The final expression is therefore

$$J_S^{tot} = \left(D_S + \frac{K_S D_C D_{CS} C^{tot}}{(K_S D_C + S_0 D_{CS})(K_S D_S + S_{\Delta x} D_{CS})} \right) \frac{\Delta S}{\Delta x} \tag{60}$$

The first term in equation (60) is that of regular transport, $D_S (\Delta S/\Delta x)$, and requires no further comment. It is in the second term that the facilitation is expressed: If it is assumed that $S_{\Delta x} = 0$, facilitated transport will be given by

$$\frac{D_{CS} C^{tot} S_0}{(K_S D_C + S_0 D_{CS}) \Delta x}$$

which for $S_0 D_{CS} \ll K_S D_C$ is linear in S_0. On the other hand, for $S_0 D_{CS} \gg K_S D_C$, the facilitated flow reaches the limiting value of $C^{tot}/\Delta x$ and becomes saturated.

To compare equations (60) and (54), we may write

$$\Delta \mu_s = (\mu_s^0 + RT \ln S_0) - (\mu_s^0 + RT \ln S_{\Delta x}) = RT \ln \frac{S_0}{S_{\Delta x}} = RT \frac{\Delta S}{\bar{S}} \tag{61}$$

where $\bar{S}$ is defined by equation (61) and approaches $\bar{S} = (S_0 + S_{\Delta x})/2$ for small values of S. Thus,

$$J_S^+ = \frac{RTL \, \Delta S}{\bar{S}} = \left(\frac{D_S \bar{S}}{\Delta x} + \frac{K_S D_C D_{CS} \, C^{tot} \bar{S}}{\Delta x \, (K_S D_C + S_0 D_{CS})(K_S D_C + S_{\Delta x} D_{CS})} \right) \frac{\Delta S}{\bar{S}}$$

and hence

$$L = \frac{RT}{\Delta x} \bar{S} \left(D_S + \frac{K_S D_C D_{CS} C^{\text{tot}}}{(K_S D_C + S_0 D_{CS})(K_S D_C + S_{\Delta x} D_{CS})} \right) \qquad (62)$$

Equation (62) concludes our treatment of facilitated transport. It shows how the phenomenological coefficient may be interpreted and demonstrates that it comprises both facilitation and saturation properties of the process.

5. COMPLEX MEMBRANES[19]

5.1. For many purposes, it is sufficient to study the behaviour of single membranes. There exist, however, industrial processes, such as desalination by electrodialysis, which are based on the utilization of membrane stacks, composed of alternating positive and negative membranes, the mastery of which requires an understanding of the operation of composite membrane systems. The evaluation of the rules of transport through an array of membrane elements is particularly important in biology, where every cellular and tissue cover is a complex system made up of several layers characterized by different permeabilities for electrolytes and nonelectrolytes. In this section, the significance of some of the basic features of complex membranes composed of elements arranged in series will be evaluated. We shall consider in some detail the well-investigated 'bilayer' composed of two perm-selective elements, one of which carries fixed positive charges and has a selective permeability to anions, and the other of which is negatively charged and has a preferred cation permeability.

It can be readily shown that when a regime of stationary flow is attained and no chemical interaction takes place either in the membranes or in the intermembrane space, the flows become constant and assume the same value throughout the system. At the same time, the overall forces acting across the composite system may be decomposed into partial forces acting on each membrane element and the total force is found to be the sum of all individual forces.

With these simple concepts, we may approach the treatment of a bilayer in the special case that no water flow accompanies the transport of salt and electrical current across the complex membrane. If we denote all parameters and coefficients for one layer by α and those for the other by β we may apply directly equations (53) with the restrictions of stationary flow described in the previous paragraph.

$$J_s^{\alpha} = \omega_{\alpha} \Delta \pi_{\alpha} + \frac{t_1^{\alpha}}{F} I_{\alpha} \qquad J_s^{\beta} = \omega_{\beta} \Delta \pi_{\beta} + \frac{t_1^{\beta}}{F} I_{\beta} \qquad (63)$$

$$I^{\alpha} = \kappa_{\alpha} \left[\Delta \psi_{\alpha} + \frac{1}{F} (t_1^{\alpha} \Delta \mu_1^{\alpha} - t_2^{\alpha} \Delta \mu_2^{\alpha}) \right];$$

$$I^{\beta} = \kappa_{\beta} \left[\Delta \psi_{\beta} + \frac{1}{F} (t_1^{\beta} \Delta \mu_1^{\beta} - t_2^{\beta} \Delta \mu_2^{\beta}) \right] \qquad (64)$$

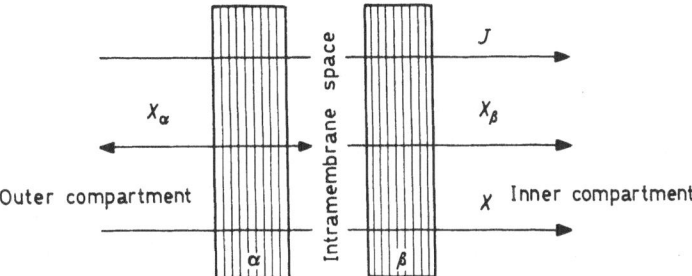

Figure 3. Scheme of a composite membrane composed of a series array of membrane elements α and β. The total force X acting across the composite membrane is additive in the forces acting on each element $X = X^\alpha + X^\beta$. In stationary cases the flows J are continuous across the system

$$J_s{}^\alpha = J_s{}^\beta; \qquad I^\alpha = I^\beta \tag{65}$$

$$\Delta\psi^{tot} = \Delta\psi_\alpha + \Delta\psi_\beta; \qquad \Delta\pi^{tot} = \Delta\pi_\alpha + \Delta\pi_\beta \tag{66}$$

$$\Delta\mu_1^{tot} = \Delta\mu_1{}^\alpha + \Delta\mu_1{}^\beta; \qquad \Delta\mu_2{}^{tot} = \Delta\mu_2{}^\alpha + \Delta\mu_2{}^\beta$$

The implications of equations (66) may be shown as follows: let us consider explicitly the intramembrane space, the volume of which may be very small but nonvanishing. Let the concentration of the permeating salt in the intermembrane space be c^*.

$$\Delta\pi_\alpha = 2RT\,(c_1 - c^*); \qquad \Delta\pi_\beta = 2RT\,(c^* - c_2)$$

and

$$\Delta\pi^{tot} = \Delta\pi_\alpha + \Delta\pi_\beta = 2RT\,(c_1 - c_2). \tag{67}$$

Similarly,

$$\Delta\mu_1{}^\alpha = RT\ln\frac{c_1}{c^*}; \qquad \Delta\mu_1{}^\beta = RT\ln\frac{c^*}{c_2}$$

and

$$\Delta\mu_1{}^{tot} = \Delta\mu_1{}^\alpha + \Delta\mu_1{}^\beta = RT\ln\frac{c_1}{c_2} \tag{68}$$

An equivalent expression holds for $\Delta\mu_2$.

Inserting equations (63) into (65) we obtain,

$$2RT\,\omega_\alpha\,(c_1 - c^*) + \frac{t_1{}^\alpha}{F}I = 2RT\,\omega_\beta\,(c^* - c_2) + \frac{t_2{}^\beta}{F}I \tag{69}$$

and hence

$$c^* = \frac{\omega_\alpha c_1 + \omega_\beta c_2}{\omega_\alpha + \omega_\beta} + \frac{(t_1{}^\alpha - t_1{}^\beta)}{2RT\,(\omega_\alpha + \omega_\beta)}\frac{I}{F} \tag{70}$$

Equation (70) shows some remarkable features which deserve consideration. The most important is that the intramembrane concentration c^* is seen to be a function of the electrical current. In the case of the two semi-infinite compartments, adjacent to the simple membrane considered

245

previously, we could assume that the external parameters of the state were independent of the flows. Now we encounter a small finite volume the parameters of which are functions of the flow pattern. When $I = 0$, $c*$ assumes the value

$$\bar{c*} = \frac{\omega_\alpha c_1 + \omega_\beta c_2}{\omega_\alpha + \omega_\beta} \tag{71}$$

which is a weighted average of the external concentrations. If $t_1^\alpha - t_1^\beta$ is positive, as is the case in *Figure 3*, $c*$ will increase with increasing I until a breakthrough of the permselectivity occurs. More interesting however, is the case in which I decreases and assumes negative values. Ultimately, a limiting value I_0 will be reached which makes $c*$ zero:

$$0 = \bar{c*} = \frac{t_1^\alpha - t_1^\beta}{2RT(\omega_\alpha + \omega_\beta)} \frac{I_0}{F}$$

or

$$\frac{I_0}{F} = \frac{2RT\,\bar{c*}\,(\omega_\alpha + \omega_\beta)}{t_1^\alpha - t_1^\beta} \tag{72}$$

Equation (72) is a simplified expression for the desalination process in a cell with walls made of oppositely charged permselective membranes. There is, however, another important aspect revealed by equation (72). Since no negative concentrations are known, I_0 is the most negative current which can flow through the membrane systems (i.e. before breakdown of the water molecules occurs). Thus positive electro-osmotic forces may increase both I and $c*$; negative electro-osmotic forces, however lead to a *limiting flow of current* which does not change with decreasing E. The presence of the intra-membrane space transforms therefore a composite membrane into a rectifier. Even if Ohm's law is expected to hold in the range of layer positive Is, it breaks down when $I \to I_0$.

Equations (70), (71) and (72) may be condensed to a single expression

$$c* = \bar{c*}\left(1 + \frac{I}{I_0}\right) \tag{73}$$

5.2. The rectification properties are cast into a quantitative form by utilizing equations (64), (65) and (68).

$$\frac{I}{\kappa_\alpha} = \Delta\psi_\alpha + \frac{1}{F}(t_1^\alpha \Delta\mu_1^\alpha - t_2^\alpha \Delta\mu_2^\alpha) = \Delta\psi_\alpha + \frac{RT}{F}(t_1^\alpha - t_2^\alpha)\ln\frac{c_1}{c*}$$

$$\frac{I}{\kappa_\beta} = \Delta\psi_\beta + \frac{1}{F}(t_1^\beta \Delta\mu_1^\beta - t_2^\beta \Delta\mu_2^\beta) = \Delta\psi_\beta + \frac{RT}{F}(t_1^\beta - t_2^\beta)\ln\frac{c*}{c_2}$$

Upon adding these expressions and noting that $t_1^\alpha + t_2^\alpha = t_1^\beta + t_2^\beta = 1$ we obtain

$$I\left(\frac{1}{\kappa_\alpha} + \frac{1}{\kappa_\beta}\right) = \Delta\psi + \frac{RT}{F}[(t_1^\alpha - t_2^\alpha)\ln c_1 - (t_1^\beta - t_2^\beta)\ln c_2]$$

$$+ \frac{2RT}{F}(t_1^\beta - t_1^\alpha)\ln c*$$

or, inserting (73),

$$\Delta\psi = -\frac{RT}{F}[(t_1{}^\alpha - t_2{}^\alpha)\ln c_1 - (t_1{}^\beta - t_2{}^\beta)\ln c_2]$$

$$+ \frac{2RT}{F}(t_1{}^\alpha - t_1{}^\beta)\ln \overline{c^*} + \frac{2RT}{F}(t_1{}^\alpha - t_1{}^\beta)\ln\left(1 + \frac{I}{I_0}\right) + I\left(\frac{1}{\kappa_\alpha} + \frac{1}{\kappa_\beta}\right) \quad (74)$$

The first two terms on the right-hand side of the equation are independent of the electrical current I and may be identified with the 'resting potential' $\Delta\psi_0$. $(1/\kappa_\alpha) + (1/\kappa_\beta) = \rho$ is the total electrical resistance of both membranes (excluding the intramembrane space). Inserting $\Delta\psi_0$ and ρ into equation (74) a lucid and suggestive equation is obtained:

$$\Delta\psi - \Delta\psi_0 = (t_1{}^\alpha - t_1{}^\beta)\frac{2RT}{F}\ln\left(1 + \frac{I}{I_0}\right) + I\cdot\rho \quad (75)$$

At higher values of I, the contribution of the logarithmic term becomes of lesser importance, so that $\Delta\psi - \Delta\psi_0 \rightarrow I\cdot\rho$, i.e., Ohm's law holds once more. On the other hand, for small values of I which tend to I_0, the term $I\cdot\rho$ becomes negligible and

$$\Delta\psi - \Delta\psi_0 = \frac{2RT}{F}\ln\left(1 + \frac{I}{I_0}\right) \quad (76)$$

In most synthetic ion-exchange membranes, ρ is sufficiently small to make $I\cdot\rho$ negligible over a wider range. It is clear that for $I\rightarrow -I_0$, $\Delta\psi$ tends to $-\infty$, or, expressed in another way, $-I_0$ is the limiting value of the current in the rectification processes of the membrane system.

5.3. Equation (76), which has a form similar to Tafel's equation for overvoltage, was found to represent adequately the experimental data found in the literature[20] and accumulated in this laboratory[21, 22]. *Figure 4*

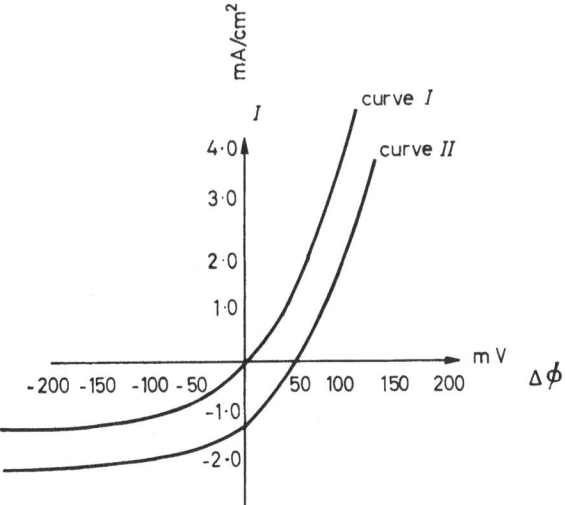

Figure 4. Rectification with a double membrane composed of cation and anion permselective membrane elements. Curve *I* for equal salt concentrations in outer and inner compartments. (Note that at $I = 0$ at $\Delta\psi = 0$.) Curve *II* for unequal salt concentrations. (Calculated by Richardson[19])

gives an example of an experimental set of $\Delta\psi$ *versus* I values and their analysis by equation (76)†.

It is rather interesting to observe that many biological membranes exhibit a rectification behaviour which closely resembles that of a synthetic bilayer. The dependence of current on potential for non-excited muscle and nerve membranes is represented in *Figure 5*.

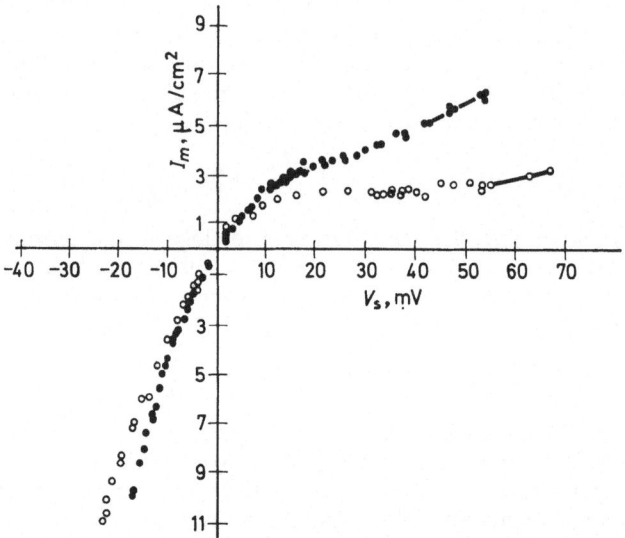

Figure 5. Relation between electrotonic membrane potential and current density for fibre 1 (○) and fibre 4 (●) [R. H. Adrian and W. H. Frey Gang, *J. Physiol.* **163**, 109 (1962)]

Although there is no reason to assume that biological membranes are structurally related to the composite permselective layer described above, it is plausible to assume that their rectification properties rest on an electrical anisotropy of cell or tissue covers. The study of membrane anisotropy is of fundamental importance for the thermodynamic grasp of active transport and will occupy our attention in the following paragraphs. The feature which emerges from the present analysis is that electrophysiological rectification studies may be a useful tool for the interpretation of anisotropic distribution of charges in complex biological structures.

An insight into the complex structure of organismic covers may be obtained also from the resting potential $\Delta\psi_0$ and its dependence on concentration. For the sake of lucidity we shall assume that both layers α and β of *Figure 3* are highly permselective. In thise case, $t_1{}^\alpha = 1$, $t_2^\alpha \simeq 0$ while $t_1{}^\beta \simeq 0$, $t_2{}^\beta \simeq 1$. Hence:

$$\Delta\psi_0 = \frac{RT}{F} [-\ln c_1 c_2 + \ln (\overline{c^*})^2] = \frac{RT}{F} \ln \frac{(\omega_\alpha c_1 + \omega_\beta c_2)^2}{c_1 c_2 (\omega_\alpha + \omega_\beta)^2} \qquad (77)$$

† In the calculation given above we neglected the dependence of the ωs, ts and κs on the average salt concentration. In certain cases this dependence is very pronounced and should be considered explicitly. A detailed analysis of these properties based on a frictional model can be found in the recent thesis of Richardson[19].

Equation (77) should be compared with the well-known equation of Nernst and Planck

$$\Delta\psi_0 = \frac{2RT}{F} \ln \frac{c_1}{c_2} \tag{78}$$

or with the equation of Taylor for a liquid junction potential. If we keep c_2 constant and vary c_1, equation (78) predicts a linear dependence of $\Delta\psi_0$ on $\ln c_1$ with a constant slope of RT/F, which was verified experimentally in numerous cases. On the other hand, $(\partial\Delta\psi/\partial \ln c_1)_{c_2}$ is neither constant nor equal to RT/F, but given by the expression

$$\left(\frac{\partial\Delta\psi}{\partial \ln c_1}\right)_{c_2,\, T} = \frac{RT}{F} \frac{\omega_\alpha c_1 - \omega_\beta c_2}{\omega_\alpha c_1 + \omega_\beta c_2} \tag{79}$$

It is clear that when $\omega_\alpha c_1 \gg \omega_\beta c_2$, $(\partial\Delta\psi/\partial \ln c_1)_{c_2} = RT/F$, corresponding to equation (78). When, however, $\omega_\beta c_2 \gg \omega_\alpha c_1$, the slope is $-RT/F$ and there is a point, $\omega_\alpha c_1 = \omega_\beta c_2$, at which $\partial\Delta\psi/\partial \ln c_1 = 0$, or the potential does not change with $\ln c_1$. This remarkable behaviour is depicted in *Figure 6* from Richardson's work.

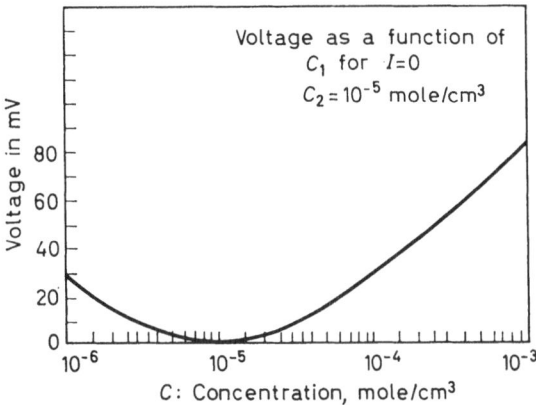

Figure 6. The resting potential $\Delta\psi$ for a complex membrane composed of two highly perm-selective elements. The concentration on side 2 is kept constant $(c_2 = 10^{-5}$ moles/cm³) while the concentration c_1 is varied. The flow of electrical current is zero $(I = 0)$. Calculated by Richardson[19]

It is rather gratifying to find that those biological membranes which show a non-linear dependence of electrical current on potential exhibit also a non-linear dependence of potential on the logarithm of ion concentration. *Figure 7* is taken from Tasaki's study of the nerve membrane of perfused axons of the squid. The data resemble those found in synthetic bilayers and described theoretically by equations (77) and (79).

6. COMPLEX MEMBRANES WITH CHEMICAL INTERACTION

6.1. Anisotropic membrane structures, stressed in the previous section, are of particular importance for interpretation of the coupling phenomena underlying active transport. As pointed out in section **2.3**, active transport

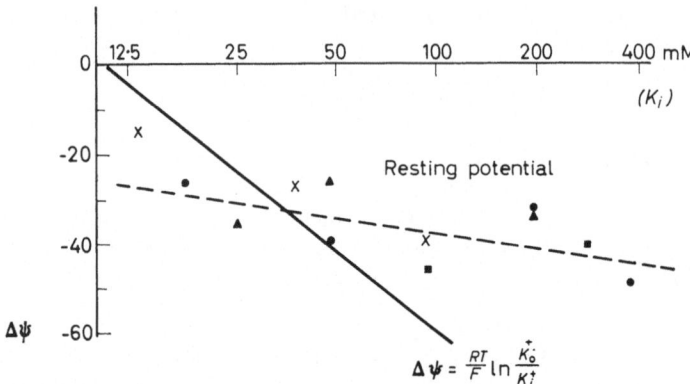

Figure 7. Effect upon the resting potential of diluting the K-perfusing fluid with isotonic sucrose solution. The ratio of Na^+/K^+ concentration was fixed at $1/10$. The abscissa represents the K^+ concentration in the perfusing solution. (Temp: 22°C.) (From Tasaki and Takenaka)

requires the possibility of coupling between diffusional and chemical flows. Some sixty years ago, Pierre Curie[23] announced, however, a principle which cast doubt on the physical possibility of such coupling. Curie's principle, introduced by Prigogine[24] into non-equilibrium thermodynamics, states that in isotropic media, coupling between flows can take place only if the flows are of the same tensorial order. Since a chemical reaction is scalar while a diffusional flow is vectorial, no active transport could occur within an isotropic medium. On the other hand, in an anisotropic space coupling is possible, and hence the anisotropic structure of biological membranes is an essential condition for the utilization of metabolic processes to drive selective diffusional flows.

Under stationary conditions there exists however another coupling possibility, denoted by Prigogine[25] as 'stationary state coupling', which does not require anisotropy and does not violate Curie's principle even in isotropic media. Stationary state coupling is not sufficient to account for the rapid, non-stationary relaxation processes going on in the living cell; it is, however, satisfactory for the treatment of chemical reactions going on in complex systems of synthetic membranes, to be discussed below.

6.2. Consider a narrow space, bounded by two permselective membranes, facing two corresponding external semi-infinite compartments. Each of the compartments contains a solution of a uni-univalent salt, with cations 1 and anions 2, and a non-electrolyte. The non-electrolyte may undergo catalytic breakdown resulting in the formation of ions 1 and 2. The catalyst is, however, confined to the narrow intramembrane space, so that a chemical process takes place only there. Such a system was studied by Blumenthal *et al.*[26], who used amides of organic acids as the non-electrolyte, which, upon hydrolytic breakdown, forms ammonium and carboxylate ions.

The change in ion concentration in the narrow space is given by the evident equation:

$$dc_1^*/dt = J_1^\alpha - J_1^\beta + J_r; \qquad dc_2^*/dt = J_2^\alpha - J_2^\beta + J_r$$

where J_1 and J_2 are the flows of cation and anion per unit area of membrane,

250

and J_r is the rate of the chemical process, calculated per unit area of the inner space.

A stationary state will be established when $dc_1{*}/dt = dc_2{*}/dt = 0$ and it is then that the ionic flows become coupled with the chemical process

$$J_1{}^\alpha - J_1{}^\beta = -J_r \qquad (80.1)$$

$$J_2{}^\alpha - J_2{}^\beta = -J_r \qquad (80.2)$$

It is worth noting that while J_1 and J_2 undergo a discontinuity, equal to $-J_r$, when passing through the inner cell, the flow of electricity remains continuous, for equations (80.1), (80.2) show that

$$I^\alpha = (J_1{}^\alpha - J_2{}^\alpha)\,F = I^\beta = (J_1{}^\beta - J_2{}^\beta)\,F = I \qquad (81)$$

We assume that ion 2 participates in the reversible electrode reaction, while the flow of ion 1, J_1, is considered as the flow of salt J_s. Equation (81) may therefore be rewritten

$$J_s{}^\alpha - J_s{}^\beta = -J_r$$

Inserting the expression J_s from equation (35), we obtain

$$\omega_\alpha \Delta\pi_\alpha + (t_1{}^\alpha/F)\,I - \omega_\beta \Delta\pi_\beta - (t_1{}^\beta/F)\,I = -J_r \qquad (82)$$

Assuming, again, ideal behaviour we may write:

$$\Delta\pi_\alpha = 2RT\,(c_1 - c{*}) \quad \text{and} \quad \Delta\pi_\beta = 2RT\,(c{*} - c_2)$$

which upon insertion into (82) gives for $c{*}$

$$c{*} = \frac{\omega_\alpha c_1 + \omega_\beta c_2}{\omega_\alpha + \omega_\beta} + \frac{(t_1{}^\alpha - t_1{}^\beta)\,I/F}{2RT\,(\omega_\alpha + \omega_\beta)} + \frac{J_r}{2RT\,(\omega_\alpha + \omega_\beta)}$$

$$= \overline{c{*}} + \frac{1}{2RT\,(\omega_\alpha + \omega_\beta)}\left[(t_1{}^\alpha - t_1{}^\beta)\frac{I}{F} + J_r\right] \tag{83}$$

Equation (83) is an evident extension of equation (70); here, too, rectification is expected and there should exist a limiting current $-I_0$ at which the salt concentration $c{*}$ in the inner cell becomes zero. The magnitude of I_0 is, however, dependent on the rate of reaction, so that variation in the activity of the catalyst will shift I_0 to higher or lower values. If the bilayer may be regarded as a rectifying diode, the intramembrane chemical reaction may be regarded as a 'grid' the catalytic activity of which may amplify electrical flows. . . .

Following the procedure of section **5.1**, we may now evaluate the dependence of I on $\Delta\psi$ for a bilayer with intramembrane chemical reaction. The

251

basic equation

$$\Delta\psi = -\frac{RT}{F}\left[(t_1{}^\alpha - t_2{}^\alpha)\ln c_1 - (t_1{}^\beta - t_2{}^\beta)\ln c_2\right]$$

$$+ \frac{2RT}{F}(t_1{}^\alpha - t_1{}^\beta)\ln c^* + I\cdot\rho \qquad (84)$$

holds in the present case, as well as in that discussed in section **5.2**. The main difference lies in the value of c^* which is now given by equation (83). Equation (84) can be cast in different useful forms: thus we may write c^* as:

$$c^* = \left[\bar{c}^* + \frac{J_r}{2RT(\omega_\alpha + \omega_\beta)}\right]\left(1 + \frac{I}{I_0}\right) \qquad (85)$$

The dependence of I on $\Delta\psi$ then becomes identical with that given in equation (75)

$$\Delta\psi - \Delta\psi_0 = \frac{2RT(t_1{}^\alpha - t_1{}^\beta)}{F}\ln\left(1 + \frac{I}{I_0}\right) + I\cdot\rho \qquad (86)$$

It should however be borne in mind that, in equation (86), the limiting current I_0 is a function of the chemical process (J_r) and that $\Delta\psi_0$—the potential at zero current—differs from that discussed above. If we denote the resting potential at zero chemical reaction by $\Delta\psi_0{}^0$ it is easily shown that

$$\Delta\psi_0 = \Delta\psi_0{}^0 + \frac{2RT}{F}\ln\left(1 + \frac{J_r}{2RT\bar{c}^*(\omega_\alpha + \omega_\beta)}\right) \qquad (87)$$

Now, if the concentrations on both sides of the composite membrane are equal, i.e., $c_1 = c_2 = c$, $\bar{c}^*$ also become c and $\Delta\psi_0{}^0 = 0$, whatever may be the values of $t_1{}^\alpha$ and $t_2{}^\beta$. On the other hand, the resting potential does not vanish but becomes

$$\Delta\psi_0{}_{(c_1=c_2)} = \frac{2RT}{F}\ln\left(1 + \frac{J_r}{2RT\bar{c}(\omega_\alpha + \omega_\beta)}\right) \qquad (88)$$

Thus the biological observation that a resting potential is maintained even when the salt concentrations are equal on both sides of the membrane is an indication that an ion-forming or ion-reducing process goes on in the membrane.

Let us finally insert c^* from equation (83) into (84) and introduce $\Delta\psi_0{}^0$. The expression obtained is

$$\Delta\psi - \Delta\psi_0{}^0 = \frac{2RT}{F}(t_1{}^\alpha - t_1{}^\beta)\ln\left[1 + \frac{(t_1{}^\alpha - t_1{}^\beta)}{2RT\bar{c}^*(\omega_\alpha + \omega_\beta)}\cdot\frac{I}{F}\right.$$

$$\left. + \frac{J_r}{2RT\bar{c}^*(\omega_\alpha + \omega_\beta)}\right] + I\cdot\rho \qquad (89)$$

There is a special interest in cases of slow flows, for which a formulation on the basis of non-equilibrium thermodynamics is applicable. In these cases, both

$$\frac{t_1{}^\alpha - t_1{}^\beta}{2RT\bar{c}^*\,(\omega_\alpha + \omega_\beta)}\,\frac{I}{F} \quad \text{and} \quad \frac{J_r}{2RT\bar{c}^*\,(\omega_\alpha + \omega_\beta)}$$

are smaller than unity and the logarithmic term may be expanded to give

$$\Delta\psi - \Delta\psi_0{}^0 = I\left[\rho + \left(\frac{t_1{}^\alpha - t_1{}^\beta}{F}\right)^2 \frac{1}{\bar{c}^*\,(\omega_\alpha + \omega_\beta)}\right]$$

$$+ J_r\,\frac{t_1{}^\alpha - t_1{}^\beta}{F\bar{c}^*\,(\omega_\alpha + \omega_\beta)} = I \cdot R_{11} + J_r \cdot R_{12} \qquad (90)$$

Equation (90) shows that if the flows are sufficiently slow, the electrical force $\Delta\psi - \Delta\psi_0{}^0$ becomes a linear function of the flows I and J_r. Moreover, it demonstrates that there exists not only a straight coefficient—the resistance, R_{11}, which relates the potential to the electric current—but that also we may expect the existence of a non-vanishing coupling coefficient R_{12}, which relates $\Delta\psi - \Delta\psi_0{}^0$ to the chemical flow, J_r.

According to the structure of the composite membrane, which determines whether $t_1{}^\alpha - t_1{}^\beta \gtrless 0$, R_{12} may be either positive or negative. On the other hand, as required by the rules of the thermodynamics of irreversible processes, R_{11} is positive definite. It will be further observed that, in accord with Curie's principle, no coupling is possible if the membrane system is isotropic, for in this case $t_1{}^\alpha = t_1{}^\beta$ and the coupling coefficient R_{12} vanishes.

Equation (90) may be regarded as a special case of an equation proposed by Kedem[27] for the description of active transport:

$$X_i = \Sigma J_k R_{ik} + J_r R_{ir} \qquad (91)$$

This expression indicates that the force X_i is related linearly to all the diffusion flows J_k passing through the system, in accordance with the phenomenological equation (18). The novel feature is the chemico-diffusional coefficient R_{ir} which makes active transport possible in anisotropic systems. In the last section we shall make use of equation (91) for some models of active transport worked out in this laboratory.

7. OBSERVATIONS ON CARRIER MEDIATED ACTIVE TRANSPORT

7.1. The discussion of active transport is furthest from the main subject of this lecture. Its inclusion is justified in that it provides the possibility of introducing additional aspects of the thermodynamic description of membrane behaviour and because of the possible relation of active transport to macromolecular contractility within active membranes.

It is not possible to sketch even in outline the scope of biological phenomena based on active transport. It embraces the operation of tissues and cells, is closely related to the metabolic transformations, and encompasses the exchange of electrolytes and non-electrolytes as well as the transport of liquids and gases. Although it is not established unequivocally whether all cases of active transport described in the literature are membrane-bound phenomena, there are several cases where chemical-diffusional coupling was clearly shown to be present in membranes. The best investigated case is that of red blood cell membranes which will serve as a model for further discussion. It is well known that red blood cells actively accumulate potassium within the cell and expel sodium into the surroundings. If the cells are haemolysed carefully, empty cells free of haemoglobin, called ghosts, may be obtained. These are essentially osmometric sacks surrounded by intact cell membranes. The remarkable property of erythrocyte ghosts is that despite their emptiness, they may accumulate K^+ and expel sodium, as long as ATP is present within the cells. This observation indicates that at least in the erythrocytes, active transport of cations is carried out by the membrane—as long as the hydrolysis of ATP provides the energy for the process.

From the point of view of our previous discussion, it is rather interesting that active transport in the erythrocyte is based on a chemical anisotropy of the membrane. Electron micrographs of the membranes do not reveal any visual anisotropy and seem to substantiate the 'unit membrane' concept, which supposes that the membrane is composed of a lipid bilayer, the external and internal surfaces of which are covered with protein layers. Physiological studies indicate, however, that the molecular composition is different on both surfaces. Several years ago, Glynn, Post and their co-workers[28] showed that the ATPase activity of the membrane requires the presence of both Na^+ and K^+ in the reaction medium. Later, Whittam[29] showed in haemolytic experiments that the site of action of the ions is different: while Na^+ and ATP must be present within the cell, K^+ must be in the external solution to enable enzymatic breakdown of ATP to take place. The evidence is rather convincing that Na^+ combines with the inner membrane surface while K^+ binds selectively to the outer surface, and it is this anisotropic binding which permits active transport to proceed.

There are indications that the operation of other biological membranes is also related to structural anisotropy. Thus the exciting technique of Baker, Hodgkin and Shaw[30] in England and of Tasaki et al.[31] in the U.S.A. led to the preparation of relatively pure nerve membranes from the axon of squids. These tubular membranes are 'active' since they are excitable and are capable of numerous responses to an external stimulus. If a proteolytic enzyme is applied externally to such membranes it has little influence on the excitability of the preparation. On the other hand, enzymatic attack on the internal membrane proteins causes rapid and irreversible damage with abolition of excitability. Another anisotropic effect is shown by the fish poison tetrodotoxin. While the application of nanomoles (10^{-9} moles) of tetradotoxin to the external surfaces of the squid axon inhibits excitability, the introduction of the poison into the intramembrane liquid leaves the membrane intact. These physiological findings, together with the electrical rectification and potential dependence on concentration discussed previously support the view that active transport takes place in complex membrane systems, with anisotropic, vectorial properties.

There is additional evidence that transport in biomembranes make use of carrier facilitation. It is not only that saturation phenomena are found to be prevalent, but recent experiments of Glynn[32] on isotope exchange in red blood cell membranes are readily explicable on the assumption that transport of sodium is based on the shuttling of a carrier back and forth in the membrane. The following discussion has therefore to make use of all the conceptual framework developed above.

7.2. The formal thermodynamic description of the sodium and potassium flows coupled with the metabolic process gives for the dissipation function

$$\Phi = J_{Na}\Delta\tilde{\mu}_{Na} + J_K\Delta\tilde{\mu}_K + J_r A \tag{92}$$

If the flows of sodium and potassium may be regarded as representing a true ion-exchange process, i.e. $J_{Na} = -J_K$, then the dissipation function reduces to the interesting form

$$\Phi = J_{Na} (\Delta\tilde{\mu}_{Na} - \Delta\tilde{\mu}_K) + J_r A$$
$$= J_{Na}\Delta\mu_{exch} + J_r A \tag{93}$$

In reality, the two flows are not exactly equal in magnitude—but for our schematic representation, we shall consider the consequences of equation (93) for the simple case of equal flows.

The new force $\Delta\mu_{exch}$ which appears in equation (93) is related to the ion distribution coefficient Γ

$$\Gamma = \frac{c_K^i/c_{Na}^i}{c_K^0/c_{Na}^0}$$

by the evident expression

$$\Delta\mu_{exch} = -RT\ln\Gamma \tag{94}$$

The phenomenological equations corresponding to equation (93) are

$$J_{Na} = L_{11}\Delta\mu_{exch} + L_{12}A$$
$$J_r = L_{21}\Delta\mu_{exch} + L_{22}A \tag{95}$$

where L_{11} and L_{22} are the straight coefficients, and L_{12} and L_{21} are the coupling coefficients expected to obey Onsager's relation $L_{12} = L_{21}$. For resting cells $J_{Na} = 0$ and hence

$$\Delta\mu_{exch} = -\frac{L_{12}}{L_{11}} A. \tag{96}$$

Inserting the ion distribution coefficient Γ from equation (94) we obtain the important relation

$$\Gamma = \exp\left(\frac{L_{12}}{L_{11}}\frac{A}{RT}\right) \tag{97}$$

which shows that if the coupling coefficient $L_{12} \neq 0$, a chemical reaction may maintain an unequal ion distribution across an anistropic membrane.

The value of Γ for human red-blood cells at 37°C is $\sim$220, which is larger by two orders of magnitude than the selectivity coefficients found with technical ion-exchangers.

When the erythrocytes are cooled to 0°C the rate of the chemical process is reduced appreciably and $\Delta\mu_{exch}$ tends to zero. On heating the cells again to 37°C, an ion flow sets in which re-establishes the original distribution. Now, the initial value of J_{Na} (at $\Delta\mu_{exch} = 0$) may be readily related to the rate of the chemical process

$$\left(\frac{J_{Na}}{J_r}\right)_{\Delta\mu_{exch}=0} = \frac{L_{12}}{L_{22}} \qquad (98)$$

Since an equilibrium mixture of ATP, ADP and inorganic phosphate could be introduced into a haemolysing cell, it is possible to test the membrane behaviours at $A = 0$. In this case

$$\left(\frac{J_r}{J_{Na}}\right)_{A=0} = \frac{L_{21}}{L_{11}} \qquad (99)$$

and if the Onsager relation holds, we should obtain from (96) and (99)

$$\left(\frac{J_r}{J_{Na}}\right)_{A=0} = -\left(\frac{\Delta\mu_{exch}}{A}\right)_{J_{Na}=0} \qquad (100)$$

Relations of the type given by equation (100) were studied by Blumenthal et al.[26] on synthetic membrane systems comprising a chemical process. A similar analysis is, however, still unavailable for the test of thermodynamic theory of active transport in biological membranes.

7.3. As pointed out previously, there is no possibility of making any thermodynamic statement on the phenomenological cofficients L_{ij} and their dependence on the parameters of state. Even the orthodox thermodynamicist, who follows precisely the commandment 'thou shalt have no graven image', is compelled to construct models which would allow quantitative correlations between flows and forces in active membranes, beyond the Onsager relation. On the basis of existing data, and following the pioneering work of several biophysical groups, we shall consider herewith an oversimplified model, which does not pretend to describe adequately all experimental findings but which is readily analysed by a physical chemist. It leans heavily on the treatment of Rosenberg and Wilbrandt, who separate the inner reaction from the outer reaction on the internal and external membrane surfaces.

It is assumed that ion transport is based on a carrier molecule C which shuttles in the membrane. The carrier may be free, and has then a selective affinity for potassium ions, which are transported as the complex CK; or it may be phosphorylated by ATP to the form CP, acquiring a strong affinity to sodium ions which are transported as a $CPNa$ complex. Figure 8 shows clearly that the phosphorylation reaction J_r^i takes place on the inner surface $(CK^i + \text{ATP} \rightarrow CPK^i + \text{ADP})$ while dephosphorylation take[s] place on the outer surface $CPK^0 \rightarrow CK^0 + \text{P}_{\text{inorganic}}$. Ion-exchange reaction[s] take place on both surfaces $(CPK^i + \text{Na}^i \rightarrow CPNa^i + \text{K}^i$ and $CPNa^0 + \text{K}^0 \rightarrow CPK^0 + \text{Na}^0)$, the overall result being transport of Na^+ to the external solution by the carrier flow J_{CPNa}, and an influx of K^+ by the carrier flow J_{CK}. The overall process is that of ion exchange, although counter gradient flow is allowed by the chemical reaction.

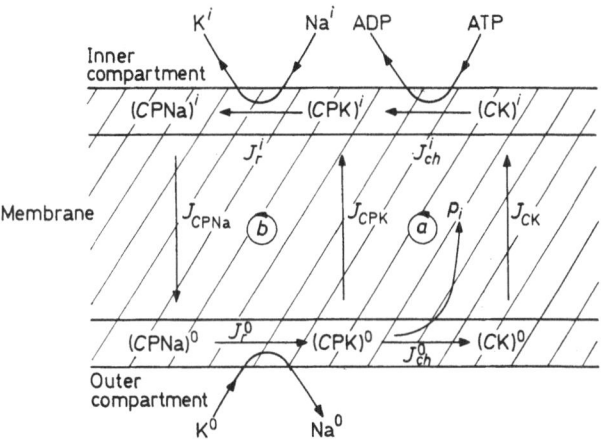

Figure 8. Schematic representation of carrier mediated, active, exchange of Na^+——K^+. J_{ch}^i is the rate of the chemical process which transforms the free carrier (CK^i) into a phos-phorylated carrier (CPK^i) $(CK^i + \text{ATP} \rightarrow CPK^i + \text{ADP})$. J_{ch}^0 represents the rate of dephosphorylation which regenerates the free carrier $(CPK^0 \rightarrow CK^0 + P_i)$. J_r^i is the rate of exchange of potassium by sodium on the inner side of the membrane $(CPK^i + \text{Na}^i \rightarrow CPNa^i + \text{K}^i)$. J_r^0 represents the rate of sodium liberation to the outer solution through exchange with external potassium $(CPNa^0 + \text{K}^0 \rightarrow CPK^0 + \text{Na}^0)$. J_{CK}, J_{CPK} and $J_{CPNa]}$ are the rates of flow of the different carrier forms across the membrane

The scheme presented here is similar to that used by Heckmann and by Hill and Kedem[13] in their treatment of facilitated and active transport by a lattice model. The common feature is the separation of the process into a chemical cycle (a) and an ion exchange cycle (b) which may be readily evaluated.

It is unnecessary to present the detailed calculation, which follows the method outlined above for facilitated-carrier mediated transport; only the final equations will be reproduced.

$$J_{\text{Na}} = L_{11}(1 - \Gamma) + L_{12}\left(K \frac{C_{\text{ATP}} \cdot C_{\text{H}_2\text{O}}}{C_{\text{ADP}} \cdot C_{\text{P}_i}} - 1\right) = L_{11}X_{\text{exch}} + L_{12}X_r$$

$$(101)$$

$$J_r = L_{21}(1 - \Gamma) + L_{22}\left(K \frac{C_{\text{ATP}} \cdot C_{\text{H}_2\text{O}}}{C_{\text{ADP}} \cdot C_{\text{P}_i}} - 1\right) = L_{21}X_{\text{exch}} + L_{22}X_r$$

The kinetic treatment underlying equations (101) makes the flows J_{Na} and J_r linearly dependent on a new pair of forces,

$$X_{\text{exch}} = 1 - \Gamma \text{ and } X_r = K \frac{C_{\text{ADP}} \cdot C_{\text{H}_2\text{O}}}{C_{\text{ADP}} \cdot C_{\text{P}_i}} - 1$$

where K is the equilibrium constant for the hydrolysis of ATP. These forces reduce, however, within a factor of RT to the thermodynamic forces $\Delta\mu_{exch}$ and A used in section **7.2** when the system approaches equilibrium. Indeed if $(\Delta\mu_{exch}/RT) \ll 1$, equation (94) gives immediately

$$\frac{\Delta\mu_{exch}}{RT} = 1 - \Gamma = X_{exch} \tag{102}$$

Similarly, we find for the affinity A

$$A = \mu_{ATP} + \mu_{H_2O} - \mu_{ADP} - \mu_{P_i} = (\mu_{ATP}^0 + \mu_{ADP}^0 - \mu_{P_i}^0) +$$

$$RT\ln \frac{c_{ATP} \cdot c_{H_2O}}{c_{ADP} \cdot c_{P_i}}$$

$$= RT\ln K + RT\ln \frac{c_{ATP} \cdot c_{H_2O}}{c_{ADP} \cdot c_{P_i}}$$

Close to equilibrium $(A/RT) \ll 1$, and hence

$$\frac{A}{RT} = K \frac{c_{ATP} \cdot c_{H_2O}}{c_{ADP} \cdot c_{P_i}} - 1 = X_r \tag{103}$$

Upon inserting equations (102) and (103) into equations (101), we regain equations (95); the model treatment provides us, however, with an explicit expression for L_{11}, L_{12} and L_{22} and verifies kinetically the validity of Onsager's theorem. As expected the coefficients are linearly proportional to the amount of carrier and its mobility, and exhibit saturation properties ascribed to facilitated transport.

The explicit dependence of the L_{ij}s on the parameters of the state permits a quantitative description of several aspects of active transport, as will be found in the paper of Blumenthal, Katchalsky and Ginzburg[33].

7.4. The model treatment leaves open the problem of the mechanism of carrier transport across the membrane. Although in the formal description of the carrier flows it was assumed that we may write $J_K = P(C_K^0 - C_K^i)$, there is little doubt that neither the free nor the phosphorylated carrier move according to the rules of free diffusion. Study of erythrocyte membranes shows that they have a tough structure displaying a viscoelastic behaviour resembling that of swollen nylon[33]. Rapid transport through such a medium would require a special mechanism which differs in essence from the random movement of small molecules. Recent studies of Post et al.[34] and of Hokin et al.[35] indicate that the carrier of erythrocyte membranes consists of protein molecules which undergo phosphorylation. It is an attractive hypothesis that the conformational change which is expected to accompany the phosphorylation would develop sufficient forces to transport the permeant across the membrane. A model demonstration of such a possibility is provided by the mechanochemical engines built in this laboratory[36] (*Figure 9*).

These engines utilize the reversible contractility of regenerated and cross-linked collagen fibres (product of the Ethicon Co., New Jersey, U.S.A.)

which contract to about half their length by interaction with strong salt solutions such as LiBr and expand reversibly upon washing with water. The original fibre is constructed of highly stretched triple helices of collagen molecules. Upon interaction with salt the helices undergo a conformational change to a random coiled structure which behaves as an ideal rubber[37]. Since the contraction process develops large forces sufficient to be utilized for the conversion of chemical energy into mechanical work, it is plausible to suppose that in the living membranes also conformational changes of biopolymers may serve as the molecular basis for carrier-mediated transport.

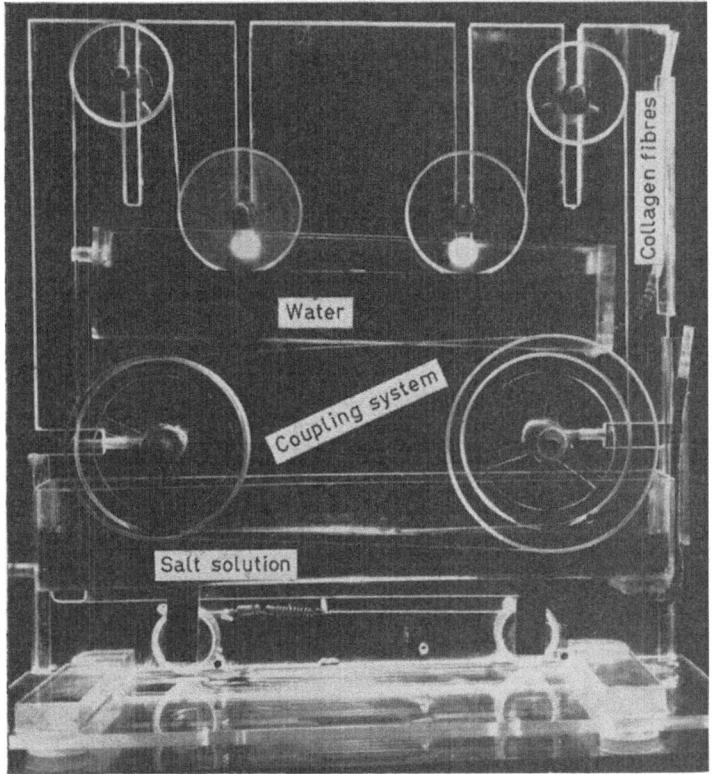

Figure 9

A tentative corroboration of this assumption is based on the following consideration of the rate of the transport in the red blood cell membrane: the sodium flow at room temperature $J_{Na} = 10^{-13}$ mole/cm²sec. Since the area of a red blood cell membrane is $1 \cdot 5 \times 10^{-6}$ cm² the number of Na⁺ ions passing across the membrane is

$$10^{-3} \cdot 1 \cdot 5 \times 10^{-6} \cdot 6 \cdot 10^{23} \simeq 10^5 \text{ ions/sec red blood cell.}$$

Various estimates lead to the conclusion that the number of sites involved in the transport is on the average $5 \cdot 10^3$. Hence, if each site comprises

E

one carrier molecule and each carrier molecule takes up a single ion per movement across the membrane, the macromolecule has to make

$$\frac{10^5}{3.10^3} = 30 \text{ cycles/sec.}$$

This is a reasonable number for macromolecular conformational change and is not far from the macroscopic contraction rates observed in collagen fibres.

Thus a deeper analysis of the performance of membranes brings us back to the study of macromolecular systems, the conformational changes and dynamic properties of which underlie the intriguing behaviour of biological systems.

Acknowledgements

This reasearch has been sponsored in part by the Air Force through the European Office of Aerospace Research (OAR) United States Air Force under contract No. F6-1052-67-C-0031.

References

[1] A. Katchalsky and P. F. Curran. *Nonequilibrium Thermodynamics in Biophysics* Harvard University Press, 1965.

[2] K. H. Meyer and J. F. Sievers. *Helv. Chim. Acta* **19**, 649 (1936).

[3] T. Teorell. *Proc. Soc. Exp. Biol. N.Y.* **33**, 282 (1935); *Trans. Faraday Soc.* **33**, 1053, 1086 (1937).

[4] F. F. Reuss. *Mem. Soc. Natur. Mose.* **2**, 327 (1809).

[5] S. R. de Groot and P. Mazur. *Non-Equilibrium Thermodynamics* North Holland Publ. Co., Amsterdam, 1961.

[6] L. Onsager. *Phys. Rev.* **37**, 405 (1931); **38**, 2265 (1931).
   H. B. G. Casimir. *Rev. Mod. Physics* **17**, 343 (1945).

[7] D. G. Miller. *Chem. Reviews* **60**, 15 (1960).

[8] O. Kedem and A. Katchalsky. *Biochim. Biophys. Acta* **27**, 229 (1958).

[9] J. Loeb. *J. Gen. Physiol.* **2**, 173 (1920); **4**, 463 (1922).
   E. Grim and K. Sollner. *J. Gen. Physiol.* **40**, 887 (1957).

[10] A. J. Staverman. *Trans. Faraday Soc.* **48**, 176 (1948); *Rec. Trav. chim.* **70**, 344 (1951).

[11] A. Katchalsky and O. Kedem. *Biophys. J.* **2**, 53 (1962);
   O. Kedem and A. Katchalsky. *Trans. Faraday Soc.* **59**, 1918 (1963).

[12] H. R. Kruyt. *Colloid Science*, Vol. I, Elsevier Publ. Co., Amsterdam, 1952.

[13] K. Heckmann. *Z. phys. Chem., N.F.* **44**, 184 (1965);
   T. L. Hill and O. Kedem. *J. Theoret. Biol.* **10**, 399 (1966).

[14] G. Eisenman, J. P. Sandblom, and J. L. Walker, Jr. *Science* **155**, 965 (1967).

[15] G. M. Shean and K. Sollner. *Ann. N.Y. Acad. Sci.* **137**, 759 (1966).

[16] J. Gabbay. *The Transport of Aminoacids Through Ion Exchange Membranes*, Ph.D. Thesis, Hebrew University, Jerusalem, 1964.

[17] W. Wilbrandt and Th. Rosenberg. *Pharmacol. Reviews* **13**, 109 (1961).

[18] R. Blumenthal and A. Katchalsky, in preparation.

[19] O. Kedem and A. Katchalsky. *Trans. Faraday Soc.* **59**, 1941 (1963);
   I. W. Richardson. *Multiple Membrane Systems as Analogues for Biological Membranes*, Ph.D. Thesis, University of California, Berkeley, 1967.

[20] H. Grundfest, in *The General Physiology of Cell Specialization*, p. 277, eds. D. Mazia and A. Tyler, McGraw Hill, New York., 1963;
   cf. also R. H. Adrian. *J. Physiol.* **175**, 134 (1964); I. W. Richardson. cf. ref. 19.

[21] P. Hirsch-Ayalon, unpublished results.

[22] Y. Katz, unpublished results.

[23] P. Curie. *Ouevres* p. 129, Gauthier-Villars, Paris, 1908.

[24] I. Prigogine. *Etude Thermodynamique des phénoménes irreversibles* Dunod, Paris and Desoer, Liège, 1947.

[25] I. Prigogine. *Thermodynamics of Irreversible Processes*, Thomas Co., Springfield, Ill., U.S.A.

[26] R. Blumenthal, S. R. Caplan, and O. Kedem. *Biophys. J.* in the press.

[27] O. Kedem, in *Membrane Transport and Metabolism*, p. 81, eds. A. Kleinzeller and A. Kotyk, Academic Press, New York, 1961.

[28] E. T. Dunham and I. M. Glynn. *J. Physiol.* **156**, 274 (1961);
I. M. Glynn. *Progress in Biophysics and Biophysical Chemistry* **8**, 242 (1957).
R. L. Post, C. R. Merritt, C. R. Kinsolving, and C. D. Albright. *J. Biol. Chem.* **235**, 1796 (1960).
J. Chr. Skou. *Biochim. Biophys. Acta* **23**, 394 (1957); **42**, 6 (1960).

[29] R. Whittam. *Transport and Diffusion in Red Blood Cells*, Arnold, London, 1964.

[30] P. F. Baker, A. L. Hodgkin, and T. I. Shaw. *Nature, Lond.* **190**, 885 (1961);
A. L. Hodgkin. *The Conduction of the Nervous Impulse*, Liverpool University Press, 1965.

[31] I. Tasaki and I. Singer, in *Biological Membranes, Ann. N.Y. Acad. Sci.* **137**, 793 (1966).

[32] P. J. Garrahan and I. M. Glynn. *Nature, Lond.* **211**, 1414 (1966).

[33] R. Blumenthal, A. Katchalsky, and B. Z. Ginzburg, p. 91, in *Proceedings of the 1st International Conference on Hemorheology*, Reykjavik, 1966, ed. A. L. Capley, Pergamon Press, Oxford/New York, 1967.

[34] R. L. Post, A. K. Sen, and A. S. Rosenthal. *J. Biol. Chem.* **240**, 1437 (1965).

[35] L. E. Hokin, P. S. Sastry, P. R. Galsworthy, and A. Yoda. *Proc. Nat. Acad. Sci.* **54**, 177 (1965);
cf. also I. M. Glynn, C. W. Stayman, J. Eichberg, and R. M. C. Dawson. *Biochem. J.* **94**, 692 (1965).

[36] I. Z. Steinberg, A. Oplatka, and A. Katchalsky. *Nature, Lond.* **210**, 568 (1966).

[37] A. Katchalsky, A. Oplatka, and A. Litan. "The Dynamics of Macromolecular Systems" in *Molecular Architecture in Cell Physiology*, p. 3, eds. T. Hayashi and A. G. Szent-Gyorgyi, Prentice-Hall, New York, 1966.

SOME FUNDAMENTAL ASPECTS OF POLYMER REACTIONS

Ichiro Sakurada

Department of Polymer Chemistry, Kyoto University, Kyoto, Japan

1. INTRODUCTION

In discussing the reactivity of macromolecules, it is generally assumed that a functional group attached to a high polymer has a chemical reactivity similar to that which would be observed for such a group in small molecules[1]. This generalization may essentially be true, but we often find reactions in which a functional group attached to a high polymer exhibits widely different reactivity to that in small molecules.

The size of a molecule is generally unimportant, but the fact that a functional group of a high polymer always has neighbouring groups is the most fundamental feature of polymer reactions.

In this lecture, based chiefly on our own experiments, I will discuss simple polymer reactions in which neighbouring groups exert great effects on the reactivity of a functional group attached to a polymer.

2. REACTIVITY OF FUNCTIONAL GROUPS OF POLYMER MOLECULES

(a) Hydrolysis of polyvinyl acetate

First let us consider an example in which the reactivity of functional groups of polymer molecules is similar to that of the corresponding low molecular weight compounds. In 1944 we carried out alkaline hydrolysis of polyvinyl acetate in acetone–water (75:25 by volume)[2]. This mixture is used because both polyvinyl acetate and alcohol are soluble in it. When methanol or ethanol is used as a solvent the reaction mechanism becomes more complicated because alcoholysis takes place simultaneously.

The rate constant k was evaluated graphically at various degrees of conversion with the following equation:

$$(dx/dt) = k (a - x) (b - x) \tag{1}$$

where a is the initial concentration of ester, b that of alkali and x the concentration of carboxylic acid at time t. The initial rate constant k_0 was extrapolated from the k conversion curve. A comparison of the initial rate constants for the hydrolyses of polyvinyl acetate and some low-molecular-weight acetates is given in *Table 1*.

Although k_0 for polyvinyl acetate is only one tenth of that for ethyl acetate it does not differ so markedly from that for isopropyl acetate, which is an ester of a secondary alcohol as polyvinyl acetate. This is a good example, demonstrating that the reactivity of functional groups of polymer molecules is similar to that of low-molecular compounds.

263

Table 1. Comparison of the rate of alkaline hydrolysis of polyvinyl acetate with those of some low-molecular-weight esters. Solvent, acetone–water (75:25); alkali, NaOH; temperature 30°C

Acetate	—CH—CH₂— | O | COCH₃	CH₂—CH₃ | O | COCH₃	CH₃—CH—CH₃ | O | COCH₃	CH₂—CH₂—CH—CH₃ | | O O | | COCH₃ COCH₃
k_0 (l.mole^{-1}. min^{-1}	0·37	3·5	0·57	4·4

Smets and his coworkers[3] carried out alkaline hydrolyses of polyvinyl-pyrrolidone derivatives and homologues to show similarity in apparent activation energy and collision frequency for the polymeric molecules and their low molecular homologous substances.

(b) Hydrolysis of polymethyl acrylate

Alkaline or acid hydrolysis of polymethyl acrylate is a good example showing the very considerable difference between the reactivity of functional groups attached to a polymer and those attached to a low-molecular-weight compound. *Table 2* gives a comparison of the initial rate constants for the hydrolysis of polymethyl acrylate and some low-molecular methyl esters of similar structure[4].

Table 2. Comparison of the rates of alkaline and acid hydrolyses of polymethyl acrylate with those of some low-molecular-weight ester (at 25°C)

Ester	—CH—CH₂— | CO | OCH₃	CH₂—CH₃ | CO | OCH₃	CH₃—CH—CH₃ | CO | OCH₃	CH₂—CH₂—CH₂ | | CO CO | | OCH₃ OCH₃
Acetone-water				
k_0 (5:2)*	0·1	3·8	1·2	4·8
(2:5)*	0·1	8·8	2·8	7·2
(6:1)†	5·2‡	—	210	—

* Alkaline hydrolysis with NaOH, k_0 is given in l.mole^{-1} min^{-1}.
† Acidic hydrolysis with HCl, k_0 is given in 10^{-3} l.mole min^{-1}.
‡ Measured at 50°C.

Not only in alkaline, but also in acid hydrolyses, k_0 for polymethyl acrylate is much smaller than that for low-molecular-weight esters. The apparent activation energy of the hydrolysis was 12 kcal/mole for polymethyl acrylate and 11–11·5 kcal/mole for methyl *iso*butyrate and dimethyl glutarate. The small k_0 for polymethyl acrylate seems to be attributed to the steric hindrance of neighbouring groups. We will return to this problem when we consider the effect of tacticity on the reaction rate.

3. CHANGE OF THE APPARENT RATE CONSTANT IN THE COURSE OF THE REACTION

(a) Esterification of polyvinyl alcohol with monochloroacetic acid and its reverse reaction

Esterification of polyvinyl alcohol $(P = 1200)$ with monochloroacetic acid and its reverse reaction were carried out in monochloroacetic acid–water mixtures[5]. The rate constants were calculated using the following equation for the reversible reaction:

$$dx/dt = k \, (a - x) \, b - k'xc \tag{2}$$

where a, b, c, and x are the initial concentrations of hydroxyl group, monochloroacetic acid, water, and ester group, respectively, and k and k' are the rate constants for the esterification and the hydrolysis, respectively.

The rate constant for the esterification may be calculated from

$$-\log (1 - p/p_e)/t = kb/2\cdot303 \, p_e \tag{3}$$

where p and p_e are the degrees of esterification at time t and at equilibrium, respectively.

Experiments were carried out at two different temperatures. As may be seen from *Figure 1*, in the range covered by the experiments, k remains constant.

The equation for hydrolysis is

$$\frac{\log \left[(p_i - p_e)/(p - p_e) \right]}{t} = \frac{k'b}{2\cdot303p_e} \tag{4}$$

where p_i is the initial degree of esterification. From *Figure 2* it may be seen that k' also remains unchanged.

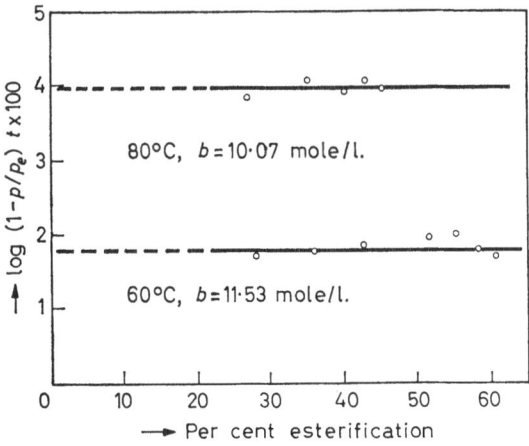

Figure 1. Esterification of polyvinyl alcohol in monochloroacetic–acid–water

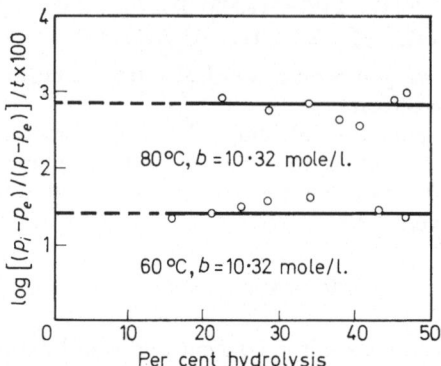

Figure 2. Hydrolysis of polyvinyl monochloroacetate in water–monochloroacetic acid

(b) Hydrolysis of polyvinyl acetate

Polymer reactions are not always as simple as in the case of polyvinyl monochloroacetate. Alkaline and acid hydrolyses of polyvinyl acetate are good examples[2]. *Figures 3* and *4* show the courses of methanolysis in pure methanol and hydrolysis in acetone–water, respectively. Both curves are sigmoidal and the course of the reaction seems to be autocatalytic. If we calculate the apparent rate constant k with equation (2), k increases linearly with increasing degree of conversion (see *Figure 5*).

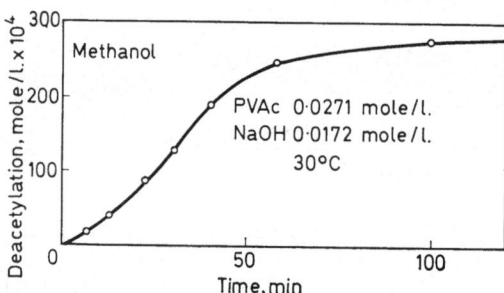

Figure 3. Methanolysis of polyvinyl acetate in pure methanol

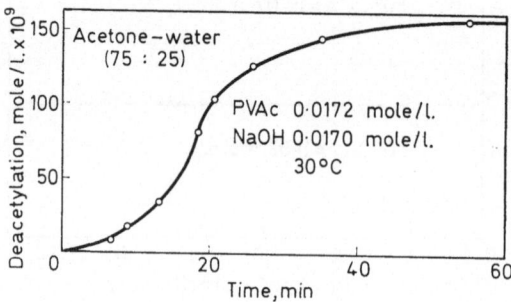

Figure 4. Hydrolysis of polyvinyl acetate in acetone–water (75 : 25)

The relation shown in *Figure 5* may be expressed by

$$dx/dt = k_0 [1 + m (x/a)] (a - x) (b - x) \qquad (5)$$

where k_0 is the initial rate constant and m a constant independent of the conversion. The calculated value of m for the reaction shown in *Figure 5* is 42; this means that the final apparent rate constant is 43 times greater than the initial one. The results shown in *Figure 5* were obtained at 30°C but the m value for the hydrolysis at 40°C was also found to be 42; it may be said, therefore, that the apparent activation energy is constant (11·8 kcal/mole) throughout the reaction.

Such an autocatalytic acceleration of the reaction may be attributed to the effect of neighbouring groups which changes in the course of the reaction. As a simplest case we can consider triads and ascribe to each triad a different rate constant[6].

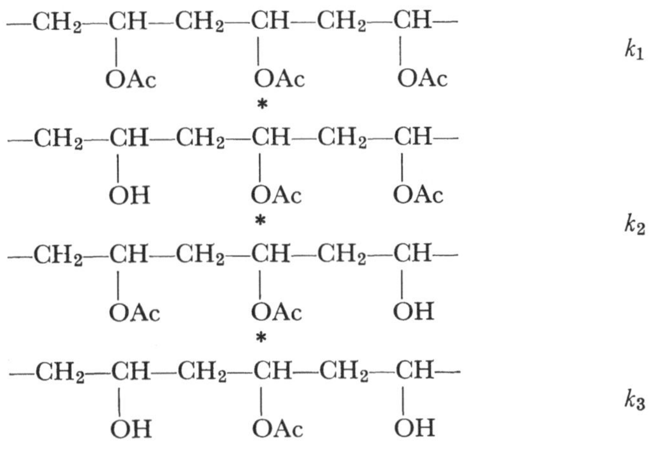

$$-CH_2-CH-CH_2-CH-CH_2-CH-$$
$$\overset{|}{OAc} \quad \overset{|}{\underset{*}{OAc}} \quad \overset{|}{OAc} \qquad\qquad k_1$$

$$-CH_2-CH-CH_2-CH-CH_2-CH-$$
$$\overset{|}{OH} \quad \overset{|}{\underset{*}{OAc}} \quad \overset{|}{OAc}$$
$$\qquad\qquad\qquad\qquad\qquad\qquad\qquad k_2$$
$$-CH_2-CH-CH_2-CH-CH_2-CH-$$
$$\overset{|}{OAc} \quad \overset{|}{\underset{*}{OAc}} \quad \overset{|}{OH}$$

$$-CH_2-CH-CH_2-CH-CH_2-CH-$$
$$\overset{|}{OH} \quad \overset{|}{OAc} \quad \overset{|}{OH} \qquad\qquad k_3$$

It is difficult to carry out an exact kinetic calculation for this case; therefore simplifying assumptions were adopted that $k_1 < k_2 < k_3$ and that the amount of the triad in which an unreacted acetyl group is surrounded by two hydroxyl neighbours is negligible at the initial stage of the reaction. The final equation is somewhat complicated but numerical calculation shows that it roughly agrees with equation (5). The approximate relation between k_2/k_1 and m is shown in *Table 3*.

As may be seen from *Table 3*, for the case $m = 42$, an acetyl group whose one neighbour is hydroxyl is hydrolysed a hundred times more easily than an acetyl which is surrounded by two unreacted groups; it follows that hydroxyl groups are formed preferably in sequence. That this conclusion is true has been shown by infrared analysis[7].

From the above argument, it would be expected that the effect of the neighbouring group would be found, even for a bifunctional low-molecular-weight compound of a similar structure[8]. The hydrolysis of ethylene glycol

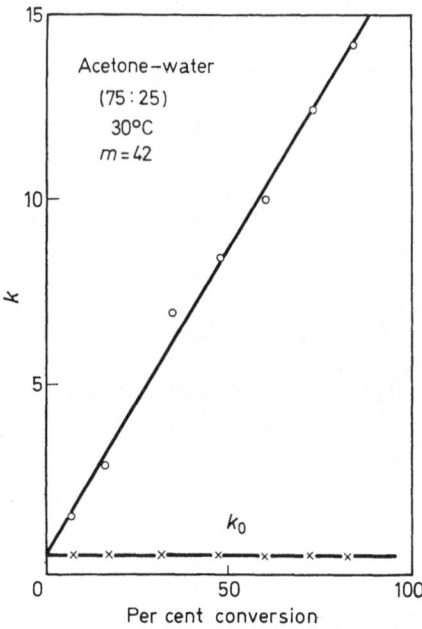

Figure 5. Hydrolysis of polyvinyl acetate in acetone–water (75:25)

Table 3. Relation between m and k_2/k_1

$a = k_2/k_1$	2	10	20	30	60	100
m	1·8	10	17	22	33	45

diacetate was carried out at first in water. The results are shown in *Figure 6*. The rate constant remains almost unchanged in the course of the reaction. When we use acetone–water (75:25) as a solvent, k increases with conversion (see *Figure 7*). Although the effect of a neighbouring group is much smaller than in the case of polyvinyl acetate, it is true that the effect exists.

It is very important to understand the mechanism of the accelerating effect of the neighbouring group. The property at first seems to be due to a steric effect, but it is not true because the effect disappears when pure water is used as a solvent. The behaviour is to be attributed to local concentration of alkali due to attraction by free hydroxyl groups. It was shown that alkali is really adsorbed by polyvinyl alcohol. It is highly probable that alkali is more easily adsorbed from acetone–water than from pure water. We are now carrying out adsorption experiments with radio-chemically crosslinked polyvinyl alcohol. We would call such an effect a hydrophilic effect because it is found when hydrophilic groups are contained in a polymer molecule.

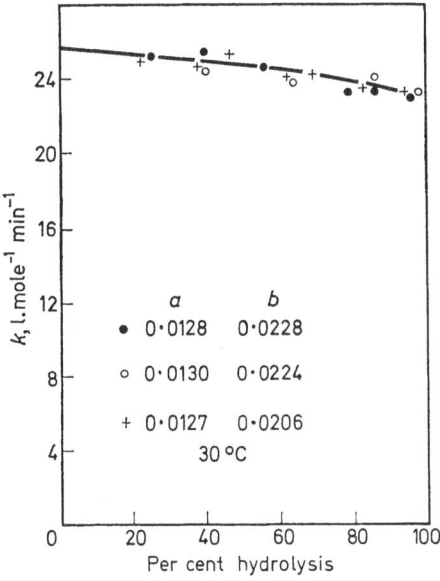

Figure 6. Hydrolysis of ethylene glycol diacetate in water with NaOH

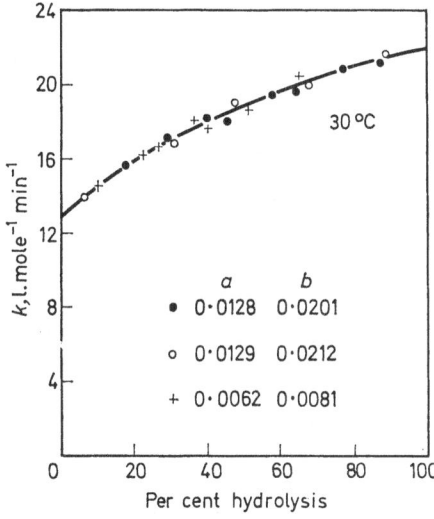

Figure 7. Hydrolysis of ethylene glycol diacetate in acetone–water (75:25)

(c) Hydrolysis of polymethyl acrylate

The course of hydrolysis of polymethyl acrylate was studied in acetone–water mixtures of different compositions[4]. The concentration of polymethyl acrylate was 0·07 mole/l., the concentration of NaOH in most cases 0·08 mole/l. and the reaction temperature 40°C. *Figure 8* shows the relations between logarithm of apparent rate constant calculated by equation (2) and percent hydrolysis. Although there is no great difference in the initial rate constant, the value of the change of k with percent conversion varies widely from one composition to another. In acetone-rich systems, k increases rapidly with percent hydrolysis, at acetone: water $= 4:3$ by volume, k is almost independent of the conversion, and in water-rich systems k decreases rapidly with increasing percent hydrolysis.

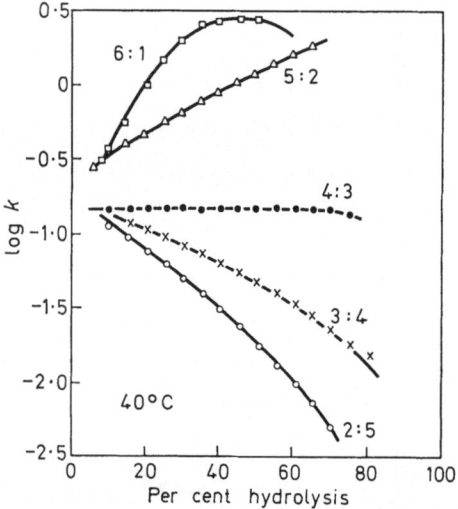

Figure 8. Hydrolysis of polymethyl acrylate in acetone–water mixtures with NaOH (Figures by the curves show the acetone–to–water ratio)

Similar experiments were carried out also at 25° and 50°C; there was no great change in the shapes of the curves. The apparent activation energy was found to be *c*. 12 kcal/mole throughout the reaction.

The rapid drop of k with increasing percent hydrolysis in water-rich systems may be attributed to the formation of carboxylate groups, which dissociate into ions in water-rich systems; polymer molecules become negatively charged and repel hydroxyl ions so that the hydrolysis reaction is depressed. This is the same electrostatic effect as Katchalsky[9] has reported for the hydrolysis of pectin.

With increasing concentration of acetone, the dissociation of carboxylate groups become less extensive and the effect of the electrostatic repulsion becomes suppressed. Undissociated carboxylate groups show a similar hydrophilic effect to hydroxyl groups in the case of hydrolysis of polyvinyl acetate, and attract NaOH, so that the catalyst concentration in the neighbourhood of polymer molecules becomes higher to increase k. On the other hand, even in acetone-rich systems it is true that the dissociation takes place and the charge number of polymer molecule increase with increasing percent hydrolysis to exert a negative effect. Therefore a maximum will

270

appear in the log k–percent hydrolysis curves. We do not intend to neglect the steric effect, but the attraction and repulsion effects of the catalyst seem to be much more important.

4. THE EFFECT OF THE DEGREE OF POLYMERIZATION ON THE REACTION RATE

(a) Hydrolysis of polyvinyl acetate

The effect of the degree of polymerization on the rate of alkaline hydrolysis was studied for polyvinyl acetates with various degrees of polymerization; the viscometrically determined degree of polymerization ($\bar{P}_v$) varied from 50 to 20,000[2]. The degree of polymerization had practically no effect on the rate of hydrolysis, but a polyvinyl acetate of $\bar{P}_v = 50$ seemed to have a slightly higher rate. Therefore experiment was undertaken again with poly-vinyl acetate of a very low degree of polymerization, which had been pre-pared by polymerization of vinyl acetate in butyraldehyde[10]; the degree of polymerization was found to be $\bar{P}_v = 11\cdot6$ and $\bar{P}_n = 9\cdot8$. The rate of alkaline hydrolysis of this polyvinyl acetate was compared in dioxane–water (3:1 by volume) at 40°C with that of a conventional polyvinyl acetate ($\bar{P}_v = 2300$). As may be seen from *Figure 9*, there is no essential difference between two samples except that the k–percent hydrolysis curve for the polyvinyl acetate of the very low degree of polymerization lies somewhat higher than the other.

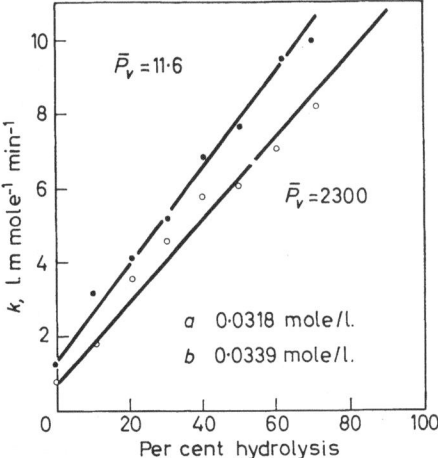

Figure 9. Hydrolysis of polyvinyl acetates of low and high degrees of polymerization

(b) Hydrolysis of polyallyl acetate

The effect of degree of polymerization on the rate of polymer reaction is not always so simple as that demonstrated in the foregoing section. According to our unpublished work[11], the behaviour of polyallyl acetate in alkaline hydrolysis is different from that of polyvinyl acetate.

Polyallyl acetates of higher degrees of polymerization were prepared by reduction of polymethyl methacrylate to polyallyl alcohol, based upon a process proposed by Schulz[12], followed by acetylation. Polyallyl acetates of lower degrees of polymerization were prepared from polyallyl alcohol or esters which had been obtained by direct polymerization of the corresponding monomer.

Polyallyl acetate of the lowest degree of polymerization used in the experiment had a molecular weight of 580. *Figure 10* shows relations between apparent rate constants at percent hydrolyses 0% (k_0) and 50% (k_{50}) and degree of polymerization. As a qualitative measure of the degree of polymerization, $[\eta]$ measured in acetone at 30°C was used as the abscissa.

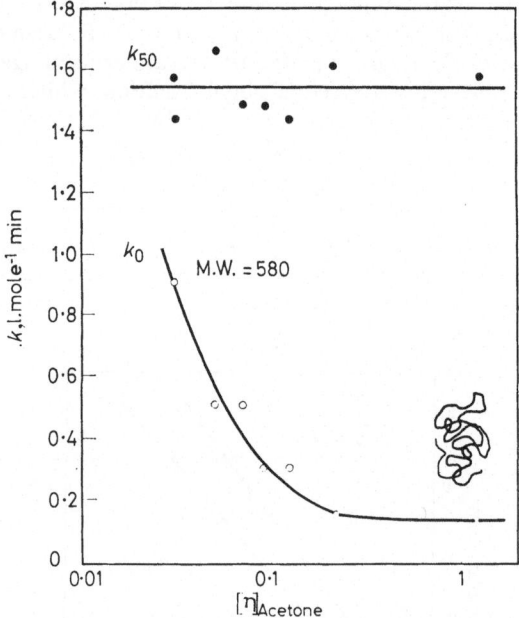

Figure 10. Rate constants of hydrolysis (at per cent hydrolysis of 0 and 50) for polyallyl acetates of various degrees of polymerization

As may be seen from the figure, k_0 decreases rapidly with increasing degree of polymerization, while k_{50} is almost independent of it. An acetone–water (8:2 by volume) mixture was used in the experiment as the solvent, because both the initial and final polymers are soluble in this mixture, but the mixture is a poor solvent for polyallyl acetate. The polymer molecule is therefore, at the initial stage of the reaction, in a state of tighter coiling, which hinders chemical reaction. With increasing degree of hydrolysis, the mixture becomes a better solvent and the coiling is loosened. This is the reason why the apparent rate constant increases with percent hydrolysis especially by polymers of higher degrees of polymerization and becomes independent of the degree of polymerization when hydrolysis has taken place to some extent.

5. NEIGHBOURING GROUP EFFECT ON THE REACTIVITY

It was recognized by Morawetz and Zimmering[13], and Zimmering et al.[14] in a study of the hydrolysis of acrylic and methacrylic copolymers containing a small portion of p-nitrophenyl methacrylate that the hydrolysis of the ester groups was much more rapid than the hydrolysis of p-nitrophenyl esters of monofunctional group. They demonstrated that the reaction does not involve hydroxyl ions but that the velocity of the process is governed by the attack of an ionized carboxyl on the ester function.

Smets and his co-workers[15] also carried out interesting experiments on the hydrolysis of methacrylic acid–methacrylate copolymer and obtained similar results.

These reactions are rather specific while the effects are observed only when an ionized carboxyl can react with an ester group.

The neighbouring-group effect which was pointed out in the course of the alkaline hydrolysis of polyvinyl acetate and polyacrylate is not specific but more general and fundamental.

We shall now consider some examples of steric and electrostatic effects not pertaining to the rate of reaction but concerning the final conversion of polymer.

The first example is the triphenylmethylation of polyvinyl alcohol[16]. It was shown experimentally that the highest degree of substitution was 42·4%. If we assume that the two direct neighbours of a trityl group are not able to undergo the reaction, the theoretical value for the maximum conversion is 43·2%; the agreement is satisfactory. The residual hydroxyl groups could be acetylated.

$$-CH_2-CH-CH_2-CH-CH_2-CH-CH_2-CH-CH_2-CH-CH_2-CH-$$
$$\quad\ \ |\qquad\quad |\qquad\quad |\qquad\quad |\qquad\quad |\qquad\quad |$$
$$\quad\ \ OTr\quad\ OH\quad\ \ OTr\quad\ OH\quad\ \ OH\quad\ OTr$$

The second example is the electrostatic effect on polymer reaction. By acetalization of polyvinyl alcohol with aldehydes which do not contain an ionizable group it is not difficult to obtain a percentage acetalization which is near to the theoretically expected value of 86·5%[17].

Aldehyde	Acetalization %
Palmitinaldehyde	85·0
Chloracetaldehyde	85·8
o-Chlorbenzaldehyde	84·6
Benzaldehyde	83·0

When acetalization is carried out with aldehydes which contain a sulphonic acid group, it is impossible to reach such a high degree of acetalization[18]. The maximum degrees of acetalization for aldehyde sulphonic acids were as follows:

Aldehyde	Acetalization %
β-Butyraldehyde sulphonic acid	57·6
o-Benzaldehyde sulphonic acid	44·0
2,4-Benzaldehyde disulphonic acid	36·0

The results seem to show that not only isolated but some other hydroxyl groups are unable to undergo the acetalization reaction due to the repulsive effect of sulphonic acid groups.

6. EFFECT OF THE TACTICITY ON THE REACTION RATE

Glavis[19] reported in 1959 that conventional and syndiotactic polymethyl methacrylates are hydrolysed in an alkaline medium relatively slowly, while the hydrolysis of isotactic polymer proceeds very rapidly and to a higher conversion.

Smets and De Loecker[20] prepared methacrylic–ester–acid copolymers from syndiotactic polymethyl methacrylate (A) and from polymethyl methacrylate which is considered as a mixture of conventional and isotactic polymers (B), and carried out hydrolysis in a buffered aqueous solution at different degrees of neutralization. Polymer A hydrolysed about four to five times faster than the conventional polymethyl methacrylate while polymer B hydrolysed very slowly.

Similar experiments were carried out by Smets and Van Humbeeck[21] on acrylic acid copolymers. In the case of acrylic–acid–methacrylate copolymers the isotactic system hydrolysed 3 to 5 times more rapidly than the conventional ones.

(a) Hydrolysis of polyvinyl acetates
We shall explain in some detail the effect of the tacticity of polyvinyl acetates on the rate of alkaline hydrolysis.

Isotactic polyvinyl acetate derived from isotactic polyvinyl ether, conventional (atactic) polyvinyl acetates, polyvinyl acetate derived from poly-(vinyl trifluoroacetate) and that derived from poly(vinyl butyral) were hydrolysed in a mixture of acetone–water (7:3 by volume) using NaOH as the catalyst[22]. Conventional ones include eight different polymers obtained by polymerization of various vinylesters at different temperatures. The percentage hydrolysis *versus* reaction time curves are shown in *Figure 11*.

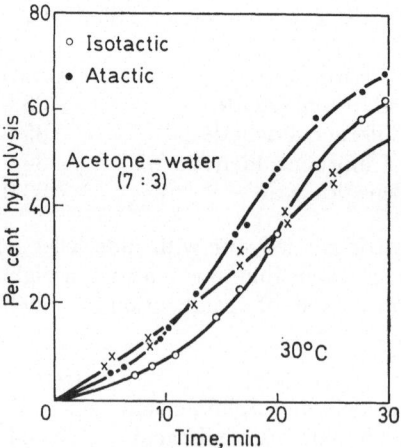

Figure 11. Alkaline hydrolysis of polyvinyl acetates of various tacticities

274

The courses of hydrolyses of all conventional polyvinyl acetates and that of polyvinyl acetate derived from poly(vinyl trifluoroacetate) may be represented by a single curve, whereas those of isotactic polyvinyl acetate and polyvinyl acetate derived from poly(vinyl butyral) differ remarkably from the first curve. The initial apparent rate constant k_0 and a measure of the autocatalytic effect m are shown in *Table 4*. Experimental data for syndiotactic polymer are not available.

Table 4. Rate of the hydrolysis of polyvinyl acetates of various tacticities

Tacticity	k_0	m
isotactic	0·14	49
atactic	0·23, 0·21	39, 38
rich in head-to-head structure	0·50	5·6

The structure of polyvinyl acetate and polyvinyl alcohol derived from poly(vinyl butyral) is not yet clear; the infrared spectrum of these polyvinyl alcohols differs slightly from that of a conventional one. It was once considered to be rich in syndiotactic structure, but according to our recent research it is more plausible that it is rich in head to head structure.

The initial rate of hydrolysis of the atactic polymer is larger than that of the isotactic one, while m of the former is smaller than that of the latter. In the case of the atactic structure, steric hindrance is smaller because the distance of acetyl groups is longer than it is in the isotactic structure, therefore an atactic polymer exhibits larger k_0. When the hydrolysis of ester groups occurs to some extent, hydroxyl groups, and hence adsorbed hydroxyl ions, are located closer to an unreacted acetyl group in the case of isotactic than in the case of syndiotactic polymer; therefore m is greater for the isotactic polymer.

Acetalization reactions of the above three kinds of polyvinyl alcohol were also compared. *Table 5* shows the equilibrium constants for the acetalization; we can see here again that there are large differences in chemical activity among these three kinds of polymer.

Table 5. Equilibrium constant for the acetalization of polyvinyl alcohols with various tacticities

Tacticity	Equilibrium const.
isotactic	3100
atactic	1200, 1400
rich in head-to-head structure	17

(b) Hydrolysis of polymethyl acrylates

In a study of the titration curve of neutralization of polyacrylic acids derived from polyacrylic esters obtained by a radical polymerization of the corresponding monomer, and polyacrylic acids obtained by polymerization

275

of acrylic acid it was found that with one exception all polyacrylic acids give a practically identical titration curve (see *Figure 12*)[23].

The exceptional polyacrylic acid was obtained by polymerization of acrylic acid under such a condition that the electrostatic repulsive forces between growing chain and monomer were very large. The polymerization condition was: solvent, water–ethanol (94·9:3·8); monomer conc., 1%; initiator, azobisisobutyronitril, pH was adjusted to 7 and the temperature was 60°C. The titration curve for this polyacrylic acid lies under that for the conventional polymers. From the polymerization condition and the comparison of titration curves, it was considered that this exceptional polymer is rich in syndiotactic structure.

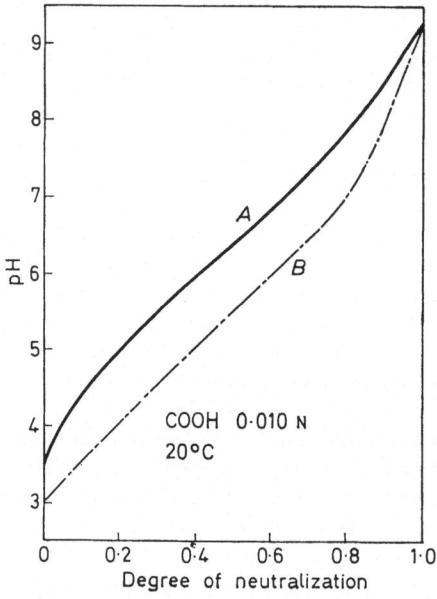

Figure 12. Potentiometric titration curves of polyacrylic acids (*A* conventional, *B* special)

The conventional polyacrylic acids and the syndiotactic one were converted to methyl ester, and the rate of alkaline hydrolysis of these esters was measured in acetone–water (7:3 by volume)[24]. The rate of hydrolysis is shown in *Figure 13*. Corresponding to titration curves, the hydrolysis rates of all conventional polymethyl methacrylates fall on a curve, whereas that of the syndiotactic polymer is widely separated from the others, k_0 for the conventional and syndiotactic polymers was 0·09 and 1·7 l. mole^{-1} min^{-1}, respectively. It is noteworthy that the latter value nearly agrees with that of methylisobutyrate which has already been given in *Table 2*. It seems that an ester group of polymethyl acrylate is more or less isolated when the polymer has a syndiotactic structure, so that steric hindrance of neighbouring groups is not very large and shows nearly the same activity as a simple monofunctional low-molecular-weight ester.

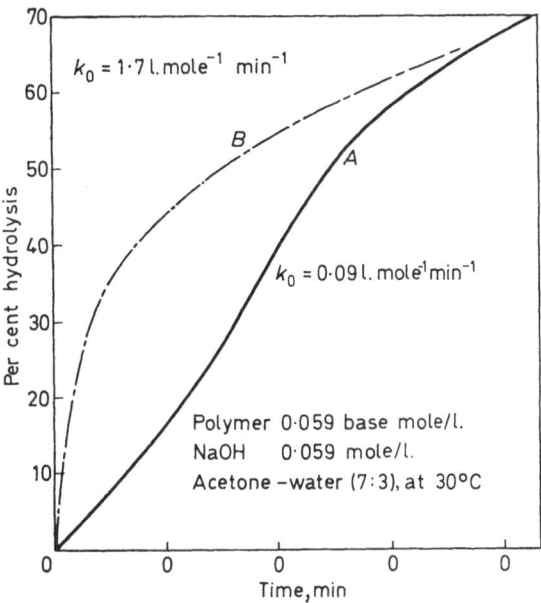

Figure 13. Hydrolysis of polymethyl acrylates (*A* conventional, *B* special)

7. HYDROPHOBIC INTERACTION IN THE POLYMER REACTIONS[25]

(a) Hydrolysis of esters with polymeric sulphonic acids

It was recently recognized that hydrophobic interaction plays an important role in the polymer reactions. The first typical example is the hydrolysis of low-molecular-weight esters with polymeric sulphonic acids such as polystyrenesulphonic acid.

In the case of the hydrolysis of ester with hydrochloric acid, it may be assumed that acid and ester molecules are distributed homogeneously throughout the reaction system and that hydrolysis takes place at any point in the system. On the other hand, in the case of hydrolysis with polymeric sulphonic acid, we may assume that hydrogen ions, counter ions of the sulphonic groups, are located exclusively in the neighbourhood of polymer molecules (*Figure 14a*). Hydrolysis can occur only in this region of the solution, because outside the region hydrogen ions capable of causing hydrolysis do not exist. If the ester molecules are distributed homogeneously in the solution, there is no apparent change in the rate of reaction, because the decrease of the volume of reaction is cancelled with the increase of the local hydrogen ion concentration. If the ester molecules are adsorbed by the polymer catalyst, the situation is somewhat different (*Figure 14b*). The concentration of the ester becomes higher just at the place where hydrogen ions are concentrated; in such a case it is expected that the polymer sulphonic acid shows a higher apparent rate constant than hydrochloric acid.

A comparison was carried out of the hydrolysis rate of ester with polystyrenesulphonic acid and hydrochloric acid.

Two types of polystyrenesulphonic acids were employed for the experiments; one is a polystyrenesulphonic acid obtained by the polymerization of monomeric styrenesulphonic acid (PSS) and the other a partially sulphonated polystyrene (PS-S), the degree of sulphonation of PS-S varying between 23 and 77 mole %.

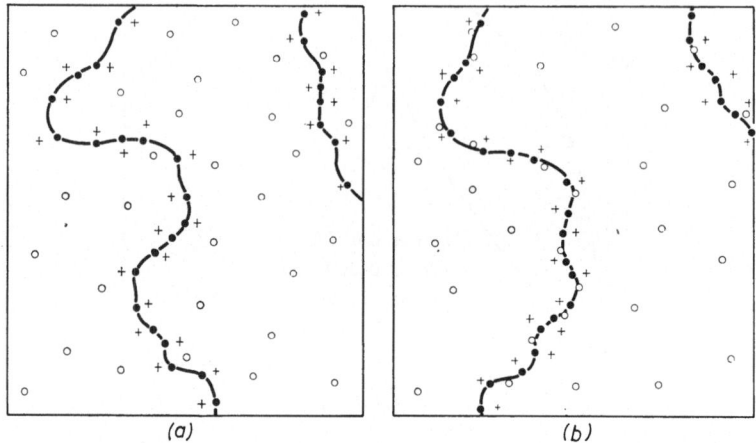

Figure 14. Hydrolysis of ester with polymeric sulphonic acids. ○ Ester, ● Acid, + Hydrogen ion. (a) No local concentration of ester. (b) Local concentration of ester

Hydrolyses of methyl- and n-butyl acetates were carried out with polystyrenesulphonic acids and hydrochloric acid. The experimental conditions and results are given in *Table 6*; r of the last column of the table being a ratio of the apparent rate constant for the polymer catalyst to that for hydrochloric acid: $r = k/k_{HCl}$.

Table 6 shows that methyl acetate is hydrolysed with polystyrenesulphonic acids with a greater rate constant than hydrochloric acid. It is

Table 6. Hydrolyses of methyl acetate and n-butyl acetate with polystyrene-sulphonic acids in water

Catalyst	Methyl acetate Ester conc.: $2 \cdot 50 \times 10^{-2}$ M catalyst conc.: $5 \cdot 00 \times 10^{-3}$ N temp.: 40°C		n-Butyl acetate Ester conc.: $2 \cdot 85 \times 10^{-2}$ M catalyst conc.: $5 \cdot 00 \times 10^{-3}$ N temp.: 40°C	
	$k. 10^2$ l.mole^{-1} min^{-1}	r	$k. 10^2$ l.mole^{-1} min^{-1}	r
HCl	2·31	1·00	2·30	1·00
PSS	2·39	1·03	4·42	1·92
PS-S(77)	2·79	1·21	5·91	2·57
PS-S(65)	3·25	1·41	6·94	3·02
PS-S(52)	3·53	1·53	10·35	4·50
PS-S(40)	3·63	1·57	14·5	6·31
PS-S(33)	4·86	2·10	18·3	7·96
PS-S(23)	5·06	2·19	23·5	10·22

remarkable that the lower the degree of sulphonation the greater the catalyst effect of the polystyrenesulphonic acid. It shows clearly that benzene rings, especially nonsulphonated benzene rings, exhibit a cooperative effect for the catalysis.

In the case of n-butyl acetate which has a longer alkyl group than methyl acetate, general aspects are similar but polystyrenesulphonic acids have a greater catalytic effect than in the case of methyl acetate.

Similar experiments had been carried out with various esters at various temperatures and we came to the conclusion that a large r value is due to the local concentration of ester in the neighbourhood of polymer molecule caused by hydrophobic interaction. If this conclusion is true, r is expected to decrease when some organic liquid is added to the reaction mixture. The hydrolysis was thus carried out in aqueous acetone instead of in pure water. As was expected, r decreased rapidly by addition of acetone to the reaction mixture (see *Table 7*).

Table 7. Effect of added acetone on the rate of hydrolysis of n-butyl acetate with polystyrene-sulphonic acids (Ester conc. $2 \cdot 85 . 10^{-2}$ M; catalyst conc. $5 \cdot 00 . 10^{-3}$ N; temperature $40°$C)

Vol. % acetone in reaction mixture	r for following catalyst	
	PS-S (40)	PSS
0	6·31	1·92
15	2·61	1·48
30	1·55	1·12

Such a hydrophobic interaction is not always observed for polymeric sulphonic acids independent of its nature, but only those polymeric sulphonic acids with strongly hydrophobic groups exhibit the effect. *Table 8* gives examples of polymeric sulphonic acids with $r \fallingdotseq 1$ and $r > 1$.

Experiments were also undertaken with cross-linked sulphonated polystyrenes such as Dowex 50 W[26], because it is possible in this case to measure the adsorbed amount of the ester on the resin, and to compare the local concentration of ester with the effectiveness of the catalyst. As it is impossible to measure the adsorbed amount with H-type resins because of the simultaneous hydrolysis of ester, the adsorption measurements were carried out with Na-type resins. When c_0 is the initial concentration of ester of the system and c_i the concentration of ester in the resin phase, then c_i/c_0 is a measure of the local concentration of ester in the neighbourhood of polymer molecules. The value of c_i/c_0 was compared with r measured with the use of H-type resins. Several examples are shown in *Table 9*. As may be seen from the table, there is a close parallelism between r and c_i/c_0, so that it may be said that our argument that larger r values are mainly due to local concentration of ester in the neighbourhood of polymer molecules is definitely proved.

Mention will now be made of the electrostatic interaction in the hydrolysis of esters with polymeric sulphonic acids.

Table 8. Catalytic effect of polymeric sulphonic acids

A. r is nearly equal to 1
PVS: Polyvinyl sulphonic acid;
PVS-VA: Copolymer of vinyl sulphonic acid and alcohol;
PVBS: Polyvinyl butyral sulphonic acid;
PVBeS: Polyvinyl benzal sulphonic acid;

B. r is larger than 1
PSS: Polystyrenesulphonic acid;
PS-S: Partially sulphonated polystyrene;
PVS-St: Copolymer of vinyl sulphonic acid and styrene;
PAN-S: Partially sulphonated polyacenaphthylene;
PMS-S: Partially sulphonated poly α-methylstyrene;
PS-Sti-S: Partially sulphonated copolymer of styrene and stilbene

Table 9. Comparison of ester adsorption (c_i/c_0) and catalyst effectiveness r (at 40°C)

	Ester	Methyl acetate	n-Butyl acetate	tert-Butyl acetate	Phenyl acetate	Hexyl acetate
Dowex W × 2	c_0^*		0·0305	0·0295		
	c_i^*		0·0863	0·0758		
	c_i/c_0		2·83	2·57		
	r		2·70	1·74		
Dowex W × 4	c_0^*	0·0504	0·0305	0·0295	0·0157	0·0136
	c_i^*	0·0814	0·0956	0·0804	0·0789	0·0721
	c_i/c_0	1·62	3·13	2·73	5·02	5·30
	r	1·52	2·90	1·84	6·30	4·70

* c_0 and c_i are given in mole/l.

When esters of an amino acid are employed, it may be expected that the local concentration of the ester in the neighbourhood of a polymer molecule occurs due to the electrostatic attraction between sulphonic acid groups of the catalyst and amino groups of the substrate to result in a higher rate of hydrolysis.

Experiments were carried out employing o- and p-aminobenzoates in 50 vol. % aqueous acetone keeping catalyst·and substrate concentration constant. The initial rate constant k_0 was determined graphically by extrapolation. The results are shown in *Figure 15*; as expected r has large values. It is noteworthy that r increases with increasing degree of sulphonation. This is a remarkable contrast to the case of hydrophobic interaction, where r decreases with increasing degree of sulphonation (see *Figure 16*). If isolated sulphonic acid groups contribute to increase r, the above result is difficult to explain, as the effect of PS-S of a lower degree of sulphonation is practically the same as that of hydrochloric acid; a certain sequence of sulphonic acid groups seem to be essential to show large r-values.

(b) Hydrolysis of polymeric esters with long-chain alkyl and alkylbenzene sulphonic acids[27]

Now we wish to treat the reverse case of the above-mentioned hydrolysis of low-molecular-weight esters with polymeric sulphonic acids, i.e., the

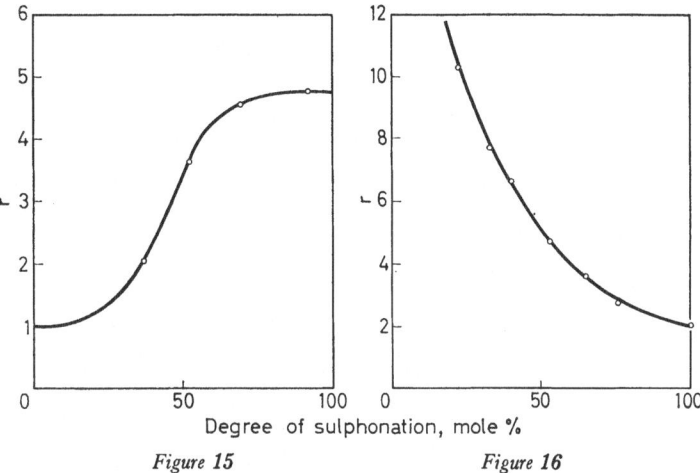

Degree of sulphonation, mole %

Figure 15   Figure 16

Figure 15. Hydrolysis of *p*-aminobenzoate with polystyrenesulphonic acids in 50% aqueous acetone (catalyst conc., 0·05N; ester conc., 0·08M; temp. 80°C)
Figure 16. Hydrolysis of butyl acetate with polystyrenesulphonic acids in water (catalyst conc., 0·05N; ester conc., 0·00285M; temp. 40°C)

hydrolysis of polymeric esters with low-molecular-weight sulphonic acids (*Figure 17*). As a polymeric ester, water-soluble partially-acetylated polyvinyl alcohol and as the catalysts, alkyl- and alkylbenzene sulphonic acids were employed. Experimental details and results are shown in *Table 10*.

As may be seen from the table, octyl sulphonic acid is found to show nearly the same rate of hydrolysis as that with hydrochloric acid, whereas, in the case of hydrolysis with alkyl and alkylbenzene sulphonic acids such as dodecyl, hexadecyl, octadecyl and dodecylbenzene sulphonic acids, much higher rates are observed. These higher rates may be attributed to

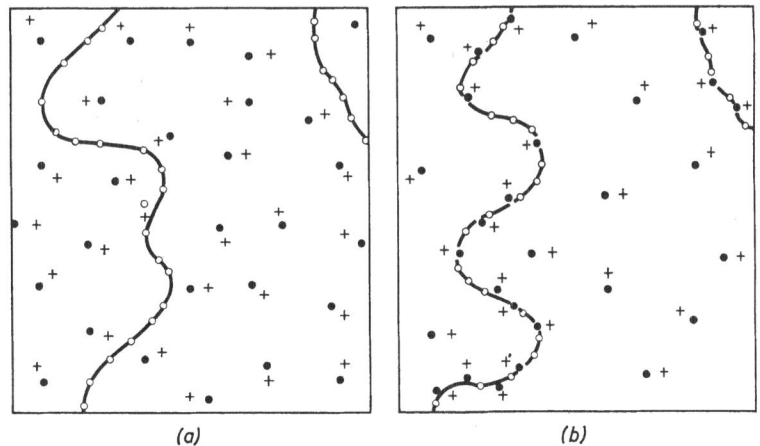

(a)   (b)

Figure 17. Hydrolysis of polymeric esters with low-molecular-weight sulphonic acid. ○ Ester, ● Acid, + Hydrogen ion. (*a*) No local concentration of sulphonic acid. (*b*) Local concentration of sulphonic acid

Table 10. Hydrolysis of acetylated polyvinyl alcohols with low-molecular-weight sulphonic acids
(Conc. of acetylated polyvinyl alcohol 0·30 g/l.)

Catalyst	Catalyst conc. N	Temp. °C	r for polyvinyl alcohol with following degree of acetylation		
			9·21	23·3	33·6
Octyl s.a.	0·010	40	1·1	1·1	1·3
Dodecyl s.a.	,,	,,	12·5	13·5	12·4
Hexadecyl s.a.	,,	,,	——	19·1	22·1
Dodecylbenzene s.a.	,,	,,	20·7	20·6	20·7
Dodecyl s.a.	0·0050	50	10·4	9·3	10·8
Hexadecyl s.a.	,,	,,	19·3	20·0	——
Octadecyl s.a.	,,	,,	——	19·1	22·1
Dodecylbenzene s.a.	,,	,,	19·3	16·7	19·9

higher local concentration of the long-chain sulphonic acid in the neighbourhood of polymer molecules as a result of the hydrophobic attraction between the substrate and catalyst molecules.

(c) Hydrolysis of polymeric esters with polymeric sulphonic acids[28]

Finally, mention will be made of the hydrolysis of polymeric esters with polymeric sulphonic acids; in this case both substrate and catalyst are polymers. If hydrophobic interaction exists, it is expected that the effect is here is more pronounced. Experiments were carried out using partially-acetylated polyvinyl alcohols as substrates and polystyrenesulphonic acids as catalysts. Experimental details and results are shown in Table 11. It is seen that for all partially-acetylated polyvinyl alcohols the polystyrene sulphonic acid gives large r-values; the largest r value is about 40. This implies that the catalyst effect of the polystyrenesulphonic acid is 40 times greater than that of hydrochloric acid under the same reaction conditions. With increasing degree of acetylation r increases, and partially-sulphonated polystyrene PS-S (31) exhibits much larger r values than pure polystyrene-sulphonic acid. Further, acetylated polyvinyl alcohols having a relatively long sequence of acetyl groups give larger r than randomly-acetylated polyvinyl alcohol. All these phenomena suggest that the hydrophobic interaction between acetyl groups of acetylated polyvinyl alcohol and benzene rings of polystyrene sulphonic acid plays an important role in the hydrolysis.

Table 11. Hydrolysis of acetylated polyvinyl alcohol with polystyrenesulphonic acids
(Substrate conc., 3·0g/l.; catalyst conc., 0·0050 N; temperature 50°C)

Degree of acetylation (mole %)	9·2	9·8*	15·4	21·5	27·0	28·2*	33·6
r { PSS	4·8	10·8	8·8	13·0	17·5	20·2	19·5
PS-S (31)	18·5	36·1	22·2	29·9	34·2	36·7	38·3

* These polyvinyl alcohols have a somewhat longer sequence of acetyl groups.

In conclusion I would say that the most important aspect of polymer reactions is not the size of molecule but the fact that a functional group of polymer molecule always has neighbours. In some cases a neighbour exhibits specific interaction with functional groups as was pointed out by Morawetz, Smets and others; but hydrophilic, electrostatic and especially hydrophobic interaction between neighbouring groups and molecules of the reaction's partner is more general and therefore fundamental.

References

[1] P. J. Flory. *J. Am. Chem. Soc.* **61**, 3334 (1939);
K. Freudenberg, W. Kuhn, W. Dürr, F. Bolz, and G. Steinbrunn. *Chem. Ber.* **63**, 1510 (1930).

[2] T. Ohsugi. Gosei-Seni-no-Kenyu **2**, 192 (1944);
I. Sakurada. Kobunshi Tembo **5**, 64 (1951).

[3] A. Conix and G. Smets. *J. Polymer Sci.* **15**, 221 (1955).

[4] I. Sakurada, Y. Sakaguchi, and S. Fukui. *Chem. High Polymers*, Tokyo (Kobunshi Kagaku) **13**, 355, 361, 408 (1956).

[5] I. Sakurada, Y. Sakaguchi, and S. Fukui. *Chem. High Polymers*, Tokyo (Kobunshi Kagaku) **17**, 83 (1960).

[6] I. Sakurada and Y. Sakaguchi. *Chem. High Polymers*, Tokyo (Kobunshi Kagaku) **13**, 441 (1956);
I. Sakurada, Y. Sakaguchi, and S. Ishiguro. *ibid.* **17**, 115 (1960).

[7] E. Nagai and N. Segane. *Chem. High Polymers*, Tokyo (Kobunshi Kagaku) **12**, 195 (1955).

[8] I. Sakurada, Y. Sakaguchi, and M. Kagau. *Chem. High Polymers*, Tokyo (Kobunshi Kagaku) **17**, 87 (1960).

[9] A. Katchalsky, N. Shavit, and H. Eisenberg. *J. Polymer Sci.* **8**, 69 (1954).

[10] I. Sakurada, K. Noma, and A. Kato. *Chem. High Polymers*, Tokyo (Kobunshi Kagaku) **15**, 799 (1958).

[11] To be published in *Chem. High Polymers*, Tokyo (Kobunshi Kagaku).

[12] R. C. Schulz. *Makromol. Chem.* **42**, 205 (1961).

[13] H. Morawetz and P. E. Zimmering. *J. Phys. Chem.* **58**, 753 (1954).

[14] P. E. Zimmering, E. W. Westhead, Jr., and H. Morawetz. *Biochim. Biophys. Acta* **25**, 376 (1957); *Chem. Abstr.* **51**, 16617d (1957).

[15] W. De Loecker and G. Smets. *J. Polymer Sci.* **40**, 203 (1959);
G. Smets and W. De Loecker. *ibid* **14**., 375 (1959); **45**, 461 (1960).

[16] K. Noma and N. Sawagashira. *Chem. High Polymers*, Tokyo (Kobunshi Kagaku) **4**, 46 (1947).

[17] K. Noma, T. Wo, and T. Tsuneda. *Chem. High Polymers*, Tokyo (Kobunshi Kagaku) **6**, 439 (1949).

[18] I. Sakurada, Y. Sakaguchi, and Y. Ohmura. *Chem. High Polymers*, Tokyo (Kobunshi Kagaku) **21**, 564 (1964).

[19] F. J. Glavis. *J. Polymer Sci.* **36**, 547 (1959).

[20] G. J. Smets and W. De Loecker. *J. Polymer Sci.* **45**, 461 (1960).

[21] G. J. Smets. *Angew. Chem.* **74**, 337 (1962).

[22] I. Sakurada, Y. Sakaguchi, Z. Shiiki, and J. Nishino. *Chem. High Polymers*, Tokyo (Kobunshi Kagaku) **21**, 241 (1964).

[23] I. Sakurada and Y. Sakaguchi. *Makromol. Chem.* **61**, 1 (1963).

[24] I. Sakurada, Y. Sakaguchi, T. Iwagaki, and Y. Mikuzu. *Chem. High Polymers*, Tokyo (Kobunshi Kagaku) **21**, 426 (1964).

[25] I. Sakurada, Y. Sakaguchi, T. Ono, and T. Ueda. *Makromol. Chem.* **91**, 243 (1966).

[26] I. Sakurada, Y. Sakaguchi, and T. Ono. *Chem. High Polymers*, Tokyo (Kobunshi Kagaku).

[27] I. Sakurada, Y. Sakaguchi, and Y. Ohmura. *Bull. Inst. Chem. Res.*, Kyoto **43**, 149 (1965).

[28] I. Sakurada, Y. Sakaguchi, and Y. Ohmura. *Bull. Inst. Chem. Res.*, Kyoto **44**, 135 (1966).

NEW METHODS OF POLYMERIZATION

C. E. H. Bawn

Donnan Laboratories, University of Liverpool, Liverpool, England

Every addition polymerization process involves two fundamental steps: initiation and propagation. Termination is also an important step but in some processes termination may be avoided by a judicious choice of experimental conditions. In the exploration of either new methods of polymerization or the development of existing methods attention may be concentrated on either the initiation process or on the propagation reaction. Sometimes both of the steps may be involved together. This paper will be largely concerned with recent progress in the study and development of new initiators and initiating processes.

The propagation process of a conventional addition polymerization of a monomer M consists of a series of reactions described by

$$P_nX + M \longrightarrow P_{n+1}X$$

The reactive moiety X attached to the chain—which may be part of the initiator—is responsible for the onset of the reaction and the monomer does not require any further activation. In such processes the initiator participates only in the step creating the growing centre and not usually in the subsequent propagation.

Three types of reactive centre may be identified: radicals, ions and coordination complexes. The ionic species may be further divided into cationic and anionic, depending on whether the growing species carries a positive charge or forms the positive end of a dipole or is negatively charged or forms the negative end of a dipole.

The most striking developments in ionic polymerization in recent years have been made in anionic polymerization and have resulted largely from the discovery of two important features of the reaction:

(1) the initiation of polymerization through an electron transfer process;
(2) the possibility of avoiding termination and transfer and therefore the feasibility of producing living polymers.

In the electron transfer initiation reaction the first step is an electron transfer from a suitable electron donor to the monomer with the formation of a radical ion

$$\text{Electron donor} + \text{monomer} \longrightarrow \text{monomer}^- \text{ (radical ion)}^+$$

The radical ion may undergo two types of reaction, namely

$$\text{Monomer}^- + \text{monomer}^- \longrightarrow \text{dimeric}^{2-}$$

285

e.g.
$$\underset{\underset{C_6H_5}{|}}{\dot{C}H-CH_2^-} + \underset{\underset{C_6H_5}{|}}{CH_2^--\dot{C}H} \longrightarrow \underset{\underset{C_6H_5}{|}}{\overline{C}H_2-CH_2}-\underset{\underset{C_6H_5}{|}}{CH_2-\overline{C}H} \tag{1}$$

or Monomer$^-$ + monomer $\longrightarrow$ dimeric radical ion

e.g.
$$\underset{\underset{C_6H_5}{|}}{\dot{C}H_2-\overline{C}H_2} + \underset{\underset{C_6H_5}{|}}{CH_2{=}CH} \dashrightarrow \underset{\underset{C_6H_5}{|}}{\dot{C}H-CH_2}-\underset{\underset{C_6H_5}{|}}{CH_2-\overline{C}H} \tag{2}$$

The dimeric species initiates the conventional chain of the polymerization growth taking place essentially on the carbanion ends since the radical end if formed in (2) disappears either by mutual termination or through another electron transfer. The concepts have been firmly established and their application to ionic polymerization has been admirably reviewed by Szwarc[1].

While anionic polymerization may be induced by transferring an electron from a suitable donor to monomer, the transfer of an electron from an electron rich monomer (donor) to a suitable acceptor may lead to cationic polymerization. For example, transfer to a cation will produce a radical and a radical cation:

$$CH_2{=}CHR + X^+ \longrightarrow \dot{X} + \dot{C}H_2-\overset{+}{C}HR$$

The process may be generalized as follows:

$$A + CH_2{=}CHR \longrightarrow \ddot{A} + \dot{C}H_2-\overset{+}{C}HR \tag{3}$$

where A the acceptor may be a neutral molecule or an ion. The process may be, and usually is, much more complex than represented by (3) and the donor D and acceptor molecule A interact to form a complex which may often be sufficiently stable to be characterized.

$$D + A \leftrightharpoons [D.A \longleftrightarrow \overset{+}{\ddot{D}}\,\overset{-}{\ddot{A}}] \leftrightharpoons \overset{+}{\ddot{D}} + \overset{-}{\ddot{A}} \tag{4}$$
$$\text{ground state}$$

In this paper we shall review the development of the general reactions (3) and (4) for the formation of cation radicals and their application as polymer-initiating reactions. Unlike the corresponding anionic reactions, cationic polymerizations are subject to monomer transfer reactions. This follows since olefinic reactivity in cationic polymerization requires the presence of strongly electron-releasing substituents. These substituents are usually polar groups or conjugated electron systems and provide in the polymer an electron-rich site of considerable reactivity. These active sites easily take part in transfer or branching reactions and the control of the propagation and molecular weight is a much more difficult operation. Isobutene, $(CH_3)_2C{=}CH_2$, is an exception to this rule and readily give much higher molecular polymers than the other commonly cationically susceptible monomers. Because of these difficulties it has not so far been possible to demonstrate the existence of the propagating dimeric cationic species as in the comparable anionic polymerization.

A further complication which arises in cationic polymerization is the ready oxidation of any free radical centre by the reagent used initially for the one electron oxidation of the olefin. This reaction is important in systems employing charge-transfer complexes as initiating agents (q.v.).

ANIONIC POLYMERIZATION IN DIMETHYL SULPHOXIDE

Before proceeding to the main topic of cationic initiation, reference will be made to some recent developments in anionic polymerization. In addition to the process of electron transfer already mentioned anionic polymerization may be initiated by the addition of a negative ion or ion pair to a suitable monomer. This mode of initiation is illustrated by the following examples:

(a) $PhC(Me)_2^-K^+ + CH_2{=}CH \longrightarrow PhC(Me)_2CH_2{-}CH\ Ph^-K^+$
$\qquad\qquad\qquad\qquad\quad |$
$\qquad\qquad\qquad\qquad\ Ph$

(b) $K^+ + NH_2^- + CH_2{=}CH \longrightarrow NH_2CH_2{\cdot}CH(Ph)^-K^+$
$\qquad\qquad\qquad\qquad\ |$
$\qquad\qquad\qquad\quad Ph$

(c) $CH_3O^- + CH_2{-}CH_2 \dashrightarrow CH_3O\ CH_2\ CH_2O^-$
with an epoxide (O bridging $CH_2{-}CH_2$)

In (a) the polymerization is initiated by an ion pair; in (b) reaction in liquid ammonia initiation is by the free NH_2^- ion. Example (c) seems to be a case where free ions initiate and propagate the reaction. A catalyst in this group of very wide applicability is the potassium derivative of dimethyl sulphoxide[2].

In recent years there has been a widespread development in the use of dimethyl sulphoxide (DMSO) as a solvent for promoting base-catalysed reaction. The most common systems involve the use of potassium *tert*-butoxide in DMSO where enhanced basic characteristics result, in part, from the equilibrium

$$K^+OBu^-(t) + CH_3{\cdot}SO{\cdot}CH_3 \rightleftharpoons CH_3SOCH_2^-K^+ + Bu(t)OH$$

The most striking feature of DMSO as solvent is its high solvating power for cations and low solvating power for anions and this leaves the dimsyl ion $CH_3SOCH_2^-$ free to take part in polymerization initiating reactions. Thus ethylene oxide, propylene oxide, acrylonitrile and methyl methacrylate are polymerized rapidly at 25°C. Molecular weights, at high conversion, were in reasonable agreement with those calculated on the basis of one polymer chain per catalyst molecule and this was confirmed by end-group analysis. With ethylene oxide and methyl methacrylate the reaction kinetics were bimolecular in the early stages of the reaction and obeyed the simple expression:

$$\text{Rate of polymerization} = k_p \, (\text{base}) \, (\text{monomer})$$

The initiation was very rapid and complete and a termination reaction is most unlikely. The measured rates of polymerization represent the propagation reaction k_p (*Table 1*). The presence of one end group per polymer indicates that initiation occurs via interaction of the dimsyl ion with the monomer:

$$CH_3 \cdot SOCH_2^- + CH_2\text{---}CH_2 \longrightarrow CH_3SO(CH_2)_2CH_2O^-$$

$$\overset{\displaystyle O}{\diagdown \diagup}$$

$$\Big| \; n \; CH_2\text{---}CH_2$$

$$\overset{\displaystyle O}{\diagdown \diagup}$$

$$\downarrow$$

$$CH_3\overset{\displaystyle }{S}OCH_2(CH_2CH_2O)_n CH_2CH_2O^-$$

The close agreement between experimental and calculated molecular weights for ethylene oxide indicates that transfer reactions are negligible. Propylene oxide, however, formed only low molecular polymers with unsaturated end groups, indicating excess transfer to solvent or monomer. These catalysts give rates of propagation for ethylene oxide about 10^4 greater than obtained by Gee and coworkers[3] using sodium methoxide in dioxane as initiator, and the molecular weights are much higher than those usually obtained in homogeneous systems.

Attempts to polymerize styrene with this catalyst were not entirely successful unless the ratio styrene : DMSO was greater than unity. At lower ratios a complex mixture of unsaturated molecules formed by reaction of the dimsyl ion and styrene, similar to those described by Walling and Bollyky[4].

Since this work was completed, there have been several preliminary reports of related polymerizations in DMSO[5,6], but few kinetic data are available for comparison. The values of k_p for propylene oxide and ethylene oxide in DMSO are much smaller than those observed in the present work[7], although the reasons for these differences are not at present clear.

CATIONIC POLYMERIZATION

Detailed studies of cationic polymerization have always lagged behind those of free radical and anionic polymerization largely because of the lack of well-defined initiator systems. The initiating substances used in this type of polymerization fall into the broad groupings: (*a*) protonic acid and acid surfaces; (*b*) Friedal–Crafts halides, and (*c*) carbonium-ion salts, all of which give carbonium ions or oxonium ions by addition of a proton or cation to an olefin or cyclic oxide

$$I^+(H^+) + CH_2\text{==}CHR \longrightarrow ICH_2\text{---}\overset{+}{C}RH \qquad (1)$$

Table 1. Polymerization in DMSO catalysed by potassium t-butoxide at 25·0°C

Monomer	Concentration (mole/litre)		$10^3 \times$ Rate (mole litre^{-1} sec^{-1})	$10^3 \times k_p$ [a] (litre mole^{-1} sec^{-1})	Degree of Poly[a][b]	
	[Monomer]	$10^3 \times$ [Initiator]			Calculated	Found
Ethylene oxide	3·22	11·6	2·32	6·2	198	139
	4·25	12·2	8·35	12·4	174	193
	4·31	2·19	1·22	12·8	985	750
	5·50	12·2	8·35	12·4	226	261
Propylene oxide	1·81	6·83	0·25	1·92		Oily liquid
Methyl methacrylate	1·30	5·30	3·40	49	123	281
	1·28	0·26	0·04	13	2430	1862
	1·32	1·18	0·52	33	550	355
	1·30	2·73	2·78	78	250	260
	1·58	7·70	7·60	63	100	140
	2·21	7·70	9·60	56	150	132
	1·28	7·64	6·00	61	90	135

a Calculated from the expression: Rate = k_p [Monomer] [Initiator]
b For ethylene oxide, polymerization was taken to completion and yields of polymer recovered were in excess of 80%.
For methyl methacrylate, polymerization was stopped at 50% conversion.
Estimation of DP was from measurements of intrinsic viscosity using the following relationships:
Polyethylene oxide $[\eta] = 1·25 \times 10^{-4} M^{0·74}$ in water at 30°C
Polymethylmethacrylate $[\eta] = 0·75 \times 10^{-4} M^{0·74}$ in benzene at 25°C.

$$I^+(H^+) + \boxed{\begin{array}{c} -(CH_2)_n- \\ \\ -O- \end{array}} \longrightarrow H-O^+ \quad \boxed{(CH_2)_n} \qquad (2)$$

Strong acids, especially sulphuric and perchloric, readily polymerize monomers such as styrene and it was thought until recently that initiation was due to carbonium intermediates formed according to (1). Recent work has established that the propagating species in this reaction is not a carbonium ion—either free or paired—but an ester formed from the monomer and the catalyst[8]. Friedal–Crafts halides which form the most important group of catalysts do not usually initiate alone but catalyse the initiation of some other substance—the cocatalyst—which provides the actual initiating fragments. The cocatalyst is usually water but acids, alcohols and other polar substances cause initiation. The species formed may have only a transitory existence and there are obvious difficulties in defining the precise initiator. Furthermore, kinetic studies of cationic polymerization are difficult to evaluate because the cocatalyst can act as a transfer agent or even as a reaction terminator. True termination processes are not often encountered in these cationic processes and the chain activity is usually destroyed by some side process.

It seems therefore clearly desirable that, in order to overcome some of the difficulties mentioned above, the catalyst and cocatalyst components should be completely utilized in the form of a well-defined, stable and easily characterizable material. The most useful substances for this purpose are stable carbonium-ion salts. They are well known and the particular salts which have been used extensively in our work are those of the triphenyl methyl and tropylium cation. These salts with the general formulae $Ph_3C^+X^-$ and $C_7H_7^+X^-$, where X is a stable anion, ClO_4^-, $SbCl_6^-$, BF_4^-, PF_6^-, etc., are easily prepared as stable crystalline materials and readily and reproducibly polymerize tetrahydrofuran, vinyl alkyl ethers, n-vinyl carbazole, acenaphthalene, styrene and other vinyl monomers[9].

Stable carbonium ions may initiate vinyl polymerization by several possible mechanisms: (1) Direct addition to the unsaturated system to give a carbonium or oxonium ion

$$I^+ + CH_2{=}CHR \longrightarrow I{-}CH_2{-}\overset{+}{C}HR$$

(2) Hydride extraction

$$I^+ + CH_2{=}CHR \longrightarrow IH + CH_2 {-}{-}{-} \overset{+}{C}H{-}R'$$

(3) The formation of cation-radical, by electron transfer or other mechanism

$$I^+ + CH_2{=}CHR \longrightarrow \overset{\cdot}{I} + \overset{\cdot}{C}H_2{=}\overset{+}{C}HR$$

The relative significance of these alternative initiating reactions will be considered in detail and in particular their relation to the initiation of the polymerization of monomers known to be readily susceptible to cationic initiation, viz. tetrahydrofuran, n-alkyl vinyl ether and n-vinyl carbazole.

The trityl hexachloroantimonate—a bright-yellow-coloured stable solid—with tetrahydrofuran gave instant discoloration of the catalyst and homogeneous polymerization of the monomer[10]. Although the initial discoloration of Ph_3C^+ cation is due to oxonium ion formation it has now been established that a subsequent reaction produces triphenyl methane before polymerization occurs, viz.

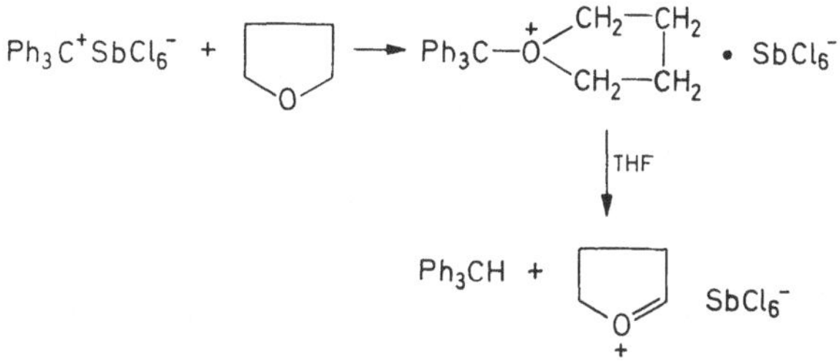

Evidence for the formation of triphenylmethane was provided initially by attempts to produce di-cations in the polymerization of tetrahydrofuran. Kuntz[11] showed that the stable dicarbonium salt

$$[Ph_2\overset{+}{C}—CH_2CH_2—\overset{+}{C}Ph_2] \; [SbCl_6^-]_2$$

produced polytetrahydrofuran having the same molecular weight as that produced by corresponding concentrations of the mono-salt $Ph_3C^+SbCl_6^-$. Later work by Kuntz[12] using n.m.r. techniques showed that triphenylmethane was rapidly formed during initiation. Independent work by Ledwith and Fitzsimmons in this laboratory has shown that triphenylmethane can be detected by gas chromatography during the polymerization initiated by $Ph_3C^+SbCl_6^-$ and can actually be isolated from the reaction mixture by chromatography on neutral alumina.

The subsequent steps in the propagation reactions are

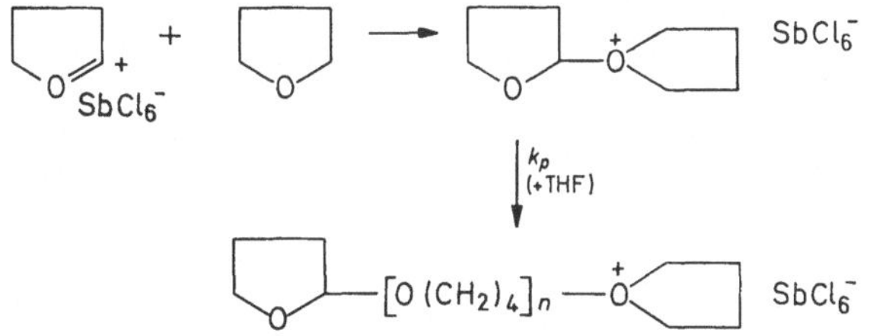

At room temperature and below, there was little termination and reaction proceeded to equilibrium conversion. Later work[13] with different anions—especially PF_6^-—showed that termination could be completely eliminated and a truly "living" polymer system established.

It was also first shown by P. Dreyfuss and Mrs Dreyfuss[14] working in this laboratory that p-chlorophenyl diazonium hexafluorophosphate was a very effective initiator for tetrahydrofuran polymerization. Extremely high molecular weight polymers were obtained and the system was free from termination reactions. The initiation was thought to be hydride abstraction, viz.

G

Cl—⟨O⟩—N₂⁺ PF₆⁻ + THF → Cl—⟨O⟩—N=N—Ö⟨ ⟩ PF₆⁻ →

→ Cl—⟨O⟩—H + N₂ + [ ], PF₆⁻

Since the publication of the use of triphenylmethane cation salts for tetra-hydrofuran polymerization, many different laboratories have used these catalysts for the polymerization of other cyclic ethers,

e.g. dioxalane ⟨ ⟩ and symtrioxane ⟨ ⟩

The effect of the anions differs for each monomer but non-terminating systems have been developed in each case. Thus for trioxane polymerization, SbF_6^- appears to be the preferred anion, whereas PF_6^- and AsF_6^- are better for dioxalane. The subject has been recently reviewed by Dreyfuss and Dreyfuss[15] and by Ledwith[16].

Parallel with the studies of the polymerization of tetrahydrofuran, it was shown that reactive olefins such as alkyl vinyl ethers, n-vinylcarbazole and alkoxy-styrenes were also readily polymerized using triphenylmethyl cation salts as catalysts. These reactions are easily carried out in a variety of non-protic solvents such as CH_2Cl_2, CH_3CN, CH_3NO_2, $ClCH_2CH_2Cl$—in which the catalyst is soluble. Many of these polymerizations are much faster than tetrahydrofuran and catalyst concentrations as low as 10^{-5} M are necessary to give measurable rates. For example, with isobutyl vinyl ether (*Table 2*) the overall rate of polymerization, k_p, estimated from the rate equation $-dM/dt = k_p(M)$ (Catalyst) or $k_p(M)^2$ (Catalyst) was of the order of 10^4 l. mole^{-1} sec^{-1}. At temperatures below 0°C, little termination occurred but the molecular weights of the polymer were less than 5000 even

Table 2. Polymerization of vinyl isobutyl ether in CH_2Cl_2 at 0°C

[CH₂=CHOBuⁱ] M	Initiator		Overall rate coefficient[a] $10^{-3} k_p$ (l.m⁻¹ sec⁻¹)	Polymer viscosity[b],[c] [η] dl/g
	[Ph₃C⁺SbCl₆] 10⁵M	[C₇H₇⁺SbCl₆—] 10⁵M		
0·079	5·64		3·0	0·071
0·19	1·93		3·6	0·087
0·19	3·75		3·7	0·083
0·19	5·57		3·6	0·084
0·19	7·42		3·4	0·084
0·19		7·1	4·1	0·085
0·08		5·32	5·0	0·072

[a] Reaction rates were followed *in vacuo* in an adiabatic calorimeter essentially as described by Biddulph and Plesch. The rate coefficient was estimated from the expression:

$$\frac{-d\,[Monomer]}{dt} = k_p\,[Monomer]\,[Catalyst]$$

[b] Measured in benzene at 25°C.
[c] The yield of polymer was always quantitative.

at $-25°C$ due to the easy occurrence of monomer transfer reactions. At the catalyst concentration of about 10^{-5} M used in these measurements equilibrium studies showed that the carbonium-ion salt was completely dissociated into ions. The detection of Ph_3C end groups in the polymer showed that initiation involved the direct reaction of the Ph_3C^+ with the double bond. It is highly probable that the free ions of the catalyst are solvated by the vinyl ether molecules to form a complex and the mechanism of the initiation and kinetics may be explained if it is assumed that this complex subsequently collapses to form a complex cation initiator:

$$R^+ + xM \longrightarrow R^+(M)_x SbCl_6^- \longrightarrow R-CH_2-\overset{+}{C}H(M)_{x+1}SbCl_6^-$$
$$\text{OR}$$

where x is the number of molecules associated with the cation in the complex.

The success achieved with triphenylmethyl cations as initiators prompted evaluation of other stable cationic salts. Thus the cycloheptatrienyl cation (tropylium ion) is well known to be of great stability and tropylium salts of anions such as ClO_4^-, PF_6^-, $SbCl_6^-$ are easily made. This salt is an effective initiator for cationic polymerization of cyclic ethers, vinyl alkyl ethers, n-vinylcarbazole, alkoxy styrene, vinylnaphthalene and acenaphylene. The polymers formed are similar to those from the trityl salts but unlike the latter salts the tropylium salts are colourless but give instant and intense colour formation on addition to the olefin and this colour disappears during the polymerization. Tropylium salts are known to form charge-transfer complexes with aromatic molecules and ethers[17] and it was evident that similar complexes were being formed as precursors of the intermediates leading to polymer formation. In these charge-transfer complexes the tropylium ion is the acceptor component and the olefin (or ether) the donor component. These complexes provided a direct method for the formation of cation-radicals and the remaining sections of this paper will deal with the general subject of cation-radicals in polymer synthesis.

Cation-radicals as initiators for polymerization

Modern theories of bonding in charge-transfer complexes assume that the donor molecule D and an acceptor molecule A interact to form a complex which may be regarded as a resonance hybrid of non-bonding and electron-transfer canonical forms, i.e.

$$D + A \leftrightharpoons [DA \longleftrightarrow D\cdot^+; A\cdot^-] \leftrightharpoons [D\cdot^+, A\cdot^- \longleftrightarrow DA]$$
$$\text{ground state} \qquad\qquad \text{excited state}$$

In the ground state the non-bonding form predominates and the charge-transfer predominates in the first excited state. The small energy separation between the ground and the first excited state gives rise to light absorption in the visible part of the spectrum and corresponds to the well established charge-transfer spectrum. When the donor molecule has a low ionization potential and the acceptor molecule a high electron affinity, electron transfer can occur to a significant extent even in the ground state of the complex. That is, mutual oxidation and reduction will occur according to

$$D + A \leftrightharpoons \text{Complex} \leftrightharpoons D \cdot^+, A \cdot^-$$

and it is therefore possible to use the three common propagating intermediates for initiation of polymerization—cation, anion and radical in the from of radical ions.

Kosower[18] has discussed in detail the conditions under which radical-ion formation from charge transfer will be significant. In polymerization reactions the most favourable situation occurs when, by suitable choice of reactants and solvent, it is possible to have a thermal equilibrium between complex and ion pairs (see *Figure 1*). Certain olefinic substances with

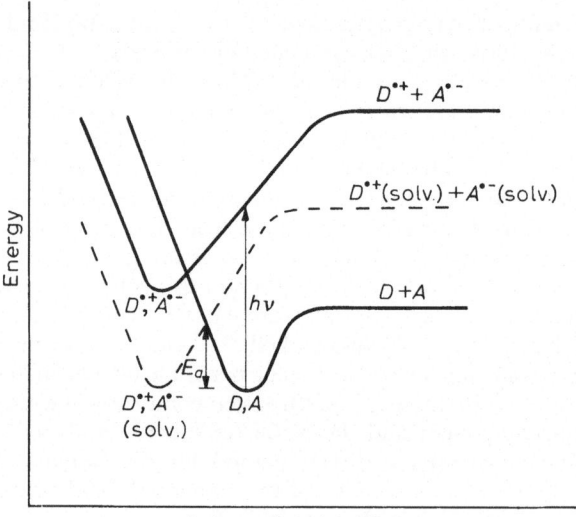

Figure 1

relatively low ionization potentials (7–9 eV) function as donors and common electron acceptors include (*a*) neutral molecules, such as quinones, anhydrides, nitrocompounds, etc., and (*b*) ionic intermediates, such as metal ions, ionic acids and carbonium ions. Both classes of acceptor are important to the polymer chemist and, by careful choice of reaction (monomer, acceptor and solvent), conditions may be attained which facilitate the radical-ion formation and the initiation of the polymerization reaction. An extremely favourable case is that when the acceptor is a cation, for example the tropylium cation[9, 19]:

$$D + C_7H_7^+ \longrightarrow \text{Complex} \longrightarrow D \cdot^+ + C_7H_7 \cdot$$

The cycloheptatrienyl radical formed may dimerize to give ditropyl

$$2 \, C_7H_7 \cdot \longrightarrow C_7H_7 - C_7H_7$$

and thus shift the equilibrium to the right in the above equation.

Using methylene chloride and acetonitrite as solvents, it has been found that tropylium chloroantimonate ($C_7H_7^+SbCl_6^-$) and tropylium tetrafluoroborate ($C_7H_7^+BF_4^-$) initiate the polymerization of n-vinylcarbazole, acenaphthylene, 1- and 2-vinylnaphthalene, vinylmesitylene and styrene. Tropylium salts are known to form stable charge-transfer complexes with aromatic molecules, and with olefinic monomers significant initiation of polymerization occurs after initial formation of coloured charge-transfer complexes between monomer and the $C_7H_7^+$ ion. The monomers used are those known to be susceptible to cationic polymerization and it seems highly probable that the $C_7H_7^+$ initiated polymerization follows a carbonium-ion mechanism. Support for this conclusion was obtained from a detailed study of the polymerization of isobutyl vinyl ether (IBVE) and vinylcarbazole (VC).

Both the catalysts $Ph_3C^+SbCl_6^-$ and $C_7H_7^+SbCl_6^-$ gave rapid and complete initiation of IBVE. Using $Ph_3C^+SbCl_6^-$ as initiator, it is certain that a classical carbonium ion is involved and the very close agreement between rate and molecular-weight data obtained with this and $C_7H_7^+SbCl_6^-$ as initiator (*Table 2*) confirms the cationic nature of the polymerizations initiated by the tropylium salt.

The tropylium-ion initiation reaction may therefore be represented as occurring *via* a radical cation:

$$C_7H_7^+SbCl_6^- + CH_2{=}CH{-}OR \longrightarrow \text{Complex} \longrightarrow$$
$$C_7H_7\cdot + CH_2{-}\overset{+}{C}H\ OR{\cdot}SbCl_6^-$$

The cation radical may dimerize directly or, after reaction with further monomer, give a di-cation as the propagating entity. Experimental confirmation of the formation of di-cations has not been achieved because of the occurrence of monomer transfer reactions. However, the proof of the occurrence of the oxidizing properties of the tropylium cation and the formation of radical-cation intermediates has been established by the use of non-polymerizable model compounds. Thus $C_7H_7^+BF_4^-$ in acetonitrile forms charge-transfer band spectra in the visible region with a large number of aromatic donors; for example, n-phenylcarbazole, carbazole, n-methylcarbazole, triphenylamine, phenothiazine and tetramethylphenylene diamine. The reaction of carbazole with the tropylium salts leads to the formation of the stable red crystalline solid:

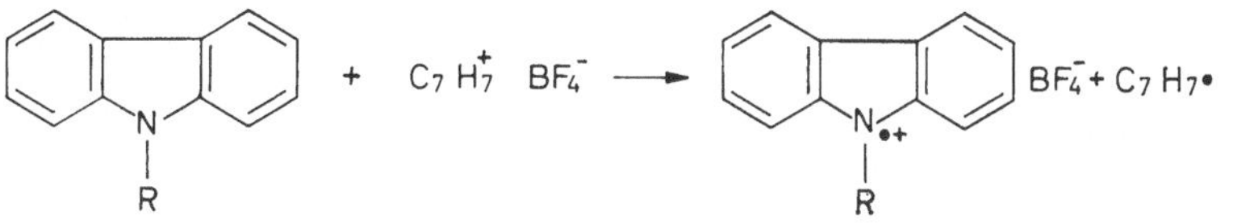

The ion-radical character has been characterized by e.s.r. and spectral measurements. The equilibrium constant for the reaction K at 25°C ($D + A \xrightarrow{K_e} DA$) was measured spectroscopically to be 3·18 l. mole^{-1}.

Other planar organic cations have also been used with equal success for the polymerization of electron-rich olefins. All of these form stable crystalline salts with many different anions and also show evidence of charge-transfer complex formation with the monomers.

There is now considerable literature information which indicates that a

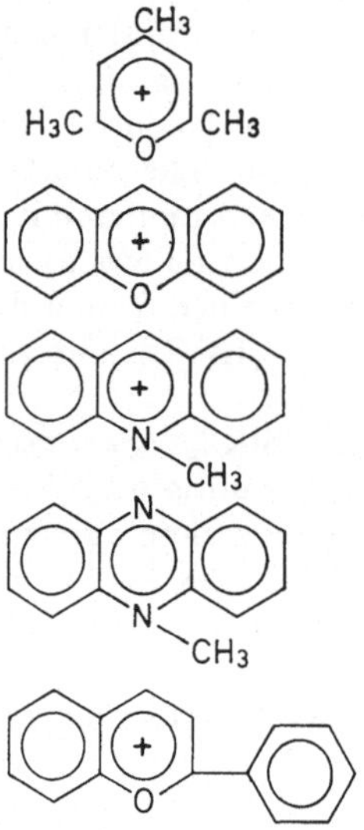

2,4,6-Trimethyl pyrylium

Xanthylium

n-Methyl acidinium

n-Methyl phenazonium

Flavylium

wide variety of neutral acceptor molecules may also be used to polymerize very reactive olefins such as n-vinylcarbazole[21]. The reactions are not so efficient as when using ionic acceptor species and each system has its own specific complications. As an example we have studied in detail the initiation of the polymerization of n-vinylcarbazole by tetracyanoethylene in methylene chloride solution. *Figure 2* shows the reactions occurring in this system and all the products shown have been confirmed and isolated.

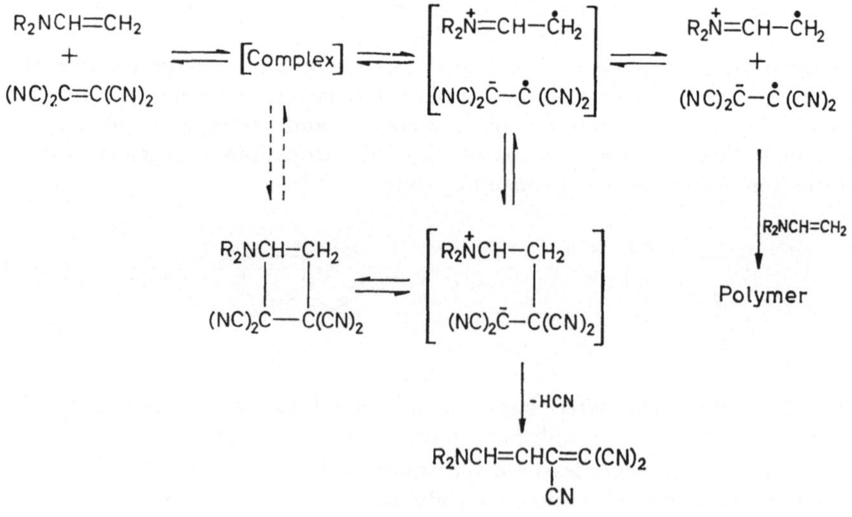

Figure 2

296

High yields of polymer may be obtained under suitable conditions and kinetic investigations *Figures 3* and *4* show that the rage of polymerization was proportional to the catalyst and monomer concentrations. The molecular weight of the polymer varied between 80 000 to 100 000 over a ten-line

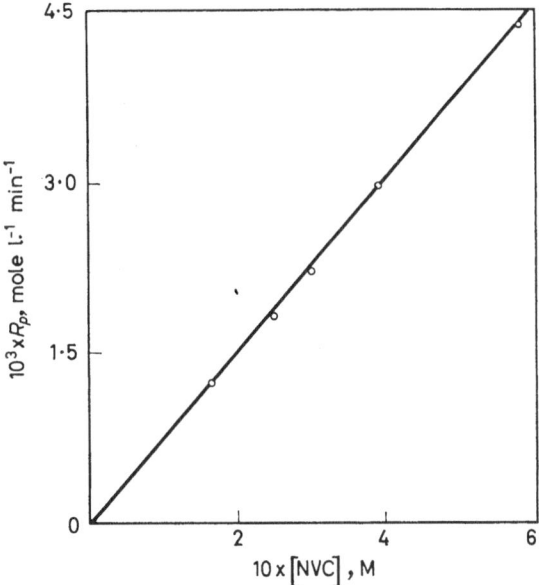

Figure 3. A plot of initial rate of polymerization (R_p) against initial [NVC]

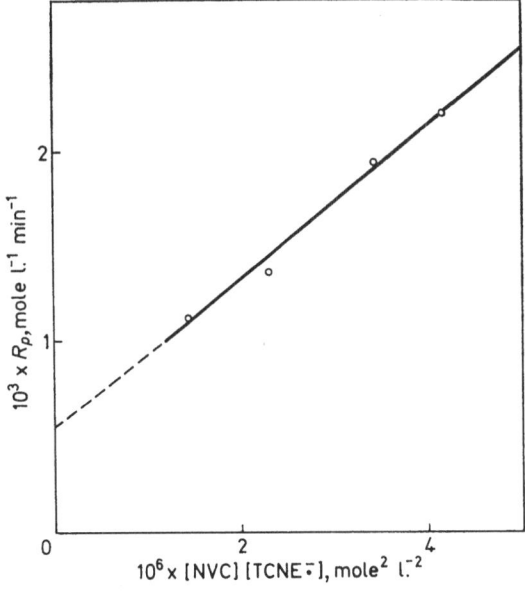

Figure 4. A plot of R_p *versus* [NVC] [TCNE⁻]

297

range of monomer concentration. The formation of the TCNE·⁻ anion-radical during the polymerization was confirmed by e.s.r. measurements. The polymerizing solution gave an eleven-line e.s.r. spectrum with a spacing of 1·60 g identical to that reported in the literature for this anion-radical. The concentration of TCNE·⁻ obtained from the area under the e.s.r. spectrum (calibrated against picryhydrazyl) was determined during the course of the polymerization and showed that the rate of polymerization R_p was given by

$$R_p = k\,[\text{NVC}]\,[\text{TCNE·}^-]$$

A particularly interesting situation arises when both donor and acceptor are polymerizable olefins. Gilbert reported[22] the spontaneous polymerization of vinyl isobutyl ether and vinylidene cyanide on mixing to give homopolymer of the two reacting monomers. A possible reaction scheme is

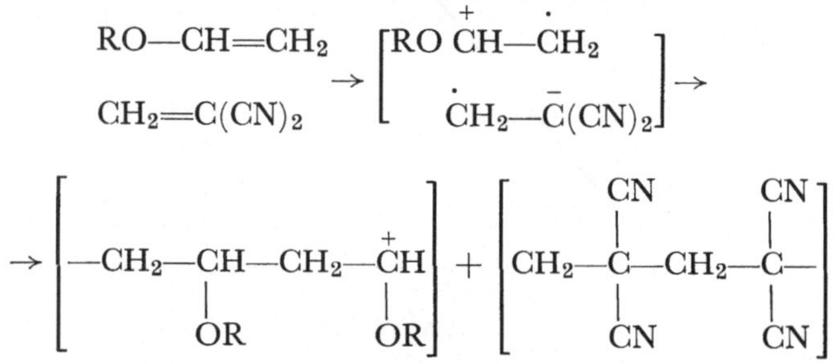

$$
\begin{array}{c}
\text{RO—CH=CH}_2 \\
\text{CH}_2\text{=C(CN)}_2
\end{array}
\rightarrow
\left[
\begin{array}{c}
\overset{+}{\text{RO CH}}\overset{\cdot}{\text{—CH}}_2 \\
\overset{\cdot}{\text{CH}}_2\text{—}\overset{-}{\text{C}}\text{(CN)}_2
\end{array}
\right]
\rightarrow
$$

$$
\rightarrow
\left[
\begin{array}{c}
\text{—CH}_2\text{—CH—CH}_2\text{—}\overset{+}{\text{CH}} \\
\quad\quad| \quad\quad\quad\quad | \\
\quad\quad\text{OR} \quad\quad\quad \text{OR}
\end{array}
\right]
+
\left[
\begin{array}{c}
\text{CN} \quad\quad \text{CN} \\
| \quad\quad\quad\quad | \\
\text{CH}_2\text{—C—CH}_2\text{—C—} \\
| \quad\quad\quad\quad | \\
\text{CN} \quad\quad \text{CN}
\end{array}
\right]
$$

Similarly Yang and Gaoni have reported[23] the spontaneous copolymerization of trinitrostyrene and vinylpyridine on mixing

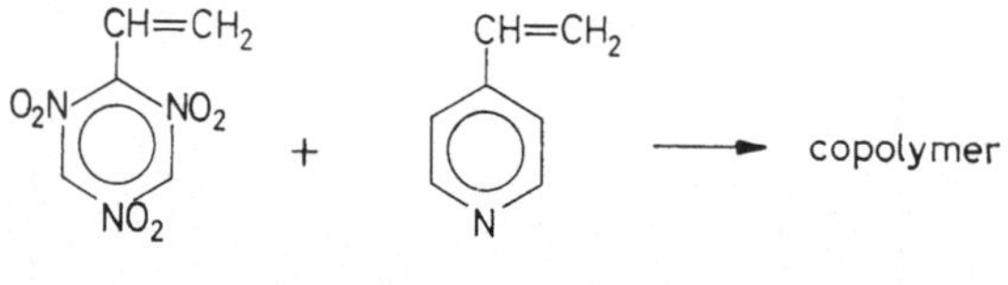

These interactants probably form charge-transfer intermediates but before generalization as to the exact mechanism of polymerization can be proposed a detailed study of the products of these and similar reactions is clearly advisable.

Formation of cation radicals by chemical oxidation

Lewis acids commonly employed for initiation of cationic polymerization can act as oxidizing agents by themselves or in conjunction with easily reducible molecules[24]. Particularly useful combinations are $\text{SbCl}_5/\text{CH}_2\text{Cl}_2$[14,25], $\text{SbCl}_5/\text{SbCl}_3$[26], SbCl_3/O_2[26], $\text{AlCl}_3/\text{CH}_3\text{NO}_2$[17], and $\text{BF}_3/\text{Pb(OAc)}_4$[28]. All these reagents have been used to prepare stable cation radicals from condensed hydrocarbons (*Figure 5*). It is evident from these observations that many common Lewis acids ($SnCl_4$, $SbCl_3$, $TiCl_4$, $AuCl_3$, $CuCl_2$ and $FeCl_3$) could function as primary oxidizing agents in their reactions with olefins. Protonic acid cocatalysts would not necessarily inhibit oxidation with Lewis acids and it has been shown conclusively by Ledwith and Woods[29] that $SbCl_6^-$ is a good electron acceptor and can be used to oxidize a wide range of substrates, for example ferrocene, arylamines and phenoxide anions. These

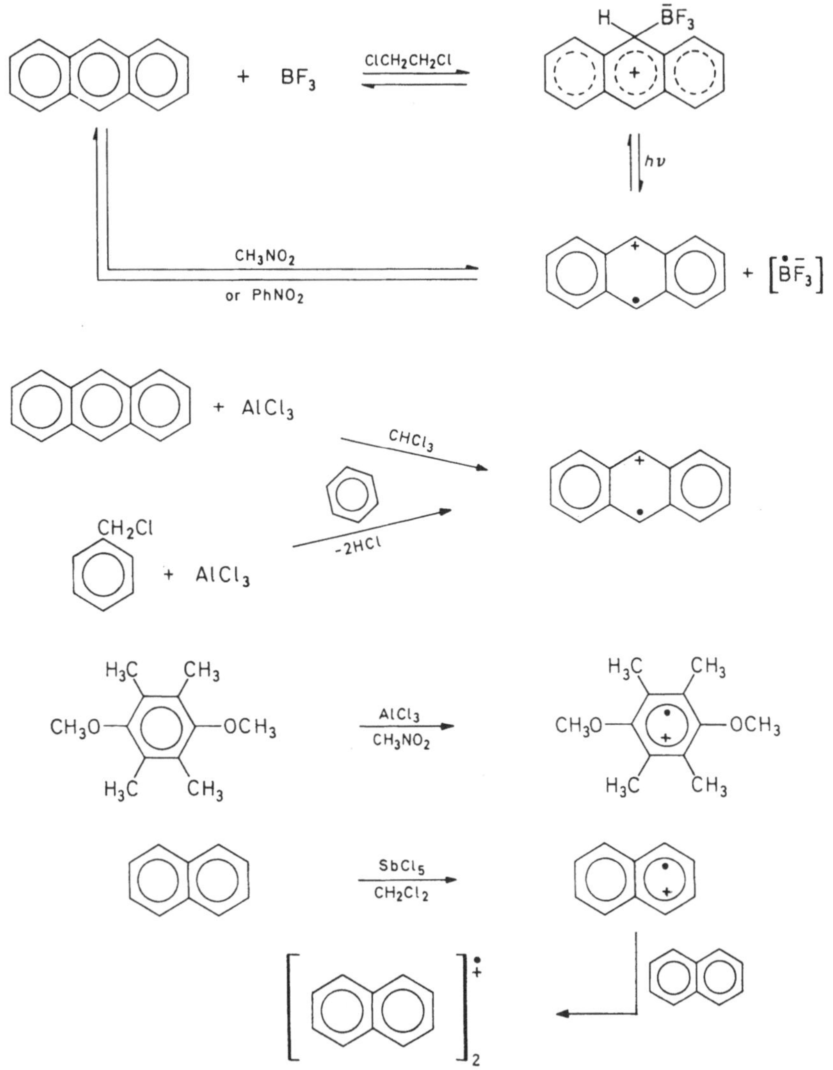

Figure 5

authors also showed that $SbCl_6^-$ used in the form of the stable quaternary salt $R_4N^+ SbCl_6^-$ is an initiating agent for the polymerization of *n*-vinylcarbazole and *p*-methoxystyrene in methylene chloride or acetonitrile solvent.

299

Typical polymerization measurements are shown in *Table 3*. Most of the initiator remains unused and it is thought that the polymerization is cationic but with a slow rate of initiation. The $SbCl_6^-$ anion will not initiate the polymerization of alkyl vinyl ethers and it therefore follows that dissociation of the anion into its components ($SbCl_5$ and Cl^-) is not the mechanism

Table 3. Polymerization of *n*-vinylcarbazole with dimethyl benzyl phenyl ammonium hexachloroantimonate in $CH_2 Cl_2$ at 20°C

[*Monomer*] mole/litre	$10^3 \times$ [*Initiator*] mole/litre	*Yield* %	*Molecular Weight*
0·24	—	—	—
0·24	9·61	78·5	5040
0·24	4·81	71·6	4270
0·24	0·96	68·8	5200
0·66	4·42	91·5	8710
0·46	4·61	87·3	5980
0·12	4·90	66·3	6490

whereby initiation occurs since $SbCl_5$ is a powerful initiator for vinyl ether polymerization. The overall oxidation reaction corresponds to

$$SbCl_6^- + 2e \rightarrow SbCl_6^{3-} \rightarrow SbCl_4^- + 2Cl^-$$

and the primary initiation step

$$M + X^+SbCl_6 \rightarrow MSbCl_6^-X^+ \rightarrow M \cdot {}^+SbCl_6^= \cdot X^+$$

followed by a rearrangement leading to a cationic propagation, e.g.

$$R_2N^+ \!\!-\!\! CH_2 \!\!-\!\! \overset{\bullet}{C}H_2SbCl_6 \cdot {}^= X^+ \rightarrow R_2N^+ \!\!-\!\! CH_2CH_2ClSbCl_4^- Cl^- X^+$$

Photo-induced formation of ion radicals

The initiating reaction discussed in the foregoing section take place in a manner which may be rationalized according to equation (4) by complex formation and electron transfer. The electron transfer is induced by thermal means as illustrated by the potential energy curves shown in *Figure 1*. The transfer process is assisted and accelerated by solvation. In non-polar solvents and even with some systems in polar solvents strong charge-transfer complexes may be formed with little or no ion-pair formation in the ground state. In such cases it is often possible to effect oxidation and reduction (initiation) by using visible or ultraviolet light. For example, tetrahydro-furan forms a donor–acceptor complex with maleic anhydride and on irradiation of a mixture of these two substances in solution a quantitative yield of the 1:1 adduct[30] may be isolated by vacuum distillation. The addition of vinyl monomers, susceptible to free-radical polymerization to the mixture of tetrahydrofuran and maleic anhydride suppresses the photochemical synthesis of (I) and results in the polymerization of the added vinyl monomer. An increase in the concentration of the maleic anhydride–THF complex functions in a manner similar to more conventional free-radical initiators in that it increases the rate of polymerization and decreases the molecular weight. In support of this view it was observed that copolymerization of the maleic anhydride occurred according to the well-known relative monomer reactivities in free radical polymerization. Thus there was no

(I)

incorporation of maleic anhydride when methyl acrylate was used as monomer but with vinyl acetate and isobutyl vinyl ether rapid and quanti-tative formation of 1:1 alternating copolymers occurred. In the latter reactions polymerization stopped as soon as all the initiator was copoly-merized.

Closely-related observations have recently been reported by Okamura and his coworkers[31] who describe the cationic polymerization of cyclic ethers and vinyl ethers by a species derived from the γ-ray and ultraviolet irradiation of ethereal solutions of maleic anhydride. Trioxane or 3,3-bis-chloromethyloxetane in the presence of maleic anhydride and oxygen poly-merized completely when irradiated and benzoyl peroxide may be used instead of oxygen. The maleic anhydride was practically unchanged at the end of the reaction and the polymerization was inhibited by hydroquinone which indicates that free radicals are involved. Addition of a small amount of the irradiated mixture to isobutyl vinyl ether caused immediate and extreme polymerization of the ether. These authors postulate that a free radical is formed from the ether by hydride extraction of an α-hydrogen atom. This is followed by a one-electron transfer from the radical to the maleic anhydride to yield the ether cation and the maleic anhydride radical anion. Cationic polymerization of the ether proceeds in the normal way. It seems reasonable to assume therefore that in our system the initial donor–acceptor complex when photoactivated gives rise to a radical ion, e.g. THF

may be replaced with other ethers and from a study of a wide series of ethers it appears that rate of polymerization of MMA parallels the ability of ethers to act as donors in complex formation. Maleic anhydrides may also be replaced by other anhydrides with considerable variation in the rates of polymerization. Clearly the nature of the donor–acceptor complexes offers a very wide scope for similar photochemical initiation reactions and other systems are being investigated.

C. E. H. BAWN

References

1 M. Szwarc. *Pure Appl. Chem.* **12**, 127 (1960).
2 C. E. H. Bawn, A. Ledwith, and N. R. McFarlane. *Chem. Commun.*, in the press.
3 G. Gee *et al. J. Chem. Soc.* 1338 (1959); 1345 (1959); 4298 (1961).
4 C. Walling and L. Bollyky. *J. Org. Chem.* **29**, 2699 (1964).
5 J. E. Mulvaney and R. L. Markham. *J. Polmyer Sci.* **4**, 343 (1966);
   L. Trossarelli, A. Priola, M. Guaita, and G. Saini. Reprint No. P578, *I.U.P.A.C. Symposium on Macromolecular Chemistry.* Prague (1965);
   G. E. Molan and J. E. Mason. *J. Polymer Sci.* A-1, **4**, 2336 (1966).
6 C. C. Price and D. D. Carmelite. *J. Amer. Chem. Soc.* **88**, 4039 (1966).
7 C. C. Price and R. Spector. *J. Amer. Chem. Soc.* **88**, 4171 (1966).
8 P. Plesch. *I.U.P.A.C. Symposium on Macromolecular Chemistry* 2, Prague (1965).
9 C. E. H. Bawn, C. Fitzsimmons, and A. Ledwith. *Proc. Chem. Soc.* **391** (1964).
10 C. E. H. Bawn, R. M. Bell, and A. Ledwith. *Polymer* **6**, 95 (1965).
11 I. Kuntz. *Amer. Chem. Soc. Polymer Preprints* **7** (1), 187 (1966).
12 I. Kuntz. *J. Poymer Sci.* **134**, 427 (1966).
13 C. Fitzsimmons and A. Ledwith, unpublished material.
14 M. P. Dreyfuss and P. Dreyfuss. *Polymer* **6**, 93 (1965); *J. Polymer Sci.* A **4**, 2179 (1966).
15 M. P. Dreyfuss and P. Dreyfuss. *Advances in Polymer Science*, 4, Springer-Verlag, p. 528 (1967).
16 A. Ledwith, in the press.
17 M. Feldman and S. Winstein. *J. Amer. Chem. Soc.* **83**, 3338 (1961).
18 E. M. Kosower. *Progress in Physical Organic Chemistry*, 3. Interscience, New York, p. 81 (1965).
19 A. Ledwith and M. Sambhi. *Chem. Commun.* 64 (1965).
20 C. E. H. Bawn, R. Carruthers, and A. Ledwith. *Chem. Commun.* 522 (1965).
21 L. P. Ellinger. *Polymer* **5**, 559 (1964); **6**, 549 (1965);
   N. Noguchi and Shu Kambara. *Polymer Letters* **3**, 271 (1965);
   K. Tsuji, K. Takakura, M. Nishi, K. Hayashi, and S. Okamura. *J. Polymer Sci.* A, **4**, 2028 (1966);
   N. Nomori, M. Hatano, and Shu Kambara. *Polymer Letters* **4**, 261 (1966); **4**, 623 (1966);
   H. Scott, G. A. Miller, and M. M. Labes. *Tetrahedron Letters* 1073 (1963).
22 H. Gilbert. *J. Amer. Chem. Soc.* **78**, 1669 (1956).
23 N. C. Yang and Y. Gaoni. *J. Amer. Chem. Soc.* **86**, 5022 (1964).
24 I. C. Lewis and L. S. Singer. *J. Chem. Phys.* **43**, 2712 (1965).
25 O. N. Howarth and G. K. Fraenkel. *J. Amer. Chem. Soc.* **88**, 4514 (1966).
26 G. B. Porter and E. C. Baughan. *J. Chem. Soc.* 744 (1958);
   J. R. Atkinson, T. P. Jones, and E. C. Baughan. *J. Chem. Soc.* 5808 (1964).
27 W. F. Forbes and P. D. Sullivan. *J. Amer. Chem. Soc.* **88**, 2862 (1966).
28 D. L. Allara, B. C. Gilbert, and R. O. C. Norman. *Chem. Commun.* 319 (1965).
29 A. Ledwith and J. Woods, unpublished material (University of Liverpool).
30 C. E. H. Bawn, A. Ledwith, and A. Parry. *Chem. Commun.* 490 (1965);
   A. Ledwith and M. Sambhi. *J. Chem. Soc.* B, 670 (1966).
31 K. Takakura, K. Hayashi, and S. Okamura. *J. Polymer Sci.* **132**, 861 (1964); **133**, 565 (1965).

STRUCTURE OF POLYMER MOLECULES AND SUPERMOLECULAR STRUCTURES

V. A. Kargin

Khimicheskoe Otdelenie, Akademia Nauk, S.S.S.R., Moskva, S.S.S.R.

From year to year we become more and more convinced of the important role played by structural phenomena in forming the properties of polymers. Mechanical properties are determined not only by changes in shape conformation and by motion of individual molecules of the polymers, but by the behaviour of larger and more complex structural formations as well. The interphase boundaries of these formations, known as supermolecular structures, are the sites where chemical reactions in polymers are most likely to begin and centres of crack formation and incipient destruction to arise. It has been found that extensive occurrence of ordered structures is typical not only of crystalline, but also of amorphous polymers, namely, glasses and rubbers.

Despite the complex morphology of structural formation in polymers it should not be forgotten that all these structures are built up of separate polymeric molecules. At first glance it seems self-evident that direct relations must exist between the properties of macromolecules and their ability to form supermolecular structures. In any case, the entire crystal chemistry of low-molecular substances is built on the assumption that such relations are inevitable.

This is all quite correct for cystals of low-molecular substances whose molecules have strictly definite shapes and sizes. But with respect to the immense molecules of polymers, in which variability of shape is inevitable, all these assumptions become much less clear. Indeed, there is always a direct relation between the properties of any structure and the properties of its structural elements. But these structural elements are not necessarily separate molecules. For example, the properties of some tactoid structure consisting of colloidal particles will, of course, be related to the geometry and size of those colloidal particles, but will have no direct relation to the properties of the molecules of the substance constituting the colloidal particles.

The shapes of most polymer molecules may vary within wide limits. When studying the simplest phenomena of structure formation quite a long time ago we came definitely to the conclusion that there are two ways by which structures can form. Sufficiently flexible molecules roll up into spherical coils or globules which form in very much the same way as the drops of a liquid under the action of surface tension. But if the macromolecules are sufficiently rigid the simplest linear structures result. No separate linear polymer molecules have been observed so far. Evidently in the great majority of cases, if not always, they aggregate into chain bunches usually containing several dozen molecules. These simple structural formations—globules and bunches—are the structural elements, the simplest supermolecular structures which combine to produce more and more complex structural forms in polymers[2]. These observations were first made 10 years ago and reported to the Symposium in Wisbaden[3]. Since then a large amount of data has accumulated and it may be asserted that many polymers can be obtained both in the linear and globular forms, because each of these forms results in an entirely different type of supermolecular structure. The very possibility of polymeric matter exisiting both in the globular and in the linear form shows that, if only the composition, structure, and size of the chain molecules are known, the supermolecular structure which the polymeric substance will form cannot be predicted. On the other hand, there is no doubt that if the formation of two forms of the simplest supermolecular structures is possible, only one of them is equilibrious. It is therefore necessary to consider in greater detail the conditions of their formation and the possibilities of their conversion into one another. The first factor to be taken into account in this connection should be the physical state of the polymer and the temperature of formation of the structures. In the elastic state, equilibrium structures may be expected. In the glassy state, structural transformations are so difficult that the polymer retains its original structural form.

The most general method is to produce polymers in the globular state from dilute solutions by evaporating the solute below the glass temperature of the polymers. Since the polymer molecules are rolled into globules in very dilute solutions, they remain fixed in this form. In this way a very great variety of polymers, ranging from polyolefins[4] and polyamides to gelatin and casein[5], can be obtained in the globular form. For many polymers, globular structures arise directly on their formation. Poly-condensation resins with high softening temperatures, such as epoxides or phenolics almost always give globular structures on preparation, though globular structures are also obtained with polyvinyl chlorides[6] and even fluorine-containing rubbers[7]. Finally, it should not be forgotten that in the globular form chain molecules are the least ordered and therefore any polymer in this form is amorphous. The rate of ordering in polymers is never high, and when a polymeric substance is precipitated from solution the rate of its separation is, as a rule, higher than the rate of ordering of its molecules. That is apparently why the first stage of precipitation of polymers as a separate phase or as separate particles is always the formation of amorphous particles, which may consist of one or many molecules. This process is typical not only for polymers, but for any substance separating as an insoluble precipitate from solutions. As far back as the twenties of this century, Haber[8] drew attention to the importance of the ratio of the rates of ordering and separation of a substance. Later Berestneva and the present author[9] showed for a large number of examples that the first stage of formation of colloid particles is always the formation of primary amorphous particles of spherical shape. This is natural, because being amorphous, they are isotropic and the rate of their growth is the same in all directions. The next stage is ordering of the molecules of the precipitated substance, which in the case of low-molecular substances results in crystallization. As a result of this the primary particle breaks up into a large number of minute, crystalline particles which form the colloid solution. *Figure 1* shows the consecutive states of formation of a colloidal solution of

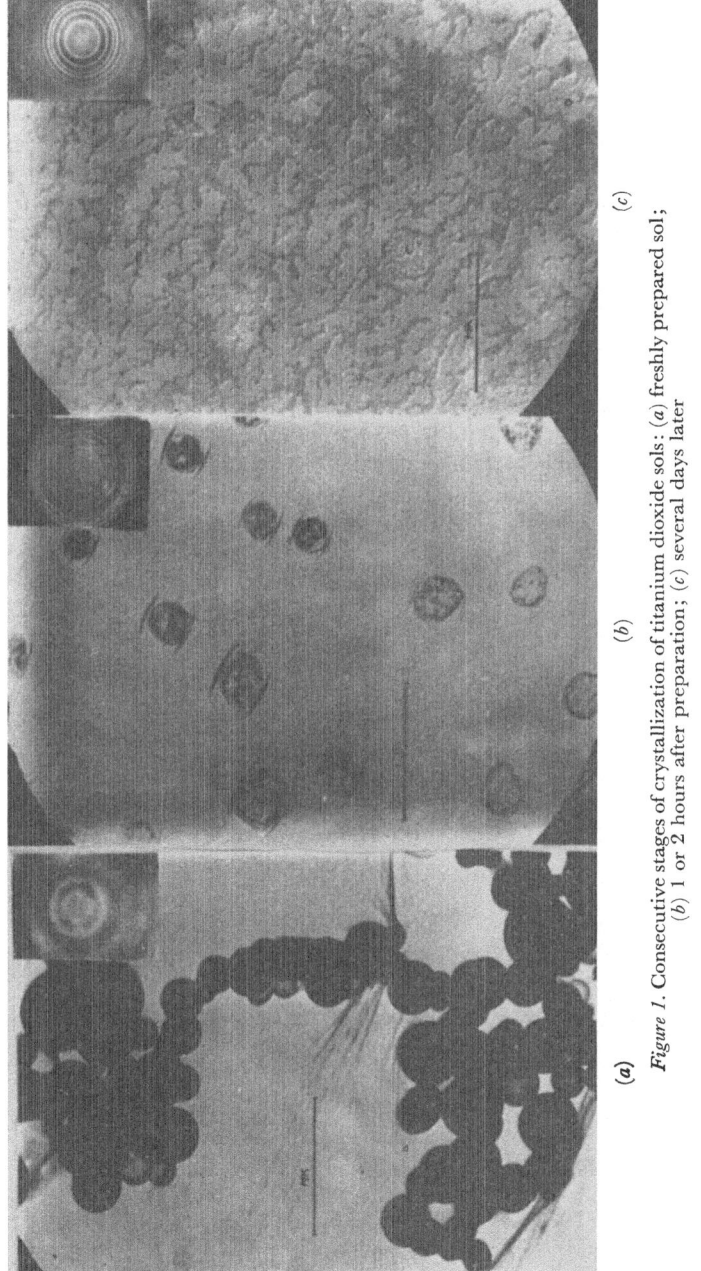

Figure 1. Consecutive stages of crystallization of titanium dioxide sols: (*a*) freshly prepared sol; (*b*) 1 or 2 hours after preparation; (*c*) several days later

TiO_2, beginning with large spherical amorphous particles which break up subsequently, due to ordering, into minute crystalline particles. All this results from the fact that the rate of separation of TiO_2 from the solution is much higher than the rate of its ordering.

This is no formal analogy. It is known that in the case of polymers also, ordered supermolecular structures form comparatively slowly both in amorphous and in crystalline polymers owing to the large size of the interacting polymer molecules. It may therefore by expected that when a polymeric substance forms, or separates from solutions, or melts, it falls out at first in the least organized form, namely, in globular form, especially if this occurs at temperatures where the molecular motion of the polymer molecules is very low. There are exceptions, when the polymer forms immediately as well organized structures, but I shall return to this later.

Hence, we should expect the simplest structures to form first; these may be unbalanced and on approaching equilibrium, will come into more perfect forms. If this transition is very slow the polymer may exist in various intermediate forms as globules, fibrils and more complex formations. In this case all perceptible relations between the composition and structure of the polymer molecules and their ability to form supermolecular structures will be obliterated. Only in the case where the structural transformations occur quickly can we expect equilibrium to be reached and direct relations to exist between the molecular build and supermolecular structure. Hence, we must examine how the transition from the simplest to the most perfect equilibrium structures occurs in polymers, and the rates of these processes for various classes of polymers and for various conditions of formation of supermolecular structures.

I have already said that below the glass temperature globular structures are quite stable in the absence of extraneous influences, and the globules

Figure 2. Replica of the spalled surface of block polyvinyl chloride

are retained as separate structural units with well-defined interfacial boundaries. The photograph of a replica of a spalled surface of polyvinyl chloride shown in *Figure 2* may serve as an example.

When the temperature rises and the polymer passes into the elastic state

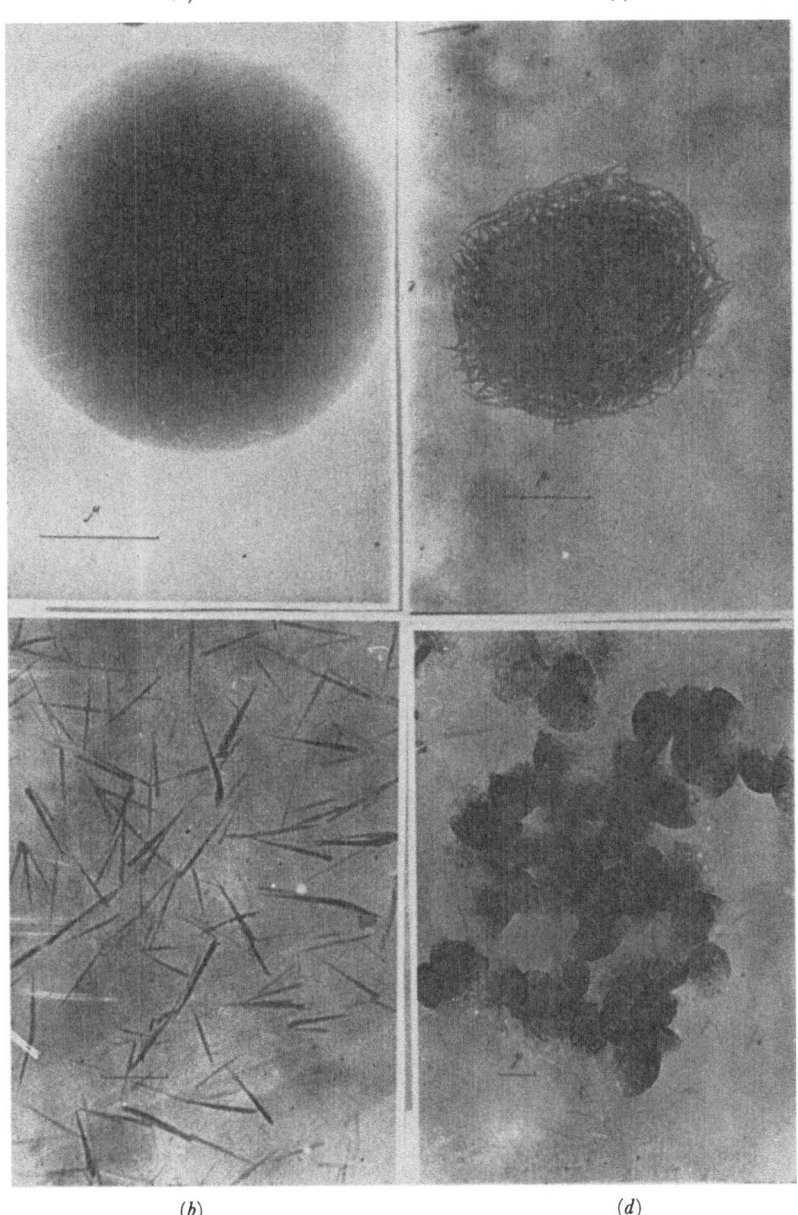

Figure 3. Isotactic polystyrene out of decaline solution: (*a*) freshly prepared; (*b*) after 3 months; (*c*) after 6 months; (*d*) after 10 months

307

H

further ordering begins. The globules merge like the drops of a liquid into larger globules which now contain numerous polymer molecules rather than one. Finally, after a critical size is reached the globules rearrange into linear or band structures typical of any polymer in the elastic state, or into bunches and fibrils characteristic of polymers with more rigid chains[10]. The rate of these transitions in pure polymeric substances is not very well known, but it is not high even in solutions. I should like to give a few examples of how the transitions from the simplest structures to more elaborate ones are accomplished when working even with very dilute solutions.

If a melt of isotactic polystyrene is cooled quickly to a low temperature it becomes amorphous and dissolves readily in hydrocarbons. If preparations for electron-microscopic photography are made out of such a solution after storage of the latter for different lengths of time, the entire sequence of structure formations can be observed beginning from the simplest structure-less species and globules down to quite perfect single crystals. It is striking that even in solutions, where conditions would seem to favour mobility of the polymer molecules the most, these ordering processes take weeks and months[11]. *Figures 3* and *4* show the structure patterns formed as a result

Figure 4. Isotactic polystyrene crystallized from a melt at 160°C

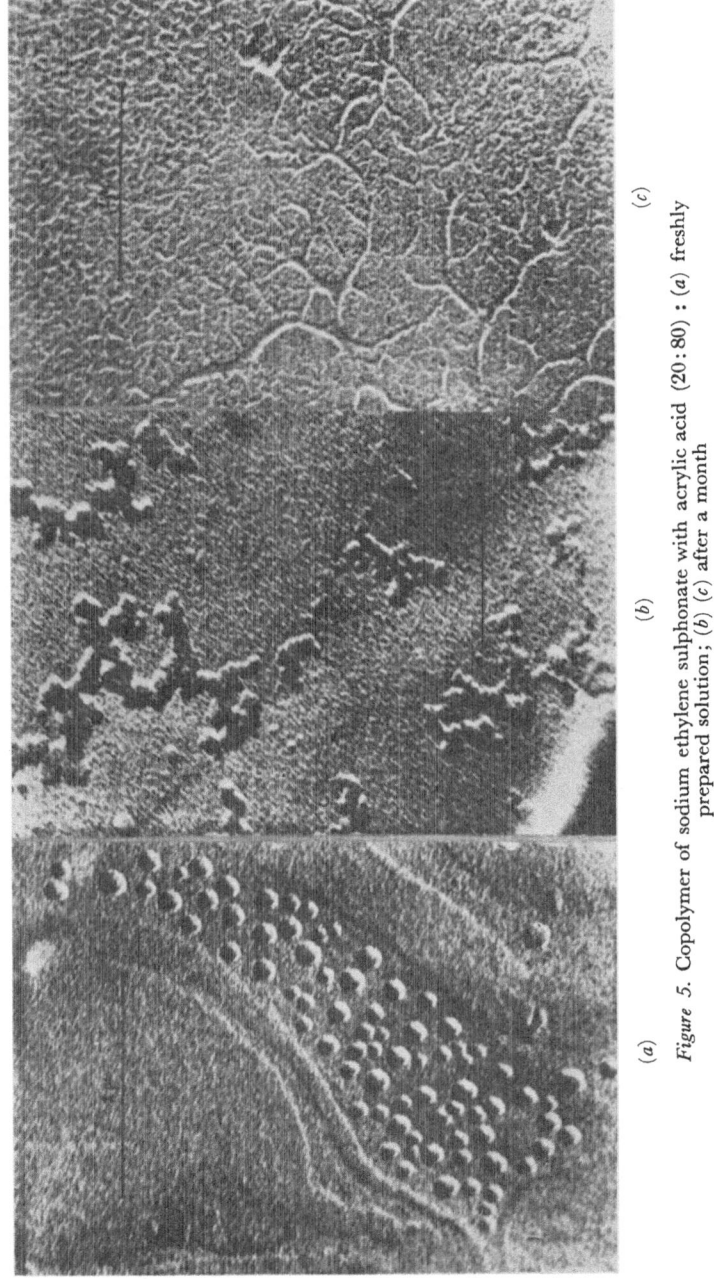

Figure 5. Copolymer of sodium ethylene sulphonate with acrylic acid (20:80) : (*a*) freshly prepared solution; (*b*) (*c*) after a month

(*a*) (*b*) (*c*)

of rapid evaporation of isotactic polystyrene solutions after various storage times. Only the last photograph shows the final state of structure formation— the formation of a crystalline substance. All the foregoing structures are amorphous, as is evidenced by the electron-diffraction pattern (taken of the most perfect structure prior to crystallization, that of long and thin fibrils).

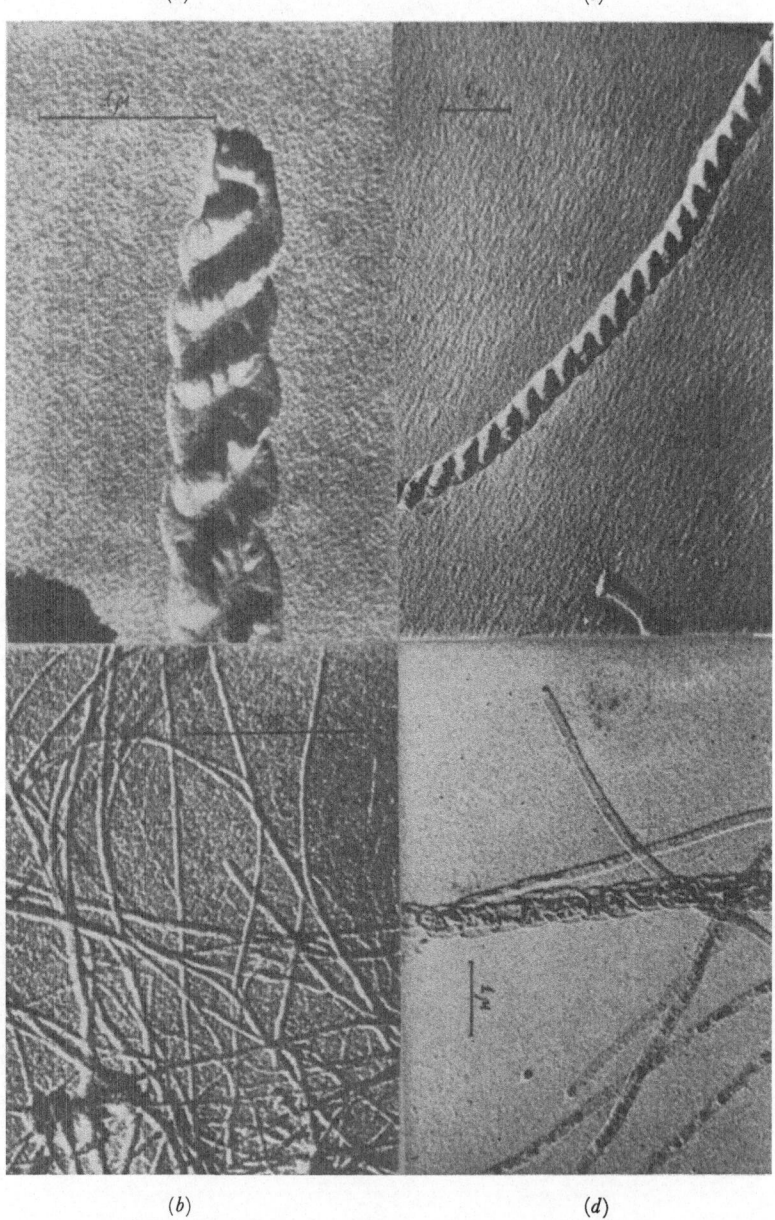

(a)   *(c)*

(b)   *(d)*

Figure 6. Copolymer of sodium ethylene sulphonate with acrylic acid (20:80): (*a*) one month after preparation; (*b*) (*c*) after 3 months; (*d*) after a year

It can be seen how slowly the transition to more perfect structures occurs and how easy it is to register intermediate, non-equilibrium stages of structure formation. Adding to this the polymorphism of the crystalline forms of isotactic polystyrene, it can readily be understood that under ordinary conditions of working with this polymer, when the time and conditions of specimen preparation do not, as a rule, favour the establishment of equilibrium, a large number of various structural forms may easily result,

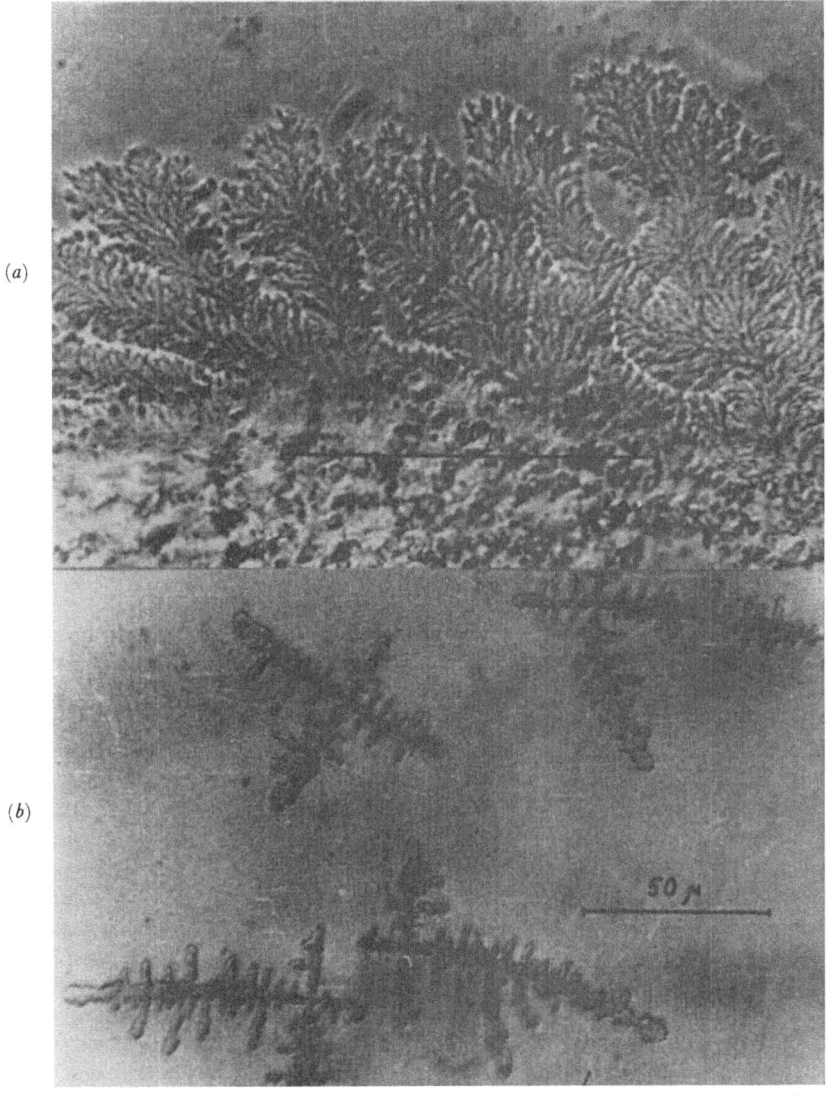

(a)

(b)

Figure 7. Optical photographs of copolymer of sodium ethylene sulphonate with acrylic acid (20:80): (a) freshly prepared solution; (b) after storing for 6 weeks

obliterating any direct relation between the properties of the initial molecules and the supermolecular structures formed.

I should like to give one more example of the sequence of various stages of structure formation, entirely free from the complications of crystallization phenomena. Perhaps the most interesting from the standpoint of studying structure formation phenomena are polyelectrolytes and their copolymers, because both the flexibility of their molecules and their intermolecular inter- action can easily be varied by changing the salt composition of the medium. At the same time they possess a striking ability to form sometimes very perfect large structures without crystallization due only to ordering of their molecules in the amorphous state (this process being accomplished in several stages[12]). *Figures 5, 6* and *7* are (as in the case of polystyrene) a series of electronmicroscopic photographs of specimens of the copolymer of sodium ethylene sulphonate and acrylic acid, obtained by rapid evaporation of the same solution after different periods of storage. These latter photographs refer to more concentrated solutions from which large structures, up to visible sizes, are formed. One can see the consecutive transition from monomolecular globules to polymolecular formations, the merging of the latter into intermediate, sometimes shapeless formations, the appearance of linear structures of a very peculiar and complex type, and finally, their transition into very perfect fibral structures, up to the formation of large dendrites.

In dilute solutions their formation is very slow. This does not mean, of course, that concentrations and temperatures cannot be found at which even very perfect structures can be obtained quite rapidly. But this series of tests again demonstrates how easily polymorphism can be achieved even in strong polyelectrolytes and their copolymers. These copolymers of sodium ethylene sulphonate and polyacrylic acid are irregular and do not crystallize completely, just like the copolymers of vinylphosphonic acid and its acrylic acid esters, described above.

The behaviour of polystyrene sulphonic acid is very much the same[13]. *Figures 8* and *9* illustrate the structure formation pattern of this polymer under conditions similar to those described above. I shall restrict myself to these examples, although at present many cases of slow structure formation could be cited both for amorphous and for crystalline polymers. I have quite deliberately been speaking of structure formation in amorphous polymers all the time because ordering in the amorphous state is a necessary prerequisite for crystallization. Therefore, the sources of polymorphism are inherent in the formation of polymeric structures prior to crystallization, and the exten- sive development of polymorphism is the main obstacle to establishment of relations between the structure of polymeric molecules and the formation of supermolecular structures. The general course of structure formation is very similar for a great variety of polymers, but depending on the conditions it may stop at various stages of the process. Naturally, in polymers with rigid chains and large intermolecular interactions, having high glass transi- tion temperatures and exisiting as glasses under ordinary conditions, the early, primitive structural forms will 'freeze'. But at the same time, under adequate conditions accelerating structure formation these polymers will form perfect structures. On the other hand, polymers with flexible and not strongly interacting molecules (rubbers) covers this path quickly, but the final and probably equilibrium structural forms of these polymers are comparatively simple.

These are the band structures characteristic of any polymer in the highly elastic state[10].

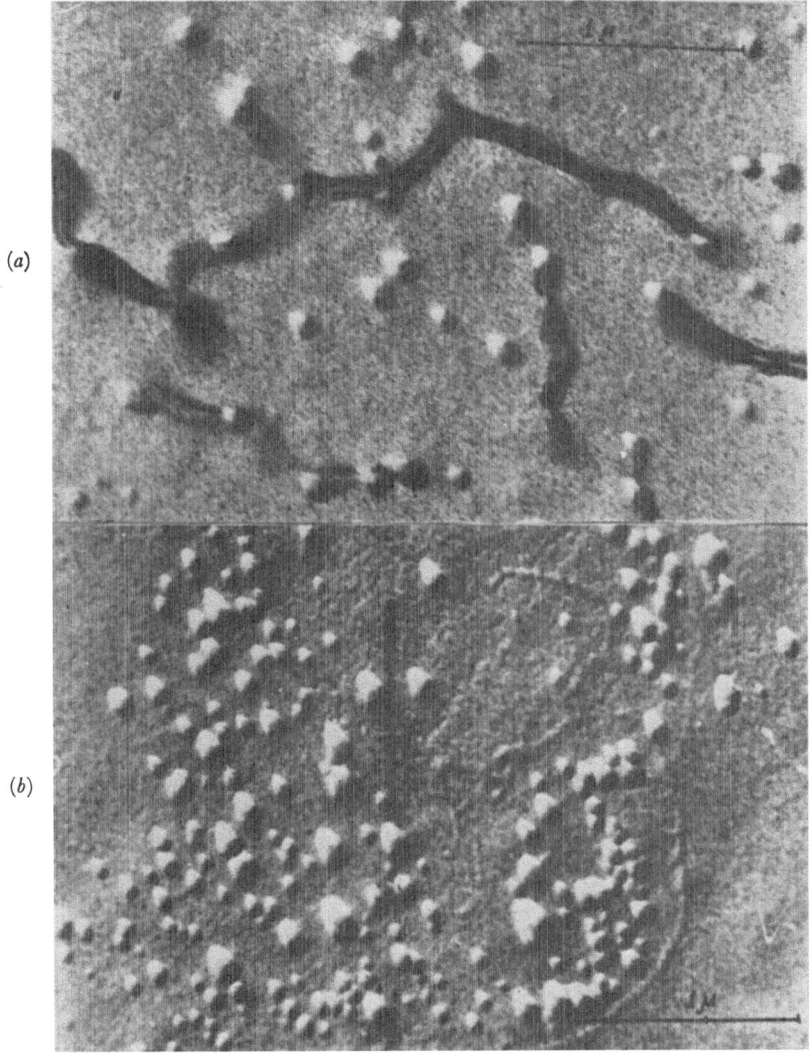

Figure 8. Sodium polystyrene sulphonate (M.W. 300,000) (*a*) freshly prepared solution, pH = 6; (*b*) 2 weeks after preparation, pH = 6

Of course, the rate of formation of the structures depends on a number of other factors as well, namely, the molecular weight, the regularity of structure and the degree of branching of the chains. For this reason different structures may not infrequently be observed to coexist within an ordinary

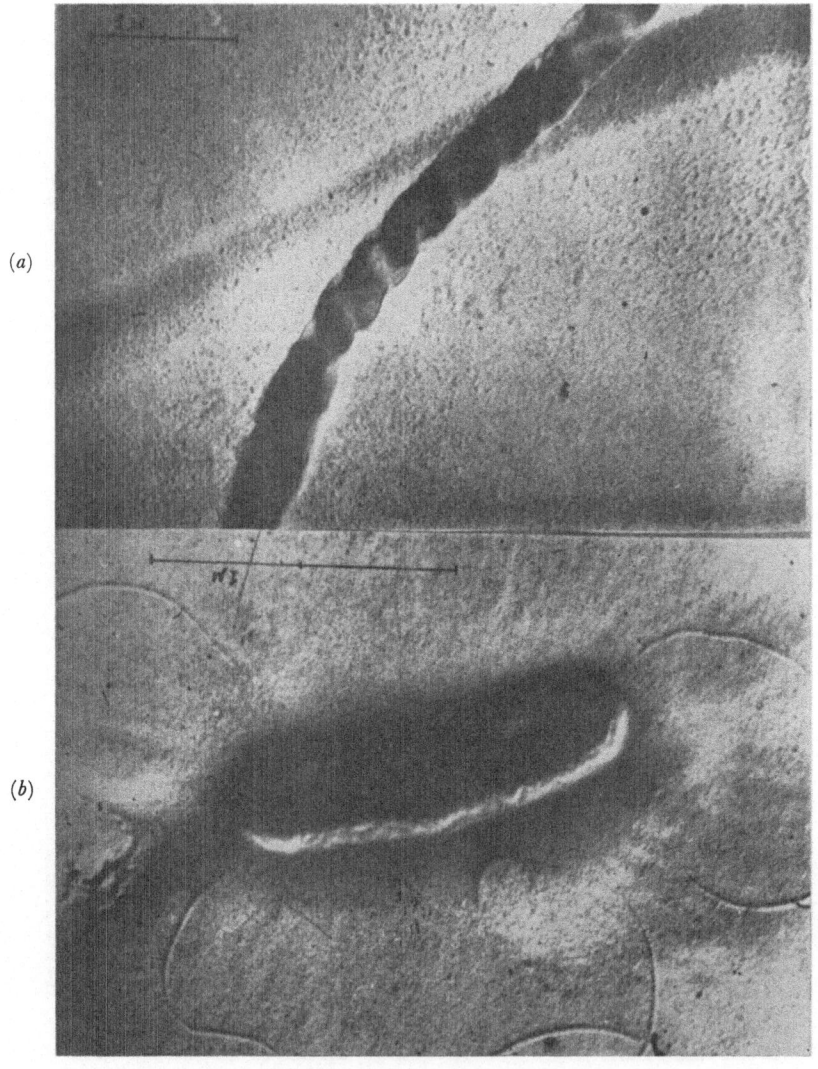

Figure 9. Sodium polystyrene sulphonate (M.W. 300,000): (*a*) 4 weeks after preparation, pH = 6; (*b*) after 6 weeks, pH = 6

specimen of a polymeric substance. An example is shown in *Figure 10* which is a photograph of partly-brominated gutta-percha, showing globules, fibrils and spherulites, i.e., the entire sequence of structures from the simplest to the highest form[14]. Hence, (owing to the stepwise, multi-stage nature of structure formation in polymers and the ready appearance and fixation of non-equilibrium structures) the conditions of formation influence the

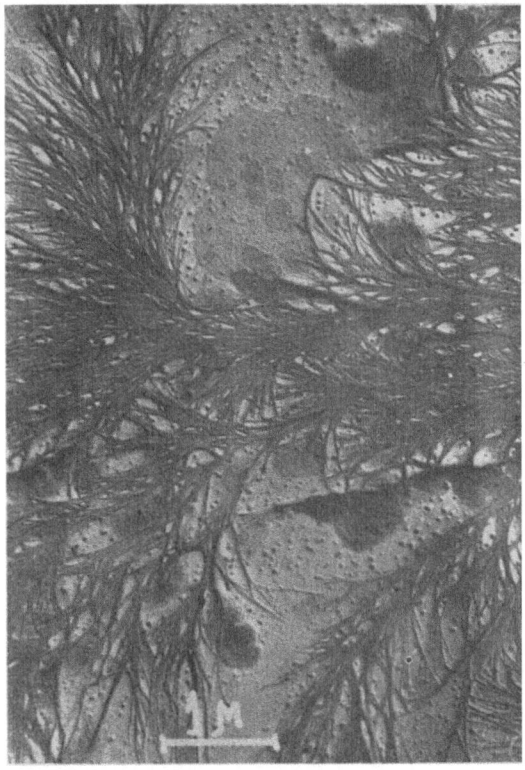

Figure 10. Partly-brominated gutta percha (16·3% Br), out of chloroform solution at 50°C

appearing supermolecular structures not less, at any rate, than the nature of the polymeric molecules themselves. The flexibility of polymeric molecules, their intermolecular interaction, and, of course, their molecular weight and degree of branching are the factors which affect structure formation in a very general way. (The regularity of the macromolecules plays an essential part only during crystallization.)

But flexibility and intermolecular interaction are constant characteristics only in the case of pure polymers. In solutions and other multicomponent systems these characteristics vary widely. Therefore on polymerization in solutions, on separation of polymers from solutions on plasticizing and in mixtures of polymeric substances, the same polymer may form various types of structures depending on the influence of the medium. Perhaps the

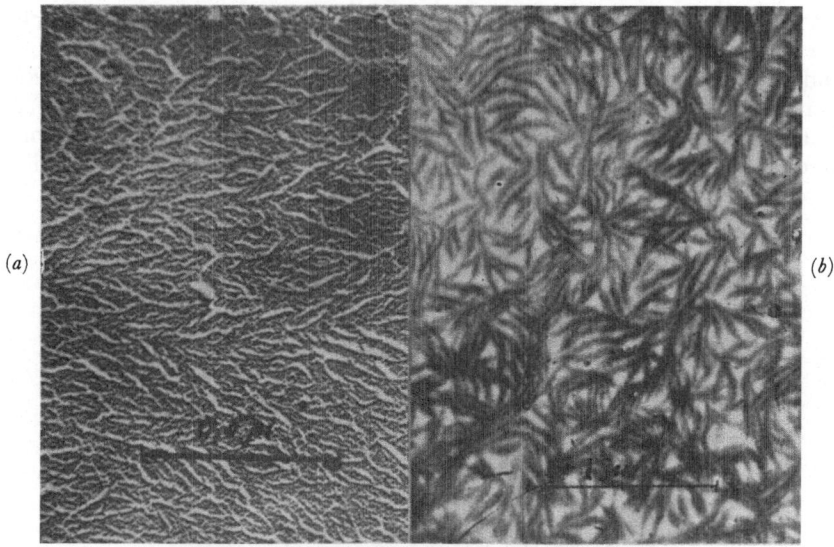

Figure 11. Conformational coil–helix transition. (a) aqueous solution of sodium salt of poly-L-glutamic acid. (b) aqueous solution of poly-L-glutamic acid, pH = 4·5

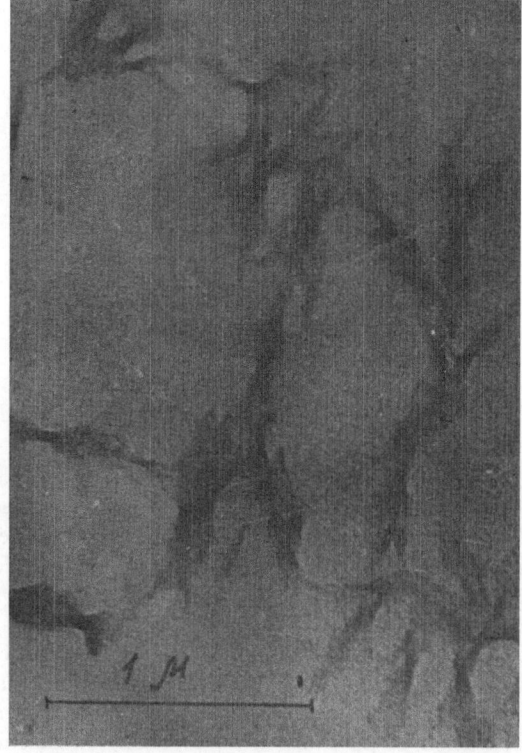

Figure 12. Poly-L-glutamic acid out of aqueous solution, pH = 3·7. Specimen prepared by thermal attachment method, (T − 45°)

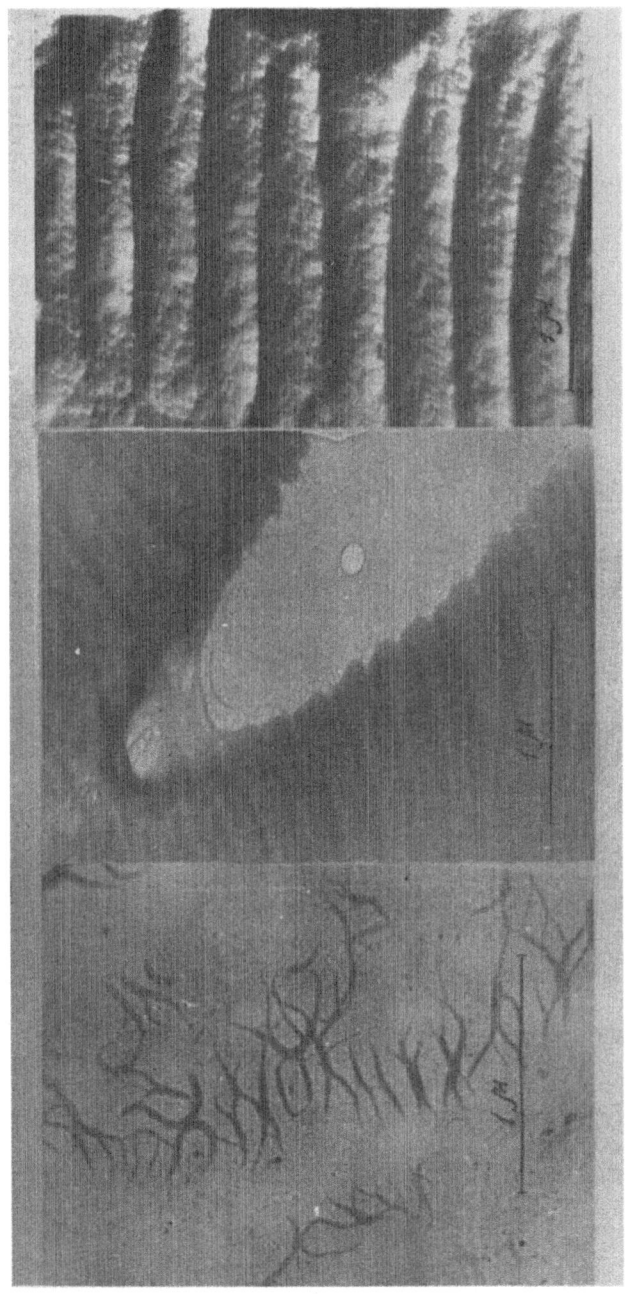

Figure 13. Poly-γ-benzyl-ʟ-glutamate out of chloroform solution

most typical are polyelectrolytes, especially weak acids, where, depending on the pH or salt composition of the solution, the undissociated flexible molecules which roll up into coils in acid media, can easily pass into drawn out charged chains in alkaline media, to form linear structures. An old and well-known example is polyacrylic acid which forms globules, and its alkali salts, which give typical linear structures[1]. Other phenomena occur in solutions of polypeptides in which conformational coil–helix transitions occur within a very narrow pH interval. For example, the sodium salt of

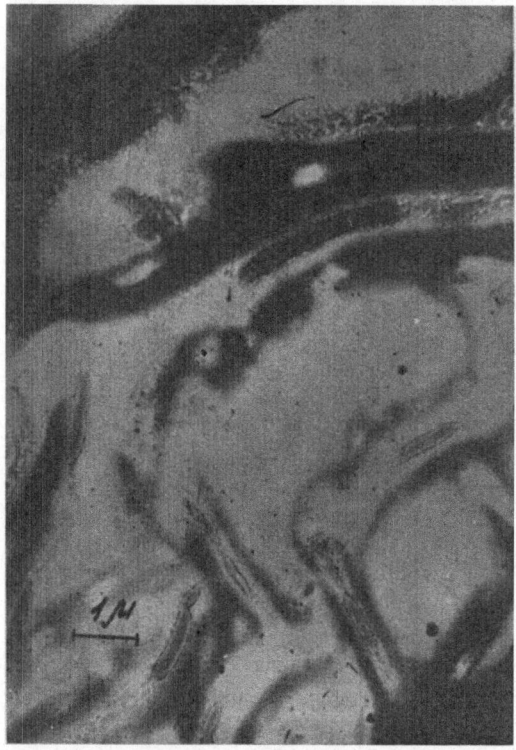

Figure 14. Poly-γ-benzyl-L-glutamate out of dichloroacetic acid solution

polyglutamic acid is a strong electrolyte whose behaviour is quite analogous to that of sodium polyacrylate and which forms the same linear structures on aggregation of drawn-out charged chains. Such a structure is shown in *Figure 11.* However, when the solution is acidified to pH = 4·5, instead of globules, as is the case for polyacrylic acid, polyglutamic acid forms a fibrillar structure, a photograph of which is given in *Figure 11b.* Well-formed fibrils can be seen to have appeared approximately of the same size, about 200 Å in cross-section and 0·5 microns long. It is curious that comparatively perfect structures not only form in the solid polymer after evaporation of the solvent, but are already existent even in dilute solutions. An investigation of solutions containing 0·03% of glutamic acid, by means of the thermal attachment method, showed the existence of well-formed aggregates containing several tens of molecules (*Figure 12*). Thus we see that conformational transitions sharply influence the formation of supermolecular structures. It is interesting that the reverse is also true, i.e., the formation of supermolecular structures promotes the transition of coils into helices.

These phenomena are typical not only of aqueous solutions. With glutamic acid esters such conformational transitions occur when polar solvents are substituted for non-polar ones and *vice versa*. *Figure 13* shows photographs of the structures of polybenzyl glutamate obtained from chloroform solution, where its molecules exist in the form of helices. The photographs in *Figure 14* are of the same polymer out of dichloroacetic acid solution, where the polymer exists in coil conformation. In the first case we see well-defined fibrillar structures, and in the second, shapeless formations. It can be seen how the order arising at the molecular level affects the formation of supermolecular structures. I would like to add that all the structural formations I have been talking about are amorphous, as is evidenced by the electron diffraction pattern[15].

I have given only a few examples of the effect of the solvent on the formation of supermolecular structures owing to changes in conformation of the polymer molecules. Many more similar examples could be given, especially for crystallizing polymers for which the importance of selecting the correct solvent is widely known.

All this reduces to the fact that by using various solvents for polymers already obtained, or during polymerization, various structures can be obtained for the same polymeric substance. Of course, after removing the solvent, some of these structures will become non-equilibrium ones, but if the polymer is under conditions of low rates of structural transformations, these non-equilibrium structures will be practically stable. We again come to the conclusion that non-equilibrium states form easily in polymers, and that this is one of the primary factors hindering the establishment of direct relations between the structure of the polymeric molecules and the formation of supermolecular structures. (These relations are only of a very general nature.)

Finally, what interests us primarily in supermolecular structures is their influence on the properties of polymeric bodies. Having outlined the basic types of polymeric structures, we may consider that each of these types will have a definite corresponding complex of properties, of which mechanical properties are of the greatest interest. However, this seemingly quite self-evident proposition is not always true. Sometimes polymers of globular structure exhibit elastic properties, although in the globular state these properties would be expected to disappear. Usually, the larger the spherulites in crystalline polymers, the more brittle the polymeric material[10]. But this rule is not always valid. (There arises the question as to the causes of these invalidities and of the unambiguity of the relations between the supermolecular structures and the mechanical properties of polymers.)

The assumption that there exists a direct relation between the type of supermolecular structure and the properties of a material is connected with a very important condition. The structure of a material should remain practically unchanged until the specimen fails. But if structural transformations occur in the specimen during deformation, the structure at the moment of breakdown will not be that of the initial specimen, but that formed as a result of the mechanical influence to which the polymer was subjected. Evidently, in this case no correspondence can even be expected to exist between the initial structure and the properties at big deformations. These phenomena of structural transformations occurring during deformation are very typical of polymers.

An old example of structural transformation is the formation of a 'neck' on

Figure 15. Electron microscopic photograph of neck of polyethylene extended to 150%

deformation, described for the case of crystalline polymers some time ago by Karothers. (It is also observed in amorphous polymers with developed structures and is formally described as a phase transformation[16].) The nature of this phenomenon remained obscure for a long time, but electron microscopy revealed that 'neck' formation is observed on a microscopic scale as well, and is actually a jumpwise transition from one supermolecular structure to another with a sharp interphasal boundary. I would like to remind you of the well-known example of polyethylene[17] (*Figure 15*) where a sharp boundary can be seen between the isotropic and the oriented parts of the specimen. Next we see the formation of a 'neck' on deformation of a large spherulite of isotactic polystyrene (*Figure 16*) where a sharp boundary can also be discerned between the unchanged and the oriented portions of the spherulite. (In addition, secondary formations can be seen which have resulted from recrystallization of the oriented parts, and these are also separated by sharp boundary lines[18].) Finally, an excellent example of jumpwise development of deformations and changes in structure is the

Figure 16. Optical photographs of necks of isotactic polystyrene crystallized from a melt and extended

extension of polyoxypropylene in the form of large spherulites[19]. *Figures 17* and *18* show the deformation pattern of this polymer in different scales, revealing a sharp transition from one structure to the other. But such phenomena are not only observed for crystalline polymers and linear structures. Comparatively recently we observed similar phenomena during the deformation of

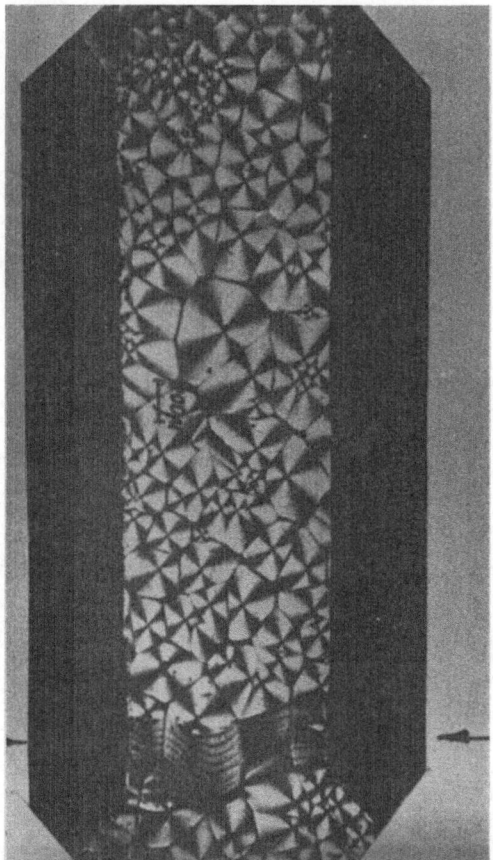

Figure 17. Extended sample of polyoxypropylene

a polymer with a globular structure[7]. Fluorine-containing rubbers possess a typical globular structure, the globules being regular in shape and close in size. The existence of a globular rubber is in itself surprising, because as long as a polymeric molecule is in globular form it should be devoid of all the typical properties of a polymer. However, when such a rubber deforms, its globules change gradually to linear conformations. In this process the globules themselves are not deformed and there is a sharp boundary between them and the newly formed oriented linear part, this boundary being just as sharp as in the deformation of the spherulites of crystalline polystyrenes. *Figure 19* is a photograph of undeformed rubber, and *Figure 20* is the same rubber stretched, at different magnifications. Here we distinctly

Figure 18. Different sections of an extended specimen of polyoxypropylene

323

see the picture of transition of the globules into oriented structureless filaments. The display of elastic properties in this case is related to destruction of the globules. We again come up against the same phenomenon, viz., breakdown of the initial structure and jumpwise transition to an entirely different type of structure.

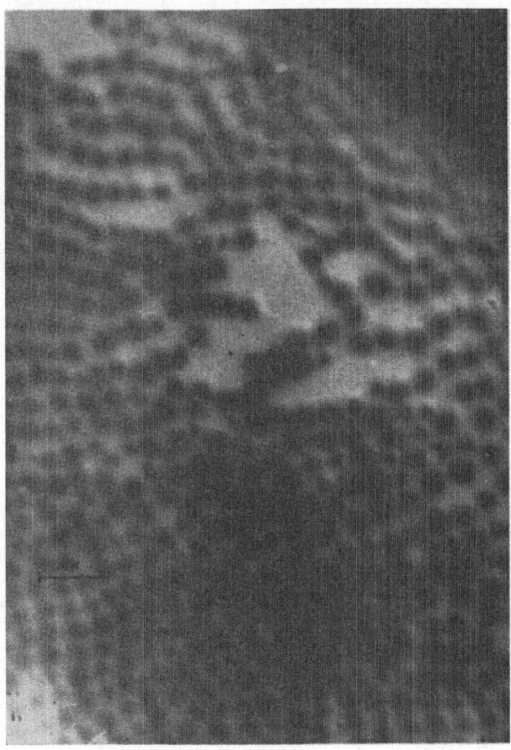

Figure 19. Fluorine-containing SKF-26 rubber out of acetone solution

Thus we must distinguish between two types of phenomena in polymer deformation. In the first, the initial structure is preserved up to the moment of failure. When this is the case the general properties of the material can be judged from its structure. In the other type the structure changes sharply during deformation and the relation between the initial structure and the properties of the material is obliterated. Evidently, structural transformations are possible during deformation only when the material posesses sufficient molecular mobility, and for this the material must be at a temperature above glass transition. Below this temperature structural transformations no longer occur and the initial structure is preserved right up to the moment of failure of the specimen. We again see how great is the part played by the transition temperatures and the physical state of the polymeric substance in the transformation of supermolecular structures. Below the glass temperature various non-equilibrium structures become fixed, and do not change even when subjected to extraneous mechanical influences. Above this temperature supermolecular structures may change into one another both spontaneously and due to external forces.

We are accustomed to the idea that major deformations in polymers are always due to changes in shape and shifting of the individual polymer molecules. That is why, even when the polymer possesses a supermolecular structure, one usually attempts to explain such deformations by the fact

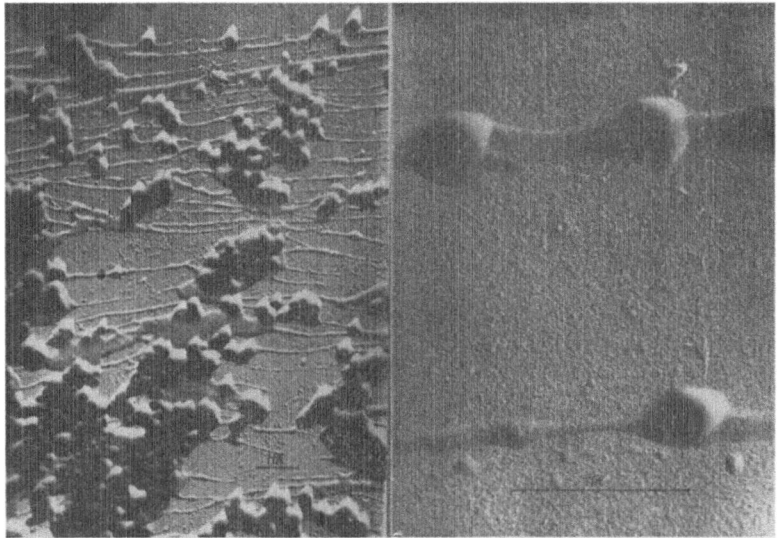

Figure 20. Extended film of fluorine-containing SKF-26 rubber out of acetone solution

that some unordered 'amorphous' portion is left in it. Meanwhile it is essential to know the contribution of the supermolecular structures themselves, especially when they consist of long, thin and sometimes also sufficiently flexible fibrillar formations. Are there cases where the deformation properties are not directly related to the properties of the separate molecules, but only to those of the supermolecular formations?

Recently we were successful in obtaining sufficiently-perfect, large-spherulite structures of polypropylene which exhibited the ability to undergo big deformations much below the glass temperature, which for polypropylene is −30°C. Already at −40° both atactic propylene and polypropylene possessing a fine spherulite structure are ruptured without perceptible elongation. Meanwhile specimens with large spherulites show ultimate elongations of 150–200% at temperatures of −40° to −60°C. Even at the temperature of liquid nitrogen the deformation may be as high as 120 or 140%[20].

It is quite natural that at such temperatures, considerably below the glass transition temperature and the brittle point (corresponding to the last dielectric loss maximum), there is no molecular mobility. The development of deformation cannot be related to the flexibility and displacement of individual macromolecules, or, which is the same, to the presence of any amorphous portions. Another proof of this is the brittleness of amorphous polypropylene at these temperatures. Hence, big deformations in crystalline propylene with a very perfect structure can be related only to the shift of large structural elements, these being well-formed supermolecular structures.

Figure 21 shows photomicrographs of the consecutive stages of extension

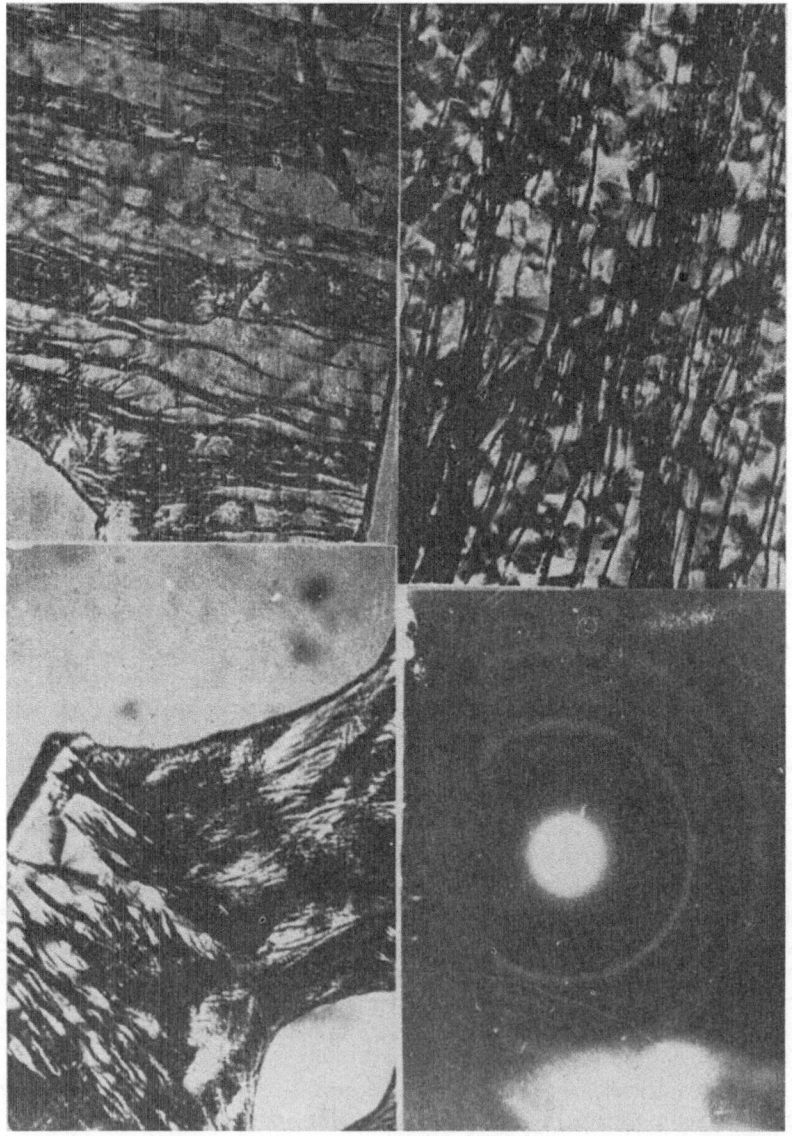

Figure 21. Consecutive stages of development of big low-temperature deformations and x-ray photograph of specimen extended to 120% at $-55°C$

of large-spherulite propylene with spherulites 80–150 microns in size. Both stretching and photographing were done at $-55°C$. The same figure shows an x-ray photograph of a specimen extended to 120% at $-55°C$ and photographed at the same temperature. Though extended to 120% the specimen showed no texture on the x-ray photograph! We found that deformation involves the displacement of blocks so large that their displacement did not even cause any texture to appear.

Thus we now have two extreme cases, namely, rubbers at high temperatures where deformations are due almost entirely to changes in conformation and displacement of macromolecules, and very-well-formed structures at low temperatures, where the deformations are due solely to the displacement of elements of supermolecular structures. The majority of real polymeric materials serve under intermediate conditions, and the current problem is to determine the contribution of each of the described phenomena to the general properties of the materials.

To conclude, it seems to me that the extraordinary simplicity of rise and fixation of non-equilibrium states in polymers excludes the existence of direct and unambiguous relations between the structure and composition of macromolecules and the appearance of supermolecular structures. Such relations may be only of a very general nature and can be considered only with reference to the physical state of the polymeric substance.

References

[1] V. A. Kargin and N. F. Bakeyev. *Koll. Zhurn.* **19**, 133 (1957).

[2] V. A. Kargin, A. I. Kitaigorodsky and G. L. Slonimsky. *Koll. Zhurn.* **19**, 131 (1957).

[3] V. A. Kargin. *Macromol. Chem.* **35A**, 77 (1960).

[4] V. A. Kargin and G. A. Koretskaya. *Vysokomolek. Soyed.* **1**, 1721 (1959).

[5] P. I. Zubov, Z. N. Zhurkin and V. A. Kargin. *Doklady Akad. Nauk SSSR* **67**, 659 (1949); *Koll. Zhurn.* **16**, 179 (1954).

[6] D. N. Bort, E. E. Rylov, N. A. Okladnov, B. P. Starkman and V. A. Kargin. *Vysokomolek. Soyed.* **7**, 50 (1965);
D. N. Bort, N. A. Okladnov, B. P. Starkman, L. I. Vidyaikina and V. A. Kargin. *Dokl. Akad. Nauk SSSR* **160**, 413 (1965).

[7] V. A. Kargin, Z. Ya. Berestneva and V. G. Kalashnikova. *Dokl. Akad. Nauk SSSR*, **166**, 874 (1966).

[8] F. Haber. *Ber. dtsch. chem. Ges.* **55**, 1717 (1922).

[9] Z. Ya. Berestneva and V. A. Kargin. *Uspekhi Khimii* **24**, 249 (1955).

[10] V. A. Kargin. *Preprinty přednaešek* **1**, 104 (1965).

[11] V. A. Kargin, T. A. Koretskaya and T. A. Bogayevskaya. *Vysokomolek. Soyed.* **6**, 441 (1964).

[12] V. A. Kargin, Z. Ya. Berestneva, E. P. Cherneva, T. D. Ignatovich and G. S. Potapova. *Dokl. Akad. Nauk SSSR*, in the press (1967).

[13] A. M. Kharlamova, E. P. Cherneva and Z. Ya. Berestneva. *Vysokomolek. Soyed.*, in the press.

[14] Tran Hyeu, N. A. Platé, V. P. Shibayev and V. A. Kargin. *Vysokomolek. Soyed.* **7**, 15 (1965).

[15] A. B. Zezin, N. F. Bakeyev, P. V. Kozlov and V. A. Kargin. *Mechanism of Processes of Film Formation from Polymer Solutions and Dispersions*, Nauka Publishers (in Russian), 1966, p. 9.

[16] V. A. Kargin and T. I. Sogolova. *Dokl. Akad. Nauk SSSR*, **88**, 867 (1953); *Zhurn. Fiz. Khim.* **27**, 1039 (1953).

[17] V. A. Kargin and T. A. Koretskaya. *Dokl. Akad. Nauk SSSR*, **110**, 1015 (1956).

[18] V. A. Kargin, T. I. Sogolova and N. Ya. *Rapoport-Molodtsova, Vysokomolek. Soyed.* **6**, 1562, 1559 (1964).

[19] V. A. Kargin, T. I. Sogolova and V. M. Rubstein. *Dokl. Akad. Nauk SSSR*, in the press (1967).

[20] V. A. Kargin, G. P. Andrianova and G. G. Kardash. *Vysokomolek. Soyed.* **9A**, 267 (1967).

ION-EXCHANGE MEMBRANES—CORRELATION
BETWEEN STRUCTURE AND FUNCTION

HARRY P. GREGOR

Polytechnic Institute of Brooklyn, New York, U.S.A.

The study of a complex system such as an ion-exchange membrane can be approached from two directions, which are not mutually exclusive. We can make use of the theoretical framework of irreversible thermodynamics as first applied by Staverman[1] to membrane systems, and as amplified by a number of other investigators, particularly Kedem and Katchalsky[2-4]. Here, we measure fluxes and generalized forces (the latter all too often involving the estimation of thermodynamic activity coefficients) and compute the phenomenological coefficients. This is essentially a "black box" approach which must give us correct and exact answers, provided the requirement of linearity between flux and force is obeyed, as they seem to be under the conditions which obtain in these systems.

The other approach is a mechanistic one, where we employ models and appropriate theoretical treatments to predict the behaviour of real systems. The two approaches are not independent of one another, but complimentary; both require the measurement of a number of different, independent parameters of the same system.

MEMBRANE STRUCTURE

In this contribution, we are concerned entirely with the so-called "tight" membranes which possess a high concentration of fixed-charge groups in a dense, porous gel. These are the ion-exchange membranes of commerce, which demonstrate a high and nearly ideal selectivity for the counter-ion species at all but the higher (>0.2 M) concentration levels.

The membranes upon which most of the experimental measurements described herein were performed are those available commercially from Asahi Chemical Industries (ACI). These are prepared by the dissolution of linear polystyrene (or a copolymer with styrene as the principal constituent) in a mixture of styrene and divinylbenzene (or another difunctional agent) which is then polymerized to form a cage polymer system, but where both the cross-linked and linear portions are principally of polystyrene. Blocks of these materials are then treated as follows: a microtome knife shaves off large sheets, which are then sulphonated to produce the cation-exchange material or treated with chloromethylether (or by an analogous reaction to form the benzyl chloride derivative), followed by treatment with trimethylamine to produce the anion-exchange membranes[5]. These membranes are "tight", and they are unusually uniform. Pieces taken from the same sheets are usually reproducible to within $\pm 10\%$, and materials obtained from the manufacturer at times separated by several months or even a year are as nearly alike, although samples obtained 4 years apart do show substantial differences in capacity but not in their intrinsic properties.

These membranes are also reasonably free of imperfections. Before all experiments they were examined for the absence of pinholes by placing the blotted membrane on a piece of dry filter paper, wetting the top of the membrane with a dye solution and examining the white filter paper for dye spots. This is a sensitive test, particularly so because the dry paper draws solvent through the membrane. While the work of Gluekauf et al.[6] has indicated that there are undoubtedly voids in most if not all ion-exchange resin and membrane systems, these usually constitute but a small fraction of their volume and do not extend through the structure; thus, these voids have no functional importance.

These ACI membranes show an exchange capacity which is the same to all of the ions studied herein, with the possible exception of a small deviation from full capacity exhibited by the cation-exchange membranes to the tetrabutylammonium (TBA) ion[7], or to the alkylbenzesulphonate (ABS) anion in the case of anion-exchange membranes[8]. Because of their high selectivity, these membranes exhibit nearly the same behaviour in dilute solutions (<0.2 M); this applies to measurements of the electrical conductivity, electro-osmotic flux, diffusion coefficient of neutral molecules or counter-ions, and the like.

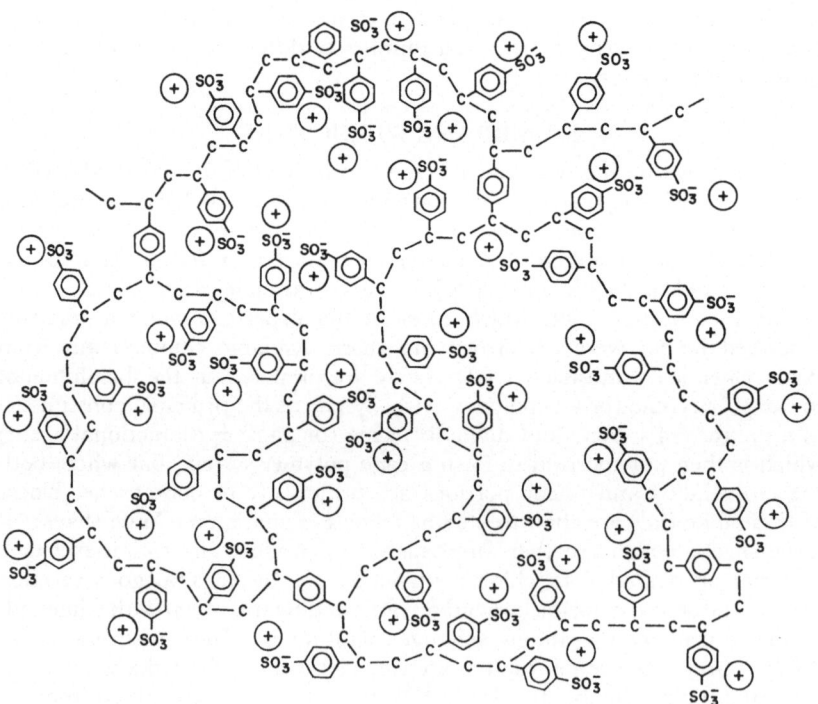

Figure 1

Figure 1 shows diagrammatically a cross-linked, completely sulphonated polystyrene polymer. The drawing is approximately to scale, for later calculations will show that the effective pore diameters are approximately 10 Å. The diagram is of a typical 5–8% DVB polystyrene sulphonic acid (PSSA) resin. Our view of the analogous membrane system is shown in *Figure 2*, where there are both cross-linked and linear polymer domains.

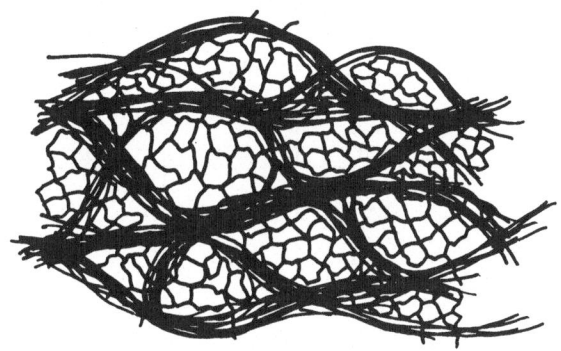

Figure 2

Figure 2 is purely conceptual: we have no direct but only indirect evidence as to the microscopic nature of the membrane. However, as will be shown later, the pore spectrum appears to be fairly narrow and the effective path length across the membrane is not too different from the actual membrane thickness (lack of tortuosity). This would suggest that the more amorphous, cross-linked portions of the membrane are preferentially sulphonated or chloro-methylated-aminated to produce ion-permeable regions.

The aromatic nature of these membranes deserves special mention. In spite of their high concentration of polar groups, these systems are still strongly aromatic and capable of strong adsorptive phenomena. This is particularly demonstrable with ions of an organic nature, particularly with the higher-molecular-weight quaternary ammonium cations and with the aromatic anions. Reference should also be made to the strong effects observed by Sakurada (see pp. 263–84) in his catalytic studies involving PSSA polymers, described elsewhere in this volume. While these adsorptive processes evidence themselves principally in thermodynamic measurements such as that of the selectivity coefficient[9, 10], they do affect the permeability of membrane systems appreciably.

A few of the properties of the ions examined in this study are listed in *Table 1*. The equivalent ionic conductance at infinite dilution is certainly a reliable although only a semi-quantitative index of the ion size because of the inapplicability of Stokes' hydrodynamics to these molecular systems. The crystal radii are well established, as are those obtained from model studies for the quaternary ammonium ions. The application of the hydrodynamic equations for the calculation of effective ionic radii in solution is discussed by Robinson and Stokes[11]. On the basis of both experimental and theoretical investigations, it appears that the numerical constant in the

Table 1. Ionic properties

	$\lambda°$	r in Å	Crystal radii Å
H	350		
Li	39		0·68
Na	50		0·98
K	74	2·1†	1·33
Me₄N	45	3·47	
Et₄N	33	4·00	
Pr₄N	23	4·52	
Bu₄N	20	4·94	
OH	195		
Cl	76	2·3*	1·81
Br	76		1·96
I	77		2·19
SCN	72		
pTS	34		
IO₃	40	3·7*	
NO₃	71		

† From distance of approach in solution. $\lambda°$ is the equivalent ionic conductance at infinite dilution.

Stokes' equation is apparently nearly 6 for large ions, decreasing to approximately 3 for ions having approximately the same radius as that of the solvent. However, the recent work of Zwanzig[12] has served to indicate that an additional correction may be required in correlating the ionic radii of the large quaternary ammonium ions with measurements of their electrical conductivity. Further, we know that the activity of water in solutions of quaternary ammonium ions is not that predicted from a consideration of ionic concentrations and radii, employing the Debye–Hückel equations. Measurements of several authors[13–15] confirm the general concept of Frank[16, 17] who views water as existing in a "normal" water state, a hydrogen-bonded "ice-like" state (which may be reinforced by the mirror effect) and another state of low entropy, the ion-hydrate state. Accordingly, ions are classified into "structure-breaking" and "structure-making" species, the structure referring to the ice-like state.

Theoretical studies have indicated that it is the radius of curvature of solvent and solute molecules which determines the extent of momentum transfer on collision and therefore the coefficient of Stokes' equations. When we later apply the classical hydrodynamic equations to the kinds of molecular structures shown in *Figure 1* and its counter-ions, we feel some confidence in this extrapolation because the radius of curvature of the molecules making up the pore walls is of the same magnitude as those of the ions passing through these pores. Under these circumstances, it appears reasonable to apply the classical hydrodynamic equations with a correction for the "hydrodynamic radius"[11].

Table 2 summarizes a number of the properties of the ACI cation-exchange membranes. Using the potassium state as the base for comparison, we see that the membrane swells slightly when larger (hydrated) exchange cations (Na and Li) are present, but appreciably and regularly as the size of the

332

Table 2. Properties of ACI cation-exchange membranes
(Capy.—2·84 meq/g)

Ion	Vol. (ml)	L (μ)	φ_p	$\bar{c}$	$\bar{m}$
H	1·42	220	0·44	2·0	4·6
Li	1·50	218	0·45	2·0	4·3
Na	1·42	217	0·41	2·0	4·9
K	1·40	221	0·41	2·0	5·1
A	1·40	216	0·42	2·0	5·1
TMA	1·78	239	0·58	1·6	3·9
TEA	1·94	247	0·63	1·5	3·8
TPA	2·07	245	0·64	1·4	4·3
TBA	1·97	244	0·68	1·4	5·1
Dowex 50-X8			0·50		5·5

Vol.—ml/g of H-form membrane.
L —thickness in microns (μ).
φ_p —pore volume fraction (including exchange ion).
$\bar{c}$ —molar concentration of fixed charge.
$\bar{m}$ —molal concentration of fixed charge.
A, TMA, TEA, TPA and TBA are the ammonium, tetramethyl-, ethyl-, n-propyl and n-butyl ammonium ions.

quaternary ammonium counter-ion is increased. There is a 40 per cent increase in membrane volume on going from the potassium to the TBA state, but only a 10 per cent increase in thickness; most of the volume increase is therefore a consequence of increases in pore diameters, if we consider these macro-pores as having their axes normal to the face of the membrane. This anisotropic swelling is one basis for our picture of these membranes as having macro-pores of cross-linked polyelectrolyte extending through the membrane, with the diameter of the macro-pores contained therein being controlled by the extent of cross-linking.

In *Table 2* we note that the pore volume fraction (which includes the counter-ion volume) also increases with (hydrated) ionic size for the in-organic ions, and more strongly as the size of the quaternary ammonium ion increases. The volume of the exchange ion, as will be seen later, consti-tutes an appreciable fraction of the total volume of the pore solution. Accurate volume and density measurements[18, 19] have shown that with most monovalent ionic species the matrix volume plus the volume of the imbibed solvent and that of the exchange ions very nearly adds up to the total, measured volume of the system. The molality or concentration in terms of imbibed solvent (*Table 2*) usually falls with increasing counter-ion size, because more water is sorbed as its concentration is reduced, due to the size of the exchange ion itself. A comparison with one of the commercial cation-exchange resins (Dowex 50-X8) shows it to be reasonably analogous with the ACI cation-exchange membranes.

The properties of ACI anion-exchange membranes, shown in *Table 3*, show a rather different correlation with the properties of exchange anions, as compared with the cation-exchange materials. Their pore volumes are substantially less, with that for the chloride state being approximately half that for the potassium state. Further, the pore volume is relatively independent of the size of the exchange anion; in fact, it is often lower with

Table 3. Properties of ACI anion-exchange membranes
(Capy.—1·35 meq/g)

Ion	Vol. (ml)	L (μ)	φ_p	$\bar{c}$	$\bar{m}$
Cl	0·94	214	0·23	1·1	5·0
I	0·93	218	0·18	1·1	6·2
SCN	0·92	211	0·18	1·2	6·4
NO₃	0·94	212	0·20	1·1	5·6
pTS	1·00	220	0·17	1·1	6·5
IO₃	0·99	216	0·32	1·1	4·3

pTS, p-toluenesulphonate.

the large organic exchange anions, similar to what is observed with the analogous anion-exchange resin systems[20]. The pore volume only increases substantially with the iodate anion. Anionic pore volumes are approximately half those for the cation-exchange membranes; the molalities for both types of membranes are approximately the same, because of the lowered water content of the anion-permeable membranes.

DIFFUSION OF NEUTRAL SPECIES

The rate of transport across these "tight" ion-exchange membranes is controlled, in large measure, by the frictional or hydrodynamic resistance which obtains. We prefer to calculate this frictional resistance in terms of a pore model. The classical work of Collander[21] first showed that pore diameters can be estimated from the relative rates of permeation of solutes of different molecular weight (and presumably size) across a porous membrane. Collander pointed out that the sorption or adsorption of solutes by the membrane could lead to erroneous conclusions, and it was therefore important to avoid solutes where such phenomena were pronounced. In dealing with "tight" membranes, we unfortunately have relatively few permeating, non-ionic solutes available and very few of these may be regarded as reasonably spherical.

The diffusive flux (J_i) of a neutral molecule across a unit area of a membrane system is given by

$$J_i = -\bar{D}_i (\bar{c}_i'' - \bar{c}_i')/L,$$

where $\bar{D}_i$ is the diffusion coefficient of the neutral species in the membrane, L is the measured thickness of the membrane, and the driving force is computed in terms of differences in the bulk (molar) concentration ($\bar{c}_i$) of the species on the left (') and right ('') sides of the membrane.

If one assumes a model wherein n tortuous paths, each of uniform radius R and length $L_p = L/h$, connect the opposite sides of a unit area of membrane, then the pore volume φ_p and the corresponding pore area a_p are related by,

$$\varphi_p = n\pi R^2 L_p/L = a_p L_p/L = a_p/h$$

where h is the "tortuosity" factor. The flux can now also be described in terms of the solution phase-diffusion coefficient D of the solute and its concentration c. If the distribution coefficient of the solute between equal pore (c_p) and solution volumes is $k = c_p/c$, then, if no membrane resistance terms are present,

$$J = -Dk\ (c'' - c')\quad a_p\ h/L = -Dk\ (c'' - c')\ h^2\varphi_p/L.$$

Here we assume that in a porous membrane it is necessary for the diffusing species to travel a path length which is tortuous on a macroscopic level and which is greater than that of the measured thickness of the membrane. The tortuosity factor has been discussed at length by several authors, particularly Carman[22]. For systems where the pore length is considerably larger than the diameter of the diffusing species, tortuosity is often taken as $\frac{1}{3}$.

The latter equation must also include two resistance factors, the so-called Ferry–Faxen terms. The former is a purely geometrical one, describing the probability that when the centre of the solute molecule of radius r strikes part of the pore area, it will pass into the pore and not be deflected. The Ferry[23] term is $F' = (1 - r/R)^2$ for a crylindrical pore and $(1 - r/R)$ for a parallel plate or rectangular pore.

The second resistance term is that due to Faxen[24], and describes the hydrodynamic resistance to the flow of a spherical particle in the centre of a pore filled with a homogeneous liquid, assuming the validity of Stokes' hydrodynamics. The employment of these latter terms has been discussed by many authors, particularly by Pappenheimer[25] and by Renken[26], and the flux equation becomes,

$$J = -Dk\ F'Fh^2\varphi_p\ (c'' - c')/L.$$

In the employment of the Ferry–Faxen equations, most authors have assumed cylindrical pores. The equations which obtain here are,

$$F'F = (1 - r/R)^2\ [1-2\cdot014\ (r/R) + 2\cdot09\ (r/R)^3 - 0\cdot95\ (r/R)^5].$$

However, with pore diameters which are of molecular dimensions it is obvious that nature has not provided us with cylindrical pores, not at least in the world of organic macromolecules. Instead, these pores must be highly irregular as indicated in *Figure 3*. It is only with pores having a diameter several times the molecular diameters that one could employ the equations for cylindrical pores with reasonable confidence. Accordingly, since the diffusing species usually is passing between two points of contact, we have preferred to employ the parallel plate or rectangular pore. Faxen has derived expressions which apply here[27],

$$F'F = (1 - r/R)\ [1-1\cdot004\ (r/R) + 0\cdot418\ (r/R)^3 -$$
$$0\cdot169\ (r/R)^5 - 0\cdot245\ (r/R)^7].$$

Because in Faxen's treatment the integration was performed starting at the centre, with insufficient terms in the expansion employed to describe

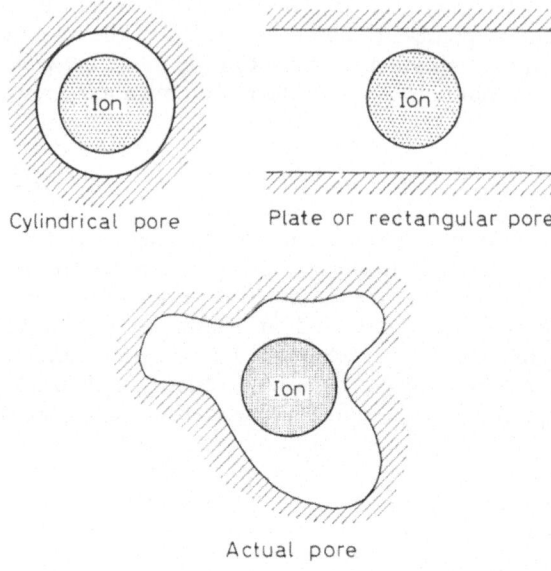

Cylindrical pore   Plate or rectangular pore

Actual pore

Figure 3

accurately the situation where the radius of the ion approaches that of the pore, we have arbitrarily corrected the last term of the function so that it goes to the proper limit at $r = a$. The difference in the two Ferry–Faxen equations is shown in *Figure 4*; the function is much more shallow in the case of the plate-like pore, and we have found that it fits the data very much better.

The diffusive flux of several low-molecular-weight, neutral species was measured across cation-exchange and anion-exchange membranes, taken from the measurements of D'Alessandro[8]. Here, Fick's law was obeyed, presumably because the concentration of diffusing species was low ($c = 0 \cdot 1$ M)

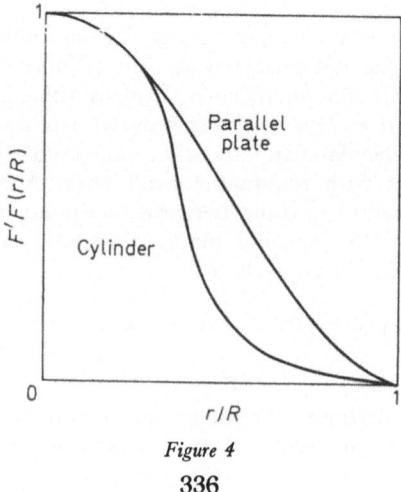

Figure 4

in all cases. *Table 4* summarizes some of these experimental results and also lists solute radii, as estimated by various authors[25, 26]. It shows too that urea and ethyleneglycol are salted-in or absorbed by these membranes, but glycerol is somewhat salted-out. We observe that while the diffusive flux decreases with increasing molecular weight, the relative diffusion coefficient $\bar{D}/D$ does not decrease regularly. There are probably two reasons for this. First, ethylene glycol is a reasonably symmetrical molecule but glycerol is not and probably diffuses in an oriented position, thus giving it an anomalously high diffusion coefficient. For example, in free solution the relative

Table 4. Diffusion of neutral molecules

Solute	Ion	$\bar{D} \times 10^7$ (cm²/sec)	$\bar{D}/D$	R (Å)	R' (Å)	$\bar{m}/m$†	r (Å)	$D \times 10^5$ (cm²/sec)
Urea	K	7·8	0·060		5·7	1·9	2·6	1·37
				3·3				
Glycol		4·3	0·043		5·5	1·4	2·8	1·00
Glycerol		4·1	(0·052)	3·8	6·4	0·8	3·1	0·78
Glucose		3·3	0·047	4·3	7·1	—	3·6	0·69
Urea	Cl	3·8	0·030		4·7	2·2		
				3·1				
Glycol		4·1	0·036		5·3	1·7		
Glycerol		3·9	(0·041)	3·5	6·1	0·7		
Glucose		0·84	0·010	3·7	5·2	—		

† $\bar{m}/m \simeq c_v/c$.

diffusion coefficient of these two species is 1·3, while in most ion-exchange membrane systems it is approximately 1·05, and this demonstrates the problem of evaluating pore diameters on the basis of the diffusion of non-symmetrical molecules. Some authors have employed hydrophilic polymers of different molecular weight for pore size determinations, with results that do not often have even qualitative significance.

We have made the assumption of uniform pores; this is obviously not correct, but difficult to subject to experimental test. If one assumes a normal, Gaussian distribution of pore radii and then applies the Ferry–Faxon equation, one finds the experimental predictions substantially unchanged, i.e., the Ferry–Faxon equations yield an average pore diameter, with little or no information on pore diameter spectra.

The relative diffusion coefficient is related to the parameters of these systems by,

$$\bar{D}_i/D_i = h^2 F_i' F_i \, (r_i/R)$$

an expression with two unknowns, the tortuosity factor h and the pore radius R. However, by using a second diffusing solute of a different radius r_j, and making the reasonable assumption that h is unchanged, we can obtain unique values of the effective pore radius R and also the tortuosity factor. A simple procedure for carrying out this computation is to calculate $F'F \, (r_i/R)/F'F \, (r_j/R)$ for different values of R and the two known values of

337

r_i and r_j; when this ratio is equal to $(\bar{D}_i/D_i)/(\bar{D}_j/D_j)$, the value of R is read from the curve. Using urea and ethylene glycol as the two solutes, and then making the other computations based upon these comparisons, we obtain the data given in *Table 4*. Here we also obtain the unexpected result that the tortuosity factor is near unity for these systems; to this point we will return later.

The diffusion coefficients of neutral species across these membranes lead us to computed pore diameters of approximately 3·3 Å for the potassium form membrane and 3·1 Å for the chloride form. The diffusion of DHO in these membranes as measured by Leszko[8] also yielded similar information on pore sizes and tortuosity.

TRANSPORT OF IONS

Since the transport of counter-ions in ion-exchange membrane systems is necessarily different, in many important respects, from that of neutral species because of the high coulombic fields present in these systems, a discussion of these transport processes is best introduced by consideration of the polyelectrolyte nature of these membranes. It was pointed out some time ago by the author[28] that ion-exchange resins and membranes are best viewed as cross-linked polyelectrolytes, with no fundamental differences in the fundamental physical parameters which obtain in both systems. Examinations of the properties of the two systems from the same point of view has shown that their behaviour is quite analogous. For example, Leszko and Gregor[8] measured the mean activity coefficient of potassium chloride in mixtures of the potassium salt of polystyrene sulphonic acid (PSSK) with potassium chloride, employing an electromotive chain containing a cation-selective glass membrane and a silver–silver chloride reference electrode. Solutions varying in total molality from 0·001 to about 3 were studied in this manner, and it was found that Harned's rule was obeyed over the entire range of concentration. Harned's rule applies for almost all simple electrolytes, but 2-1 electrolytes do not obey the rule in many cases; this makes the conformity of the PSSK–potassium chloride system the more surprising.

In relatively concentrated polyelectrolyte solutions (1–3 M) and with low concentrations of potassium chloride, the mean activity coefficient of the latter species is approximately 0·45; this value should also obtain in calculating the concentration of diffusible electrolyte in comparable ion-exchange resin and membrane systems. Unfortunately, most of these systems contain voids[6] which produce an anomolously high content of diffusible electrolyte and therefore a low, calculated mean activity coefficient.

Leszko and Gregor[29, 8] also have reported the diffusion coefficients of counter-ions, co-ions and water in the PSSK–KCl system. *Table 5* summarizes these results, wherein the parameters for the polymer solution are those measured by direct experiment, divided by the solution volume fraction to yield the "effective" value, one which should be the basis for comparison with the free solution values. The presence of the fixed anionic sites has the effect of decreasing the mobility of counter-ions, with a much smaller effect upon the co-ions, as is observed with ion-exchange resin and membrane systems[30]. From the measurements of Leszko and Gregor[29] we

Table 5. Diffusion coefficients
(3M PSSK, 0·1M KCl, 25°C)

	Polymer solution ×10⁵	Free solution ×10⁵
D+	0·76	2·2
D−	1·1	2·2
D_{HDO}	1·94	2·45
φ†	0·78	1·00

† φ is the solution volume fraction.

find that the diffusion coefficient for water in the polymer solution is sub-stantially less than for water in free solution, probably because of solvation of the highly-polar polyelectrolyte. Thus, in general we can apply the pore or "free solution" volume correction term to membrane, resin and dissolved polyelectrolyte systems, alike.

The selective uptake of one ionic species over another by an ion-exchange resin or membrane gives us a further insight into the system and the pro-cesses occurring therein. Selectivity in ion-exchange systems is determined by a number of factors. First, there is the straightforward coulombic inter-action between the fixed-charge group and the mobile ion species. Selective uptake under these circumstances is determined by differences in the distance of closest approach of ions to the charged rod. This has been investigated theoretically by Miller and Gregor[10] following an earlier study by Kagawa and Gregor[31] based upon the charged rod model (Fuoss, Katchalsky and Alfrey). Here, one must employ (hydrated) ionic radii, so under these circumstances the interactions between ions and solvent come into play. Figure 5 shows the model assumed by Miller and Gregor and their

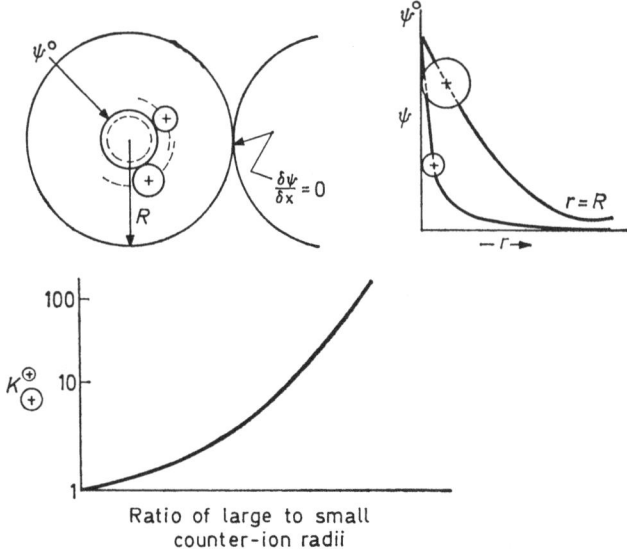

Ratio of large to small
counter-ion radii

Figure 5

calculated results. This theory was compared with the data of Gregor and Bernstein[10, 32]; an excellent correspondence between theory and experiment was obtained for a number of polymethacrylic acid polymers and the exchange of potassium against a number of quaternary ammonium cations. This system was studied at complete neutralization to produce a highly polar polymer and accordingly, in minimum of adsorptive interactions. In these polymer systems, which were cross-linked by divinylbenzene (DVB), a substantial deviation from the theory was encountered at high degrees of cross-linking, undoubtedly due to adsorptive forces between the apolar cations and the aromatic ring. Later experiments by Gregor and Greff[8] on a copolymer of methacrylic acid and ethyleneglycoldimethacrylate do not demonstrate this adsorptive effect.

The role of adsorption in ion-exchange interactions has been recognized for some time. In general, ion-exchange polymers of low capacity (and low polarity) show a strong preference towards organic ions. The study by Bregman and Gregor[9] on the selective uptake of potassium against a number of quaternary ammonium cations with PSSA resins of different degrees of

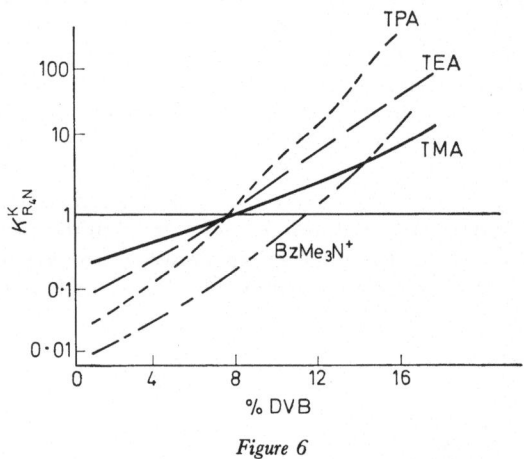

Figure 6

cross-linking (see *Figure 6*) showed that at low degrees of cross-linking (when pressure–volume effects are minimal), the organic cations are preferred, approximately in the order of their increasing molecular weight. Much greater adsorption is encountered with the aromatic benzyltrimethyl-ammonium cation, as expected. Some recent studies have shown that the changes in enthalpy accompanying these reactions are small, consistent with our knowledge of hydrophobic bonding. These adsorptive effects are different in character from the coulombic effects, but a clear-cut distinction between site-binding and non-specific coulombic binding is not easy to obtain because in many systems they differ only in degree but not in kind, and are difficult to demonstrate by reasonably absolute and independent means. In general, we may say that when the selective uptake of an ion is well in excess of that predicted from considerations of size, one may consider ion-pair formation or site-binding. *Figure 7* shows the selective uptake of a number of anions by the ACI anion-exchange membranes and a cross-linked aliphatic polyethyleneimine (PEI) resin, taken from the study of Konrad, Saber and Gregor[8]. Here it is seen that the order of selective uptake is SCN > pTS > I > Cl > IO_3 (the latter not shown) with the ACI membrane. This is the familiar Hofmeister or Lyotropic series. It is apparent that the polarizability of the halogens anions is an important factor here, as was also shown by Gregor, Belle and Marcus[20]. Adsorptive forces un-doubtedly come to play in exchange with the aromatic p-toluenesulphonate anion. It is also evident that the selective uptake of adsorbable anions increases with the increasing chain potential. For a system of two anions of the same size differing only in polarizability, the theory of Miller and Gregor[10] predicts selectivity (through different degrees of closest approach) and also an increase in the selectivity of the more polarizable anion as the mole fraction of the less polarizable anion increases; this is also confirmed by the study of Konrad *et al.*[8].

Figure 7 also presents data on the selectivity of aliphatic anion-exchange polymers for similar pairs of ions, and here we see a marked decrease in

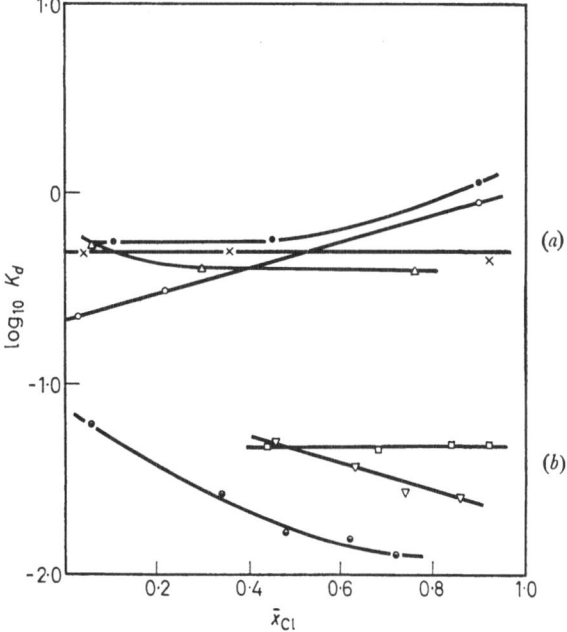

Figure 7. Selectivity coefficients measured against chloride for the following ions
(a) PEI-ECH resin — iodide ($\times$); thiocyanate ($\bigcirc$); perchlorate ($\triangle$); p-toluenesulphonate ($\bullet$)
(b) Asahi anion-exchange membrane — iodide ($\square$); thiocyanate ($\ominus$), p-toluenesulphonate ($\triangledown$)

preference. With iodide, its selectivity over chloride is about 50 with the benzyltrimethylammonium polymer, and only about 5-fold with the (analogous) diethylammonium group. This result is unexpected in its magnitude and suggests that a charge-transfer mechanism may be involved with the former. With the thiocyanate and pTS anions it is evident that hydrophobic bonding is the most significant causative factor in their selective uptake.

Self-diffusion coefficients in chloride–iodide from anion-exchange membranes (also of aromatic character) were measured by radio-tracer techniques by Andelman and Gregor[33] who found that the individual diffusion coefficients did not vary significantly over the entire range of composition varying from iodide to chloride forms, and that the self-diffusion coefficient of chloride was greater than that of iodide by a factor of 4·5. Employing the Nernst–Einstein theory which relates self-diffusion coefficients to limiting ionic conductivities (at infinite dilution) one calculates the same equivalent conductivity ratio for these ions. This composition-independent behaviour may be explicable on the basis of a chain mechanism, as postulated by Belgovskiy[8].

The measurement of the diffusion coefficient of an ion in a membrane system is not as simple and unequivocal as one would desire, because there is always resistance due to the solution boundary layer adjacent to the membrane face. One can make measurements on "sandwiches" of membranes with one pressed firmly to the other to minimize the solution resistance term or make appropriate computations. In general, the self-diffusion coefficient of counter-ions in "tight" membranes is relatively independent of concentration in solutions more dilute than 0·1 M, as was shown by Kawabe et al.[7].

The conductances of ACI cation-exchange and anion-exchange membranes are listed in Table 6. Here we find that the relative conductivity ratio λ/λ^0 does not vary greatly for the univalent, inorganic ions but does fall off sharply as the molecular weights of the quaternary ammonium ions increase. The energies of activation for conduction are not very different from those in free solution (3·3–3·7 kcal/mole) with the exception of

Table 6. Conductance of ACI ion-exchange membranes

Ion	RA (ohm cm^2)	$\bar{\kappa} \times 10^3$ (ohm^{-1} cm^{-1})	λ (cm^2/ohm eq)	λ/λ°	E_a (kcal/mole)
H	0·3	60	30	0·085	—
Li	2·3	9·3	4·6	0·12	4·1
Na	1·5	16	8·0	0·16	4·6
K	0·9	24	12	0·16	3·8
A	0·9	24	12	0·16	—
TMA	4·3	5·6	3·5	0·078	—
TEA	13·2	1·7	1·2	0·036	3·4
TPA	49·5	0·41	0·30	0·013	4·7
TBA	469	0·047	0·032	0·0017	5·9
Cl	4·7	4·6	4·2	0·055	4·8
I	19·7	1·1	0·98	0·013	—
SCN	26·8	0·79	0·69	0·0096	6·8
NO$_3$	6·8	3·1	2·8	0·040	—
pTS	717	0·031	0·031	0·00090	—
IO$_3$	11·2	1·9	1·9	0·046	5·5

RA—areal resistance, equal to $L/\bar{\kappa}$.
$\bar{\kappa}$ —specific conductivity.
λ —equivalent conductance.

those for the larger quaternary ammonium cations. In the case of the anion-exchange membranes, the equivalent conductance falls with increasing atomic weight of the halide anions, but several anomalies appear and will be discussed later. A comparison of the relative equivalent conductivities of cation and anion-exchange membranes show that the former have much larger relative equivalent conductivity, with that for potassium being approximately three times that for chloride. The energy of activation for conductivity in the anion-exchange membranes is appreciably higher than in free solution for all anions measured, particularly with the strongly adsorbed thiocyanate ion. This suggests that a different energetic process is involved in transport in anion-exchange membranes.

With the cation-permeable membranes, the Ferry–Faxen equation can be employed to estimate effective pore radii from the conductivity of ions of different size. As discussed by Kawabe[7], there are two ways of calculating pore radii. One may make the assumption that the pore volume of the system is directly proportional to the square of the pore radius (because swelling is anisotropic), and calculate individual pore radii which pertain to those for the smaller ion, given as R' values in *Table 7*. One can also compare

Table 7. Parameters of ACI cation-exchange membranes

	R' (Å)	R'' (Å)	R''' (Å)	r of ion (Å)
K	3·6		3·5	2·1
Me$_4$N	4·4		4·1	3·5
		4·9		
Et$_4$N	5·1		4·6	4·0
		5·1		
Pr$_4$N	4·9		4·8	4·5
		5·2		
Bu$_4$N	5·1		5·0	4·94

$\lambda + EO$.

ions of different size and calculate mean pore radii (shown R'' in *Table 7*). The two methods naturally yield comparable results, with radii from the latter method favouring those for the larger ion.

ELECTRO-OSMOTIC WATER TRANSPORT

The electro-osmotic (EO) permeability of membranes was measured by D'Alessandro, Bagner and Breslau[8]; *Table 8* summarizes some of these results. It is usual to express electro-osmotic transport in terms of moles of water transported per Faraday of ions, but here it is more correct to report electro-osmotic volume transport because such a large fraction of the solution volume transported consists of the exchange ion itself. *Table 8* shows that in the case of cations, the electro-osmotic volume transport increases with (hydrated) ionic size. Turning first to the non-hydrated quaternary ammonium ions, one finds that the amount of water transported per ion begins to level off as the size of the cation increases, and with the TPA and TBA ions equals, within experimental error, the water content of the membrane.

Table 8. Electro-osmosis

Ion	$\lambda/\lambda°$	φ_p	EO (ml/**F**)	EO' (H_2O/ion)	$(18\,\overline{m})^{-1}$ (H_2O/ion)	V_{ion} (ml)
H			45			
Li	0·12	0·45	163			
Na	0·16	0·41	116	6		
K	0·16	0·41	105	4	11	33
TMA	0·078	0·58	267	8·6	15	116
TEA	0·036	0·63	368	10	15	172
TPA	0·013	0·64	463	12	13	255
TBA	0·0017	0·68	632	17	14	326
F	—	—	170	9		—
Cl	0·055	0·23	90	3		30
Br	—	—	90	3		30
I	0·013	0·18	95	3		30
IO_3	0·046	0·22	159	2		140
ClO_4	—	—	105	—		—
pTS	0·001	—	224	—		—

ml/**F** —ml of water transported/Faraday of current.
EO' —moles of water per ion.
$(18\,\overline{m})^{-1}$—moles of water present/counter-ion.
V_{ion} —molar volume of counter-ion.

In other words, these ions sweep *all* of the water contained in the membrane along with it under electro-osmotic flow, equalling the "plug-flow" of engineers. This result indeed suggests that these ions very nearly approximate the size of the pores through which they move, and also suggests that there is little solvent contained in pores not penetrated by the exchange ions. It suggests, then, a remarkably homoporous system, but this conclusion may readily be erroneous. The stress placed upon the membranes by the electro-osmosis of the TBA ion was considerable; several split during the experiment and all swelled, so that the apparent discrepancy in *Table 8* wherein more water was transported than present is the result of increased swelling (and water content) during this procedure.

The electro-osmotic transport shown by the lithium and sodium forms of the membrane are not nearly as high as one would predict from their apparent hydrated diameters as estimated from their conductivity in solution[11]. For example, one should compare Li with the mean of the TMA–TEA ions on the basis of solution sizes, but it has only half the predicted EO value. The obvious conclusion is that it has lost half its water of hydration, and that the same process has occurred with Na, to a lesser extent. This conclusion reinforces the work of Kawabe[7] who calculated pore radii for the alkali metal cations by extrapolation (using φ_p data) from R values for the TMA ion, and came to the conclusion that there was considerable dehydration of the Li ion, less for Na and little for the K or A ions, as they move from the solution to the membrane phase. The constancy of λ/λ^0 values with these ions is then the result of a decrease in size and a lowered F'F drag factor.

It is interesting to note that while in exchange and other equilibrium phenomena (such as hydration[34, 35]) the hydrogen ion has apparently the same volume as does the lithium ion, presumably because its organization of water structure gives it the same effective size under these circumstances; under electro-osmosis the different transport mechanism for the hydrogen ion results in a low EO value.

Breslau[8] also measured the temperature coefficient for EO with Li and K-form membranes, and obtained the most interesting result, that the energy of activation almost exactly corresponded with that for the viscosity change of water. In other words, no significant change in hydrated size took place, unless different effects compensated for one another. Under equilibrium conditions hydration numbers are larger and more influenced by temperature changes, as was observed by Gregor and Bregman[9].

Electro-osmotic fluxes across anion-exchange membranes demonstrate a rather different behaviour than is observed with cation-exchange membranes. First, EO for the halide anions (except fluoride) are nearly the same, while that for the iodate ion is approximately doubled. A comparison with the behaviour of cation-exchange membranes shows the unexpected result that the potassium and chloride ions demonstrate approximately the same electro-osmotic transport, but their relative equivalent conductivities differ by a factor of 3. We have seen earlier that their pore diameters, as measured by the diffusion of neutral species, are but slightly different, from the magnitude which one would expect from their relative electro-osmotic water transports, but quite different from what one would anticipate from their relative equivalent conductivities. In other words, a large difference in pore diameters would be required to account for the different conductivities. On the other hand, the comparison between iodate and the TMA ion is "normal" in that their relative equivalent conductances are in approximately the same ratio as their electro-osmotic water transport terms. We must conclude that the halide ions show yet another effect.

The fluoride ion shows an EO which is very much higher than for the other halides, undoubtedly the result of changes in water structure. This is classified as a "structure breaking" ion, and its EO transport reflects this.

DISCUSSION

One can attempt to correlate these data on the basis of model studies reported in the literature. In particular, Schmid (see reference 30 for an excellent discussion of his papers) has correlated self-diffusion coefficients, equivalent ionic conductivities and hydraulic permeabilities for fine-pore membrane systems—ones where the Debye radius is large compared with the pore radius. This is obviously the case for the tight membranes with which we are concerned here. Schmid derived his equations by considering the relative velocities of the ions, the solvent and the membrane; essentially, his treatment is based upon the choice of a stationary frame of reference, in this case the membrane. Under these circumstances, the moving ion carries along with it electro-osmotic water, which in turn contributes to the conductivity of other ions in the system. The total conductivity $\bar{\kappa}_t$ is then equal to the intrinsic conductivity (velocity of water zero) plus the contribution to the specific conductivity of the electro-osmotic flow, $\bar{\kappa}_{EO}$. The intrinsic conductivity is assumed to be described by the Nernst–Einstein equation, for here we neglect the time of relaxation effect because at the usual frequency of measurement the ion traverses several fixed-charge sites in the membrane. Schmid defines a specific flow resistance term, and the ratio of the electro-osmotic specific conductivity to the "true" conductivity can be calculated in terms of the electro-osmotic flux or the ratio of the electro-osmotic and hydraulic permeability fluxes, $\bar{\kappa}_{EO}/\bar{\kappa}_t = 1 - \alpha = 1 - F\bar{c}\,(EO)\,\varphi_p$. One can determine α directly by self-diffusion coefficient measurements and then employ the Nernst–Einstein equation for comparison with the conductivity value. It should be emphasized that the theory of Schmid does not allow for specific interaction; it treats all ions as moving at the same velocity with the same electro-osmotic flow.

A test of the Schmid relationships, both from electro-osmotic fluxes and also from a combination of electro-osmotic and hydraulic permeability fluxes has been made[36]. For the potassium ion the discrepancy between theory and experiment is substantial (0·48 cf. 0·61) but not impossible. However, the correction term α falls sharply as the size of the ions increases, until for the TPA ion we find that 0·99 of the specific conductivity is that due to electro-osmosis, a most unlikely event. Further, the equation gives negative values of the coefficient with the TBA ion. The model upon which the Schmid theory is based obviously cannot be employed with these very tight membranes, particularly where the diameter of the counter-ion is an appreciable fraction of the total pore diameter.

Table 9. Comparative values

	Cation-permeable* $\bar{c}_K = 2\cdot0, \bar{m}_K = 5\cdot1$			Anion-permeable† $\bar{c}_{Cl} = 1\cdot1, \bar{m}_{Cl} = 5\cdot0$		
	K	Li	TMA	Cl	I	IO$_3$
φ_p	0·41	0·45	0·58	0·23	0·21	0·35
λ/λ°	0·16	0·12	0·078	0·055	0·013	0·046
EO	105	259	267	90	90	169
r	2·1	2·5	3·5	2·1	2·1	3·5
R_λ	3·3	3·6	4·5	3·0	3·0	4·2
%Dissoc.	100	100	100	46	10	100

* $\bar{c}_K = 2\cdot0$; $\bar{m}_K = 5\cdot1$.
† $\bar{c}_{Cl} = 1\cdot1$; $\bar{m}_{Cl} = 5\cdot0$.
R_λ^- for TEA is 4·7 Å; TPA is 5·0 Å; TBA is 5·1 Å.

Table 9 summarizes the salient experimental results obtained herein. The potassium, TMA and iodate ions show "normal" behaviour. However, with the halide anions we observe reasonably normal electro-osmotic (and hydraulic permeability[36]) behaviour, but an abnormally-low electrical conductivity. The most reasonable conclusion we can draw is that the halide (and the SCN, pTS and ClO$_4$ anions) are bound at fixed-charge sites and cannot contribute to the conductivity, although under the influence of an electric current they are able to move electro-osmotically and exhibit "normal" behaviour. Accordingly, taking reasonable values for the radii of the halide anions, from their R values we compute that the membrane in the chloride form is 46 per cent dissociated and in iodide form is 10 per cent dissociated.

Should site-binding occur, this will be evident by the selectivity coefficient measurement, as indeed it is. However, as was shown by Miller and Gregor[10], coulombic binding and site-binding show the same general characteristics and are difficult to distinguish one from another. From selectivity coefficient data, the chloride ion is preferred approximately 8 times as strongly as the iodate anion, while from conductivity data the apparent ratio is 2:1. We may then conclude that the difference is due to ordinary coulombic binding, or that *both* occur here. By the same token, the relative binding of iodide to chloride from conductivity data is approximately 5:1, while from selectivity coefficient data the ratio is 25:1. Again, the difference is assumed to be that due to coulombic binding. One may raise the question as to whether there is not a retardation of ionic motion under the influence of an electric field or a concentration gradient as the result of ordinary coulombic binding. One could explain this on the basis that the ion was not travelling through the *centre* of the pore but rather along its periphery, with a correspondingly greater "drag". Unfortunately, the hydrodynamic equations for this case are rather involved and no solution has been published. Further, the simplifications involved are rather severe and we have already made too many approximations.

CORRELATION WITH MEMBRANE TECHNOLOGY

On the basis of our extensive industrial experience in the employment of ion-exchange membranes in electrodialysis, the results which we have just described lead us to certain conclusions. First, consider the all-important problem of the "fouling" of membranes. When natural waters containing high-molecular-weight, adsorbable anions such as the humic acids or ABS (alkylbenzenesulphonate) are electrodialysed, one observes an increasing ohmic resistance of the anion-exchange membrane, such that it ultimately must be replaced. Reversal of the current does not help. This problem is particularly exacerbated when solutions of biological extracts such as fermentation liquors or sugar solutions are treated by electrodialysis; the cation-exchange membranes are quite undamaged even after several weeks or months of use, while the anion-exchange membranes fail rapidly under these circumstances, often in a matter of hours or even minutes. Some of the commercially available membranes show some relatively minor advantage over others with respect to these fouling phenomena, but no membrane is reasonably free of fouling.

Since fouling is the result of site-binding, we can approach its alleviation either by employing solutions not containing strongly polarizable anions, which is obviously impossible, or by making anion-exchange membranes of a less adsorbent character. This obviously calls for elimination of the aromatic ring system. Membranes prepared from polyethyleneimine should show lowered fouling, unless other nonpolar moieties are present in excess.

There is yet another good reason why one should avoid the aromatic ring. The benzyltrimethylammonium compound is relatively unstable; its thermal decomposition occurs rapidly at relatively low temperatures, particularly in the presence of oxygen. Some systematic investigations of the relative stability of different commercially available membranes under different conditions of temperature, acidity and basicity have been carried out by Gregor[8]. Wichterle *et al.*[37] have shown by model studies that only the symmetrical *n*-alkyl ammonium compounds would show the desired stability.

In our search for a stable quaternary ammonium polymer we may look to the work of Okomoto[38] who synthesized 2,6-isopropylmethylpyridinium iodide and found that this quaternary ammonium compound could be distilled at 250°C under reduced pressure without decomposition. Whether this particular compound can be made in the form of a polymer remains to be seen, but pyridinium compounds are highly aromatic in nature and therefore the fouling problem would still persist. The recent report by Gianinni[39] who has prepared and polymerized a variety of aliphatic olefinic tertiary and quaternary ammonium compounds suggests that it may be possible to have a quaternary ammonium polybase of high stability and non-aromatic character.

The availability of a thermally stable anion-exchange membrane would allow electrodialysis to be carried out at elevated temperatures with lowered power costs and at higher allowable current densities; and would thus give further impetus to its application to this particular technology.

It is possible also that another deleterious effect in electrodialysis may be reduced by the availability of non-adsorbing anion-exchange membranes. The Bethe–Toropoff effect is very much in evidence with ion-exchange membranes, as evidenced by a water-splitting phenomenon with hydrogen ions passing in one direction and hydroxide ions in the other. This effect has been studied by several investigators[40]; and Gregor and Miller[41] have proposed that it is a result of the unusual boundary conditions which obtain at the membrane surface, where the electric field acts to retard the recombination of hydrogen and hydroxide ions and increases markedly, in effect, the degree of dissociation of water. Different membrane types, particularly the anion-permeable membranes, show different degrees of polarization, suggesting that the water structure at the surface of these membranes may be different. Almost all anion-exchange membranes show considerably greater water-splitting than their cation-permeable counterparts, and one can only speculate that the availability of an aliphatic technology of anion-exchange membrane may alter this preferred water-splitting and thus add another dimension to the technology of electrodialysis.

ACKNOWLEDGEMENT

Much of the work described herein was supported by the Office of Saline Water of the United States Department of the Interior; the author expresses his real appreciation. The collaboration of several individuals deserves special mention: H. Kawabe of the Institute of Physical Chemistry in Tokyo; A. Leszko of the University of Krakow; S. D'Alessandro of the University of Palermo; J. Andelman now at the University of Pittsburg; Z. Konrad of the Institut Rudjer Boskovic in Zagreb; I. Belgovskiy of the Institute of Chemical Physics in Moscow; and I. F. Miller of the Polytechnic Institute of Brooklyn. It is a pleasure also to acknowledge the inspiration which Dean Herman F. Mark has given to all of us.

References

[1] A. J. Staverman. *Trans. Faraday Soc.* **48**, 176 (1952).
[2] O. Kedem and A. Katchalsky. *Biochim. Biophys. Acta* **27**, 229 (1958).
[3] O. Kedem and A. Katchalsky. *J. Gen. Physiol.* **45**, 143 (1961).
[4] O. Kedem and A. Katchalsky. *Trans. Faraday Soc.* **59**, 1918, 1931 (1963).
[5] Y. Tsunoda, M. Seko, M. Watnabe, T. Mikado, and Y. Yamagoshi. Jap. Patents 7290 and 7489.
[6] E. Glueckauf and R. E. Watts. *Proc. Roy. Soc.* **A268**, 339 (1962).
[7] H. Kawabe, H. Jacobson, I. F. Miller, and H. P. Gregor. *J. Colloid Sci.* **21**, 79 (1966).
[8] H. P. Gregor. *et al.*, in preparation.
[9] H. P. Gregor and J. I. Bregman. *J. Colloid Sci.* **6**, 323 (1951).
[10] I. F. Miller and H. P. Gregor. *J. Chem. Phys.* **43**, 1783 (1965).
[11] R. A. Robinson and R. H. Stokes. *Electrolyte Solutions*, 2nd Ed., Butterworths, London, 1959, p. 331.
[12] R. Zwanzig. *J. Chem. Phys.*
[13] D. F. Evans and R. L. Kay. *J. Phys. Chem.* **70**, 366 (1966).
[14] W. Y. Wen and S. Saito. *J. Phys. Chem.* **68**, 2639 (1964).
[15] H. P. Gregor, M. Rothenberg, and N. Fine. *J. Phys. Chem.* **67**, 1110 (1963).
[16] H. S. Frank and M. W. Evans. *J. Chem. Phys.* **13**, 507 (1945).
[17] H. S. Frank and W. Y. Wen. *Disc. Faraday Soc.* **24**, 133 (1957).
[18] H. J. Chaya. B.S. Thesis, Polytechnic Institute of Brooklyn, June, 1947.
[19] H. P. Gregor, F. Gutoff, and J. I. Bregman. *J. Colloid Sci.* **6**, 245 (1951).
[20] H. P. Gregor, J. Belle, and R. A. Marcus. *J. Amer. Chem. Soc.* **76**, 1984 (1954).
[21] R. Collander. *Soc. Sci. Fenn. Biol.* **2**, 6 (1926).
[22] P. C. Carman. *Flow of Gases Through Porous Media*, Academic Press, New York, 1956.
[23] J. D. Ferry. *Chem. Reviews* **18**, 373 (1936).
[24] H. Faxen. *Arkiv Math. Astr. Fys.* **17**, No. 27, (1922).
[25] J. R. Pappenheimer. *Physiol. Reviews* **33**, 387 (1953); see also J. R. Pappenheimer, E. M. Renkin, and L. M. Borrero. *Amer. J. Physiol.* **167**, 13 (1951).
[26] E. M. Renkin. *J. Gen. Physiol.* **38**, 225 (1954).
[27] H. Faxen. *Ann. Physik IV*, **68**, 89 (1922).
[28] H. P. Gregor. *J. Amer. Chem. Soc.* **70**, 1293 (1948); **73**, 642 (1951).
[29] M. Leszko and H. P. Gregor. *Roczn. Chem.* **40**, (7/8) 1281 (1966).
[30] F. Helfferich. *Ion Exchange*, McGraw-Hill, New York, 1962.
[31] I. Kagawa and H. P. Gregor. *J. Polymer Sci.* **23**, 477 (1957).
[32] F. Bernstein. Dissertation, Polytechnic Institute of Brooklyn, February, 1952.
[33] H. P. Gregor, J. Belle, and R. A. Marcus. *J. Amer. Chem. Soc.* **76**, 1984 (1954).
[34] M. H. Waxman, B. R. Sundheim, and H. P. Gregor. *J. Phys. Chem.* **57**, 969 (1953).
[35] B. R. Sundheim, M. H. Waxman, H. P. Gregor. *J. Phys. Chem.* **57**, 947 (1953).
[36] I. Belgovskiy and H. P. Gregor, to be published.
[37] O. Wichterle. I.U.P.A.C. Symposium on Macromolecular Chemistry, Prague, 1965.
[38] Y. Okamoto and Y. Shimagawa. *Tetrahedron Letters* No **3**, 317 (1966).
[39] U. Gianinni. I.U.P.A.C. Symposium on Macromolecular Chemistry, Brussels, 1967.
[40] H. P. Gregor and M. A. Peterson. *J. Phys. Chem.* **68**, 2201 (1964).
[41] H. P. Gregor and I. F. Miller. *J. Amer. Chem. Soc.* **86**, 5689 (1964).

THERMALLY STABLE POLYMERS

CARL S. MARVEL

Department of Chemistry, University of Arizona, Tucson, Arizona, U.S.A.

Most organic polymeric materials melt below 200°C and most of them begin to degrade rapidly at temperatures only slightly above 200°C. Thermally stable polymers are generally considered to be those which will withstand much higher temperatures without loss of strength or change of structure. In general we expect these materials to withstand at least 300°C in air and up to 500°C or higher in inert atmospheres. Polymers which show these properties are usually highly aromatic in structure, often with heterocyclic units, high melting, sometimes infusible and usually with low solubility in all solvents. This makes their fabrication very difficult and as a consequence limits their usefulness.

There are a relatively few polymers which are available commercially as plastics, films, wire-coating polymers, etc. which are stable in the temperature ranges indicated. There are other polymers which have been synthesized and tested in pilot-plant scale which show promise but are still very expensive and not generally used. Finally there are other classes which have been studied in the laboratory and have not yet reached the development stage. In this report I shall mention examples of these various classes of polymers.

There is a need for thermally stable rubbers but at the present time I know of none that will stand up to 500°C for any appreciable length of time. The best of the commercially available materials are the fluoro-rubbers but even these decompose rather rapidly in a thermogravimetric test at 400°C and are useful only at considerably lower temperatures. Viton (I) which is a perfluoropropylene–vinylidene fluoride copolymer can be used continuously at temperatures of 200°C in air for months without complete loss of properties. At that elevated temperature its tensile strength is much less than at room temperature. It is recommended for continuous service at 260°C for up to 1000 hours. For fleeting use it can give temporary protection at temperatures up to about 500°C.

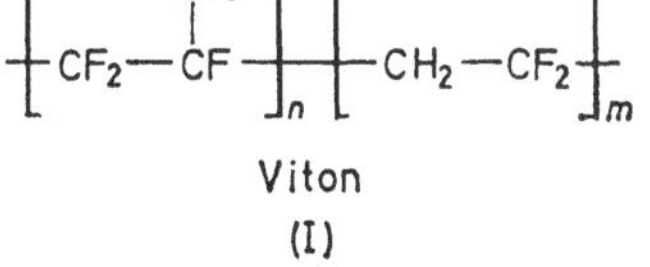

$$\left[CF_2 - \underset{\underset{CF_3}{|}}{CF} \right]_n \left[CH_2 - CF_2 \right]_m$$

Viton
(I)

I know of no rubber under development which is more promising than Viton. Although the work of Brown[1] has shown that the triazine unit joined by perfluoroalkylene units (II) has some promise, information on the current state of development is not available in the literature.

351

The polyimides are the most thermally stable of the polymers available for use as plastics and films. They have also been spun into fibres in experimental tests but are not generally available as fibres. A variety of polyimides have been synthesized and described in various papers but two commercial names, Vespel and Kapton, have been designated names for the plastic and films which are marketed. These are believed to be the polyimide (III) of pyromellitic anhydride and p,p'-diaminodiphenyl ether[2].

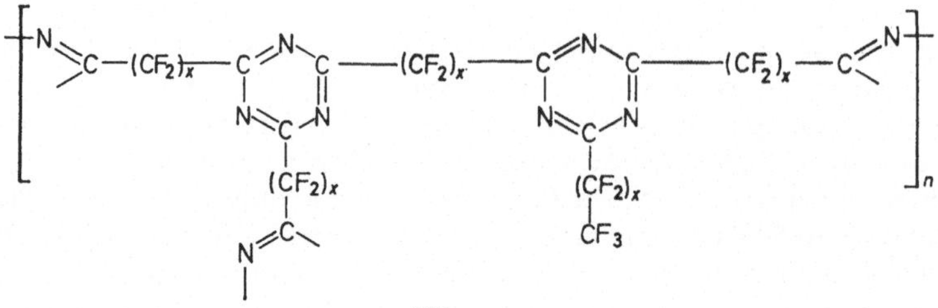

(II)

At 400°C in air, it loses less than 10% of its weight in 100 hours. It can be cooled to liquid nitrogen temperature without loss of strength. As a film it has been used successfully in air over the range of -269°C to 400°C. It begins to char at 800°C but it does not melt up to 900°C. It can be used in air at 300°C for a month and at 400°C for a day. It retains a tensile strength of 4000 psi at 500°C. At 500°C in helium its weight loss in 1000 minutes is about $7\frac{1}{2}\%$.

The polyimides must be fabricated at the amic acid stage before the final cyclization to the imide is performed. In the amic acid stage, the polymers are soluble in aprotic solvents and can be cast or spun. Then the imide ring is closed to give an insoluble infusible polymer. The polyimides are somewhat

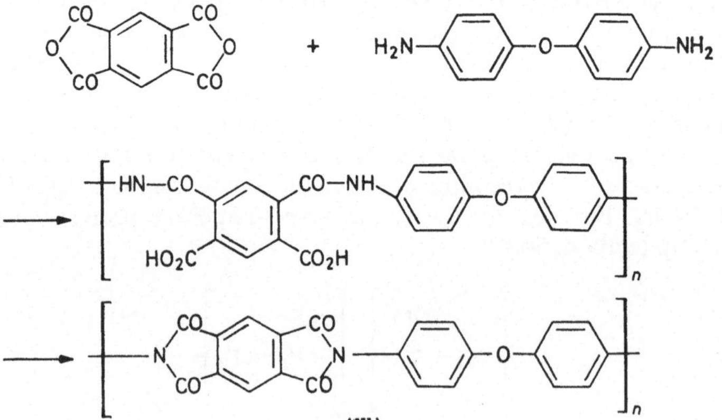

(III)

sensitive to base-catalysed hydrolysis, but this has not limited their utility.

There is an aromatic polyamide called Nomex which is used as a fibre and a paper at elevated temperatures. It does not melt but degrades rapidly above 370°C. Preston and his coworkers[3] have become interested in wholly-ordered aromatic copolyamides which show excellent high-temperature properties and better flexibility than do random polyamides. For example, the amine N,N'-m-phenylene-bis-(m-aminobenzamide) and isophthaloyl chloride in interfacial polymerization give a polymer (IV) with a melt temperature of 410°C which loses about 10% of its weight in thermogravimetric analysis at 450°C. The same amine with terephthaloyl chloride

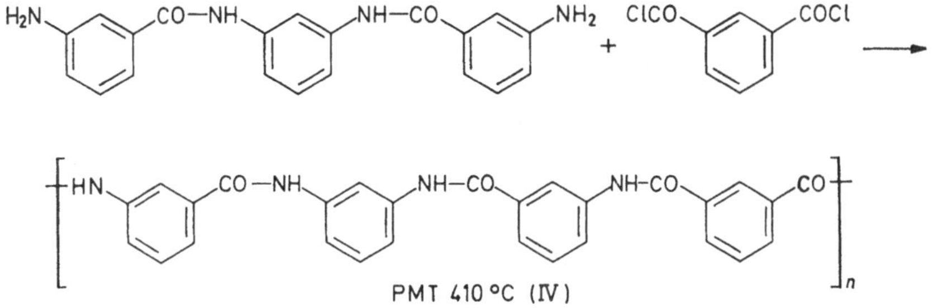

PMT 410 °C (IV)

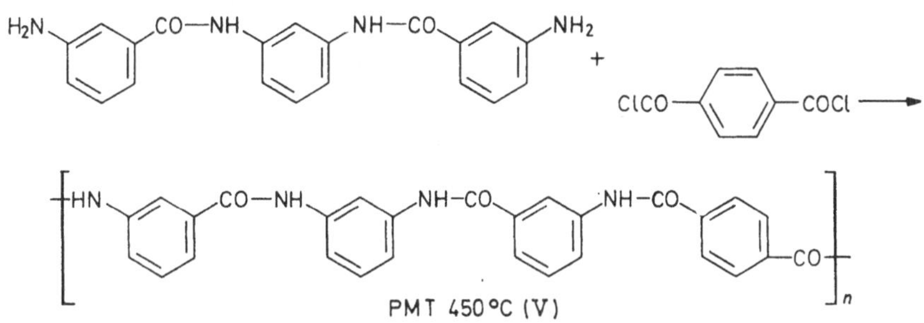

gives an ordered copolyamide (V) with a melt temperature of 450°C which loses only 10% of its weight at 500°C. A random *meta*-copolymer (VI) of the same units softened at 300°C and melted to a clear melt at 350°C. The above ordered-copolymers gave tough films and strong fibres.

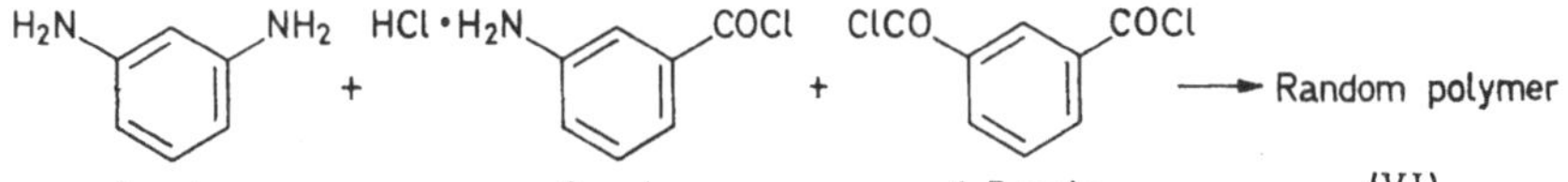

1 mole   2 mole   1·7 mole   (VI)

Softens at 300°C
Clear melt at 350°C

These ordered copolyamides have excellent radiation resistance. They have also been made with a variety of fused and multiple ring systems. Also heterocyclic units have been successfully introduced into experimental polymers to give materials which produce excellent fibres which are heat and radiation resistant. These ordered copolyamides are now available industrially on a selective evaluation basis.

Another group of experimental thermally stable polymers are the polyphenyls (VII). Some of these have reached the stage of experimental testing but as far as I am aware, none are commercially produced. Polyphenyls have been made by dehydrogenation of poly-1,3–cyclohexadiene[4] and by the Friedel and Crafts condensation of benzene[5]. The polyphenyls are deep brown to black, insoluble, infusible polymers which are extremely difficult to fabricate. Some success has been achieved by use of powdered metallurgy techniques to attain compacted materials for use as ablative materials in aerospace applications[6]. When o-, m- or p-terphenyls were used in the Friedel and Crafts condensation, a less regular structure resulted and the polymers were fusible and more tractable than the more simple polymers[7].

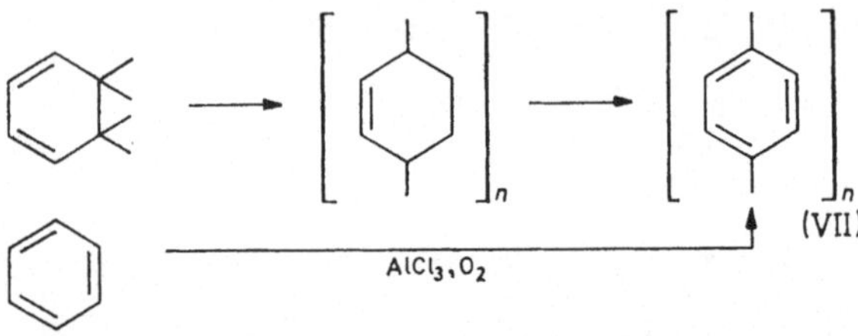

(VII)

In the last few years a large amount of work on the study of polyaromatic heterocycles has been carried out. In 1961 Vogel and Marvel[8] described the aromatic polybenzimidazoles (VIII and IX) which were obtained by the condensation of aromatic tetraamines with diphenyl esters of dibasic aromatic acids. The condensation reaction which produced the heterocyclic rings is the reaction which produces the polymer and this general type of condensation has been extended to produce a wide variety of polyaromatic heterocycles, all of which have excellent thermal stability. Brinker and Robinson[9] had previously made polybenzimidazoles with aliphatic recurring units, but had not reported any unusual properties in these materials which were prepared by condensing aliphatic dibasic acids with aromatic tetra-amines. This reaction was not useful for making all aromatic polybenzimi-dazoles since the aromatic acids seem to decarboxylate to some extent at the temperature required for the condensation. After investigating a variety of

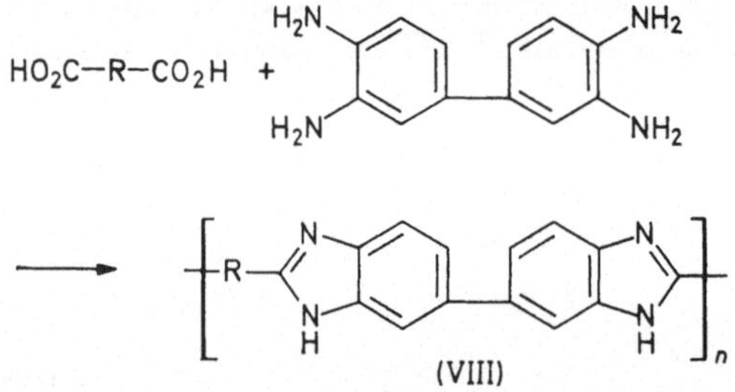

(VIII)

conditions it was found that the condensation of aromatic tetraamines with the diphenyl esters of aromatic dibasic acids gave good yields of high-molecular-weight polymers. Many such polymers have been made by this general type of condensation[10] in my laboratory. Korshak and his co-workers[11] have also made a number of polymers of this same general type.

Iwakura and his co-workers have found that this type of polymer can be made by condensing aromatic dibasic acids, their methyl esters, their amides or the corresponding nitriles with aromatic tetraamines in poly-phosphoric acid[12]. This method avoids the use of the highly-oxidatively-sensitive free aromatic tetraamines.

The polybenzimidazoles are all coloured polymers varying from deep golden yellow to black, usually without a melting point below 400°C. They

354

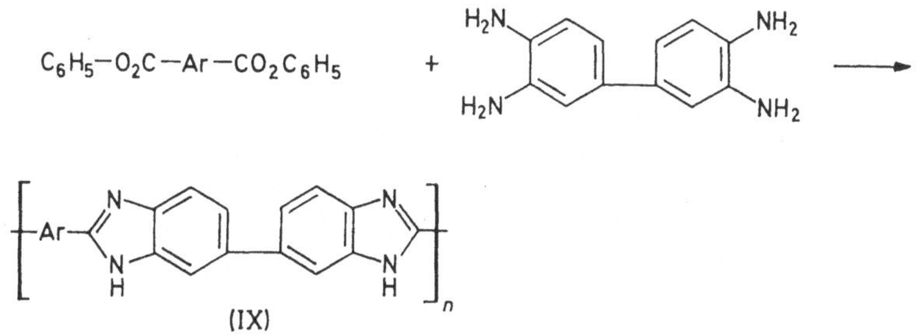

$$C_6H_5{-}O_2C{-}Ar{-}CO_2C_6H_5 \quad + $$

(IX)

vary in crystallinity and solubility. All have been found to be soluble in sulphuric acid, a few in formic acid and a few in trifluoroacetic acid. Those polymers which are not crystalline as shown by x-ray patterns, are soluble in such aprotic solvents as dimethylsulphoxide, dimethylformamide, dimethyl-acetamide, N-methylpyrrolidone and hexamethylphosphoramide. All of the materials are nonconductors of electricity. When heated to about 400°C for a short time, they become insoluble in all solvents, even sulphuric acid. In a nitrogen atmosphere they do not lose weight up to 500°C. In air they begin to oxidize rather rapidly at about 300°C. They are extremely stable to hydrolysis and are not attacked by hot strong sulphuric acid solutions or hot 25% potassium hydroxide solutions.

The polybenzimidazole (X) which has been most completely evaluated is the one derived from the condensation of 3,3′-diaminobenzidine and

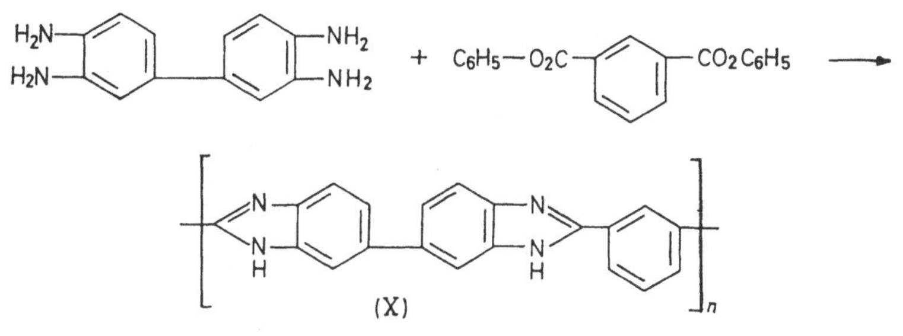

(X)

diphenyl isophthalate. The polymerization is carried out by mixing the two solid monomers and heating for an hour or two in an inert atmosphere at about 250°C until a solid foamy mass is produced. This mass is cooled to room temperature and broken up to give fine granules so that heat transfer to the interior of the mass is better and then the solid is again heated out of contact with air for several hours at 350–400°C to complete the polymeriza-tion which proceeds in the solid state. The polymer that is produced is a golden yellow solid, soluble in such solvents as dimethylsulphoxide and dimethylacetamide. It has an inherent viscosity in dimethylacetamide solution of 0·5 to 1+ depending on the time of polymerization and the exact balance of reagents. A sample which had an inherent viscosity of 0·8 in dimethylsulphoxide had an inherent viscosity of about 3·3 in sulphuric acid solution and by the light-scattering technique had a molecular weight of 54 000[13].

This polymer has been converted into films and fibres by casting or

P.A.C.—L

spinning from dimethylacetamide solution. These films and fibres show good mechanical properties up to a temperature of 300°C. In air, films lose their strength rather rapidly at temperatures above 300°C. Fibres from this polymer with tensile strength of about 7 g/denier are undergoing extensive industrial tests for high temperature use[14].

This polybenzimidazole has been shown to have merit as a glass laminating resin and as a metal adhesive when properly used. It must be applied as a low-molecular-weight polymer which can melt and wet the surfaces involved. Then polymerization must be completed *in situ* to obtain good adhesive strength[15].

Some structurally related polymers (XI) have been prepared by substituting esters of diboronic acids for the aromatic dibasic acids[16] in the polybenzimidazole synthesis. In this case it was necessary to use butyl rather

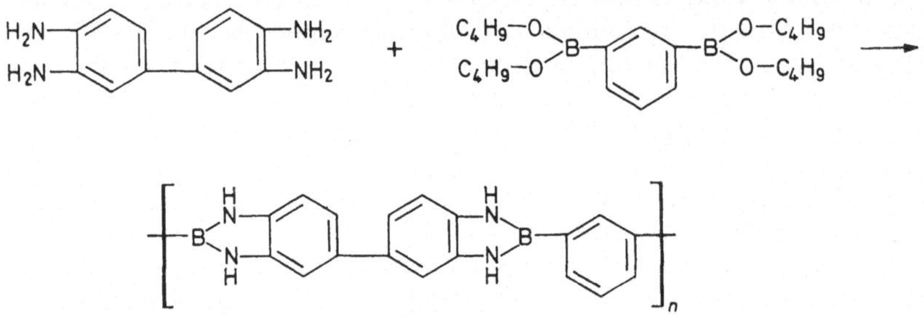

(XI)

than phenyl esters since in the latter case the liberated phenol was a sufficiently strong acid to cause cleavage of the carbon–boron linkages. These polymers possess good molecular weight, solubility and heat stability. Their sensitivity to hydrolysis was so great that they were not extensively studied.

By applying the phenyl-ester condensation reaction with aromatic amines, two new groups of polymers, the polybenzothiazoles and polybenzoxazoles, have been prepared. Hergenrother, Wrasidlo and Levine[17] condensed 3,3'-dithiobenzidine and diphenyl isophthalate to obtain a polybenzothiazole (XII) which is remarkably stable in air up to 600°C. Under these conditions the polymers lose only 1% of their weight. The polymers are soluble only in sulphuric acid and have had little application as yet.

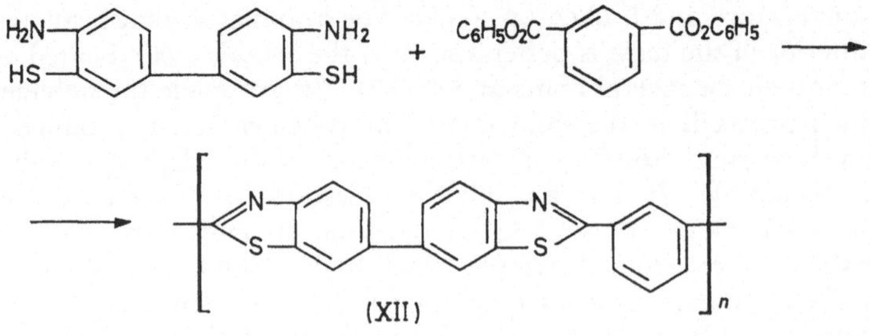

(XII)

Polybenzoxazoles (XIII) have been prepared from 3,3'-dihydroxybenzidine by its reaction with isophthaloyl chloride[18] and by condensation with

356

diphenyl isophthalate[19]. These polymers are stable to 500°C in nitrogen but are soluble only in sulphuric acid and are generally quite intractable.

Longone and Un[20] have synthesized a number of polybisthiazoles (XIV) from bisbromomethyl ketones and dithioxamide. These polymers had unusual thermal stability with weight losses of only 20% up to 500°C and

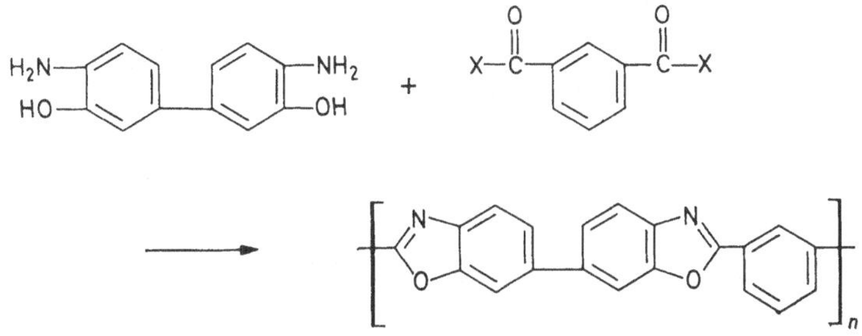

(XIII)

only 50% up to 900°C. They were yellow-brown and turned pink on long exposure to light. There was some evidence of photo-oxidation. The polymers did not melt and were generally insoluble.

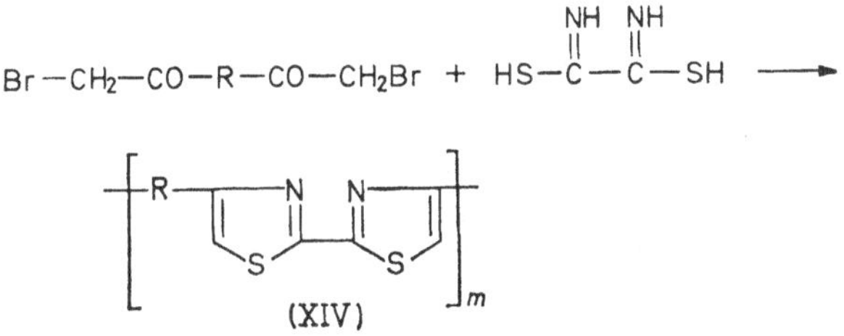

(XIV)

Another group of heat stable polymers has been developed which contain oxadiazole and triazole recurring units. By using bifunctional reagents, Abshire[21] in my laboratory was able to adapt Huisgen's[22] dipolar addition reaction for the preparation of oxadiazoles so that polymers with alternate aryl and oxadiazole recurring units (XV) could be obtained. These polymers

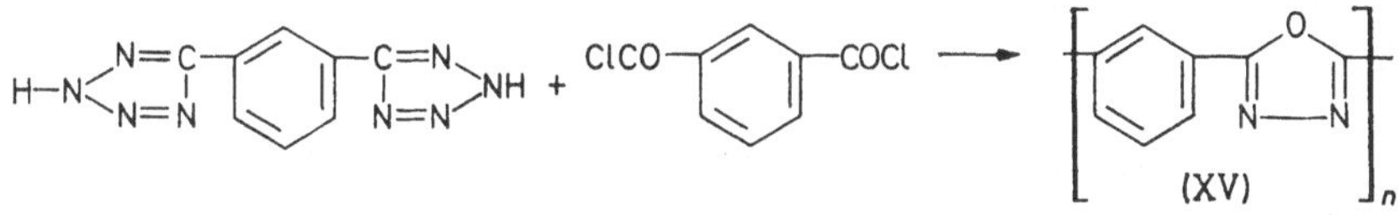

(XV)

were obtained in the 6000 molecular weight range. They were coloured, did not melt below 350°C and were soluble only in sulphuric acid. Hence, they were not useful for processing. They were, however, very thermally stable and only lost weight when heated above 520°C in nitrogen.

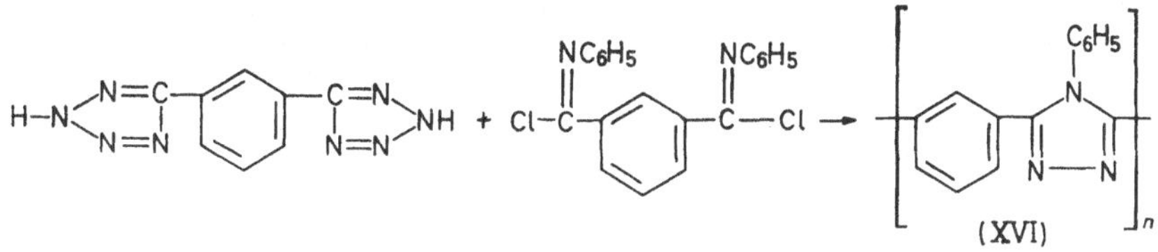

(XVI)

357

Abshire[21] used an analogous reaction of Huisgen[23] to prepare polymers with alternate aryl and triazole recurring units. These polymers (XVI) were slightly higher in molecular weight, slightly more soluble and slightly less heat stable than the corresponding oxadiazole polymers.

New procedures have been devised for synthesizing the polymers with

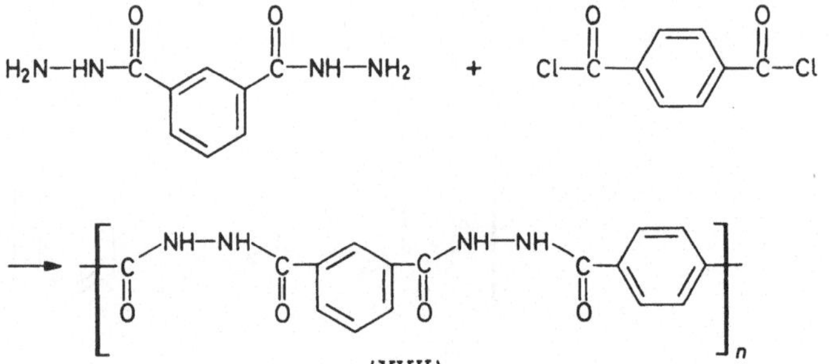

(XVII)

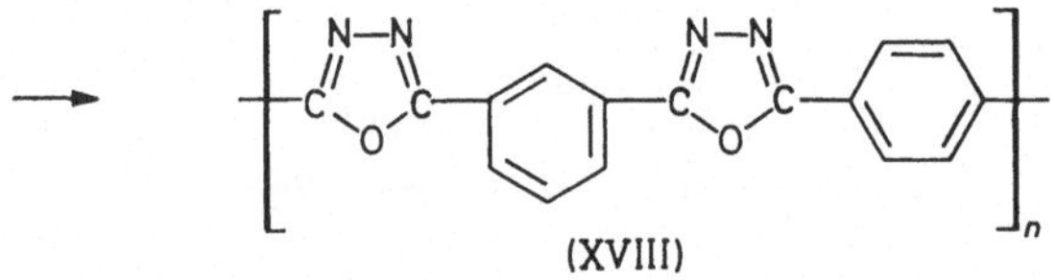

(XVIII)

oxadiazole[24] and triazole[25] recurring units. The corresponding polyhydrazides (XVII) were prepared and then converted to either the oxadiazole (XVIII) or triazole structure. This has permitted the synthesis of high-molecular-weight products which can be fabricated into fibres at the polyhydrazide stage and then converted to the fully aromatic structure. Both classes of polymers have yielded potentially useful fibres which are stable at high temperatures and which have particularly good resistance to oxidation. The fibres which have been prepared from the polyoxadiazoles have good tensile strength (3 g/denier) and good knot strength[26]. They maintain at least half of their strength after 30 hours of ageing at 400°C in air. At 300°C in air they maintain half strength for 700 hours. When tested at 300°C their strength is at least 2 g/denier.

Frazer and Fitzgerald[27] have prepared the corresponding thiadiazoles from the thiahydrazides. In this case the cyclization reaction is much simpler to carry out. These fibres have good tensile properties which are retained to the extent of 92% after 114 hours heating in air at 300°C. At 400°C, 60% of their tensile strength is retained after 32 hours in air which is even better than shown by the oxadiazole fibres.

Preston and Black[28] have used their ordered copolymer idea to develop a series of thermally stable ordered heterocyclic polymers (XIX–XXIV). The thermal properties of the series are listed in *Table* 1. The decomposition temperatures noted are the temperatures at which the weight loss in nitrogen is approximately 10%. For comparison, the figure for polybenzimidazole is 600°C, for polyimides 570°C, for oxadiazoles 500°C and for triazole 490°C, in the same test.

THERMALLY STABLE POLYMERS

Table 1. Thermal properties of ordered heterocyclic copolymers

Polymer No.	Melt transition temp. (°C)	Decomposition temp. (°C)
XIX	530	—
XX	525	520
XXI	525	550
XXII	530	510
XXIII	565	550
XXIV	450	510

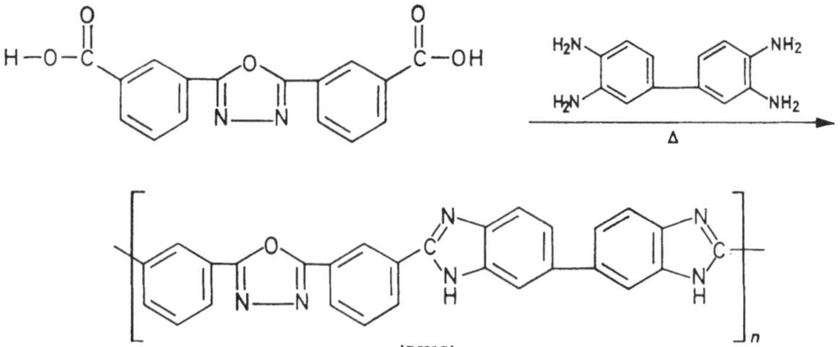

(XIX)

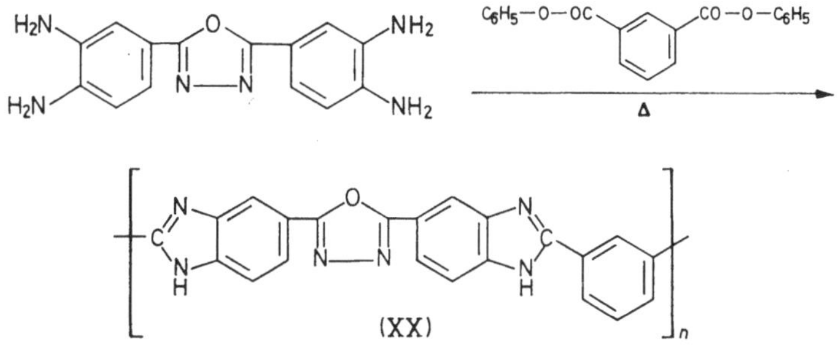

(XX)

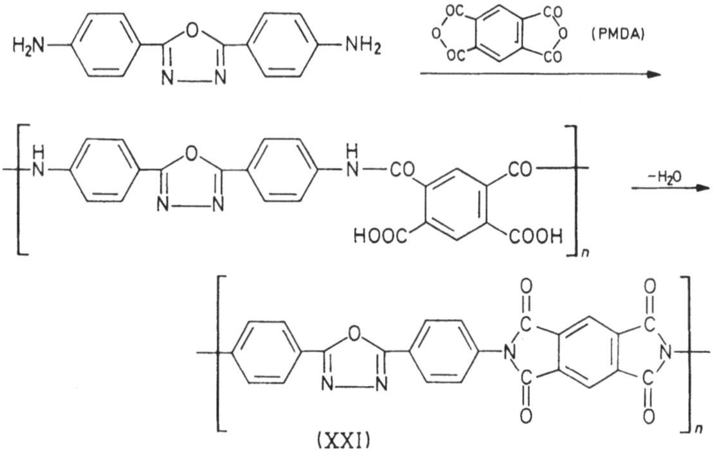

(XXI)

A number of other polyaromatic heterocycles have been briefly described. Schaefer and Bertram[29] have prepared some polymers (XXV) with alternate phenyl and pyrazole recurring units by the reaction of aromatic dihydrazines with tetraacetoethane. This polymer is remarkably stable up to 400°C in nitrogen but decomposes rapidly above that temperature. Similar products have been described by Korshak and coworkers[30]. Stille and Williamson[31] and de Gaudemaris and Sillion[32] have prepared polymers with quinoxaline

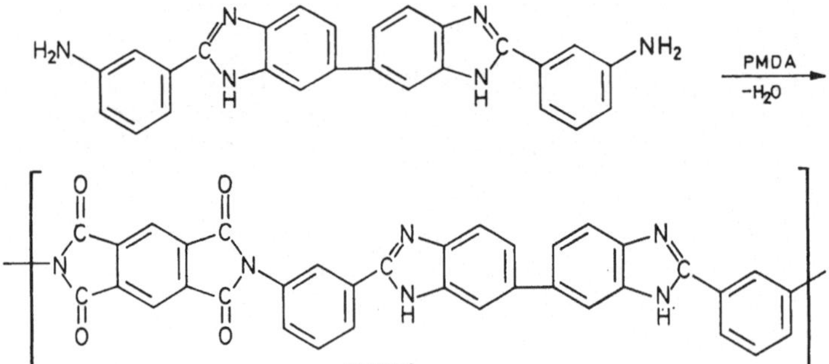

(XXII)

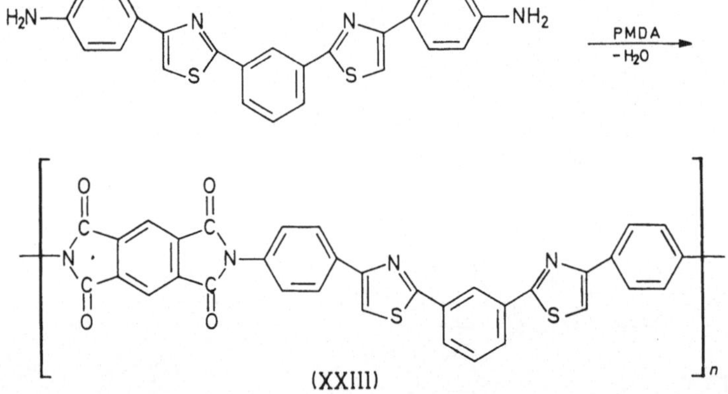

(XXIII)

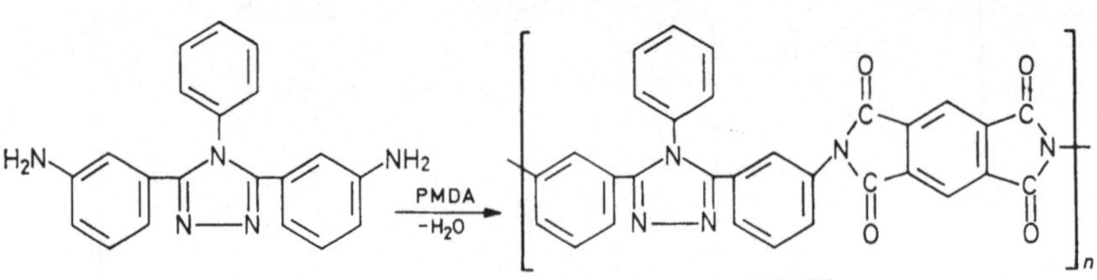

(XXIV)

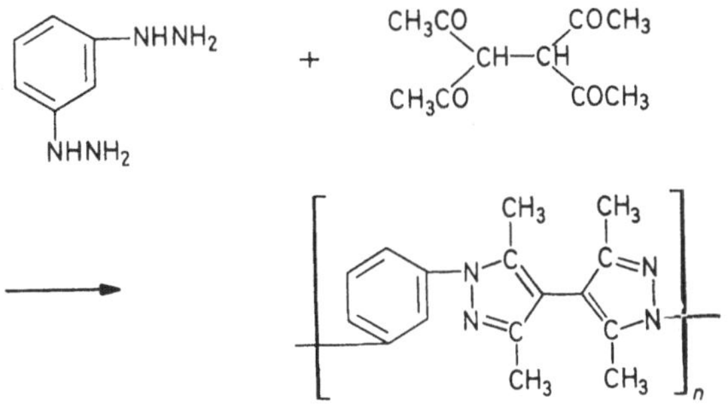

recurring units (**XXVI**) by condensing 3,3'-diaminobenzidine with 1,4-diglyoxalylbenzene. These polymers are reported to be stable in air up to 500°C and to nearly 800°C in nitrogen. Thus far they have not been fabricated into useful objects.

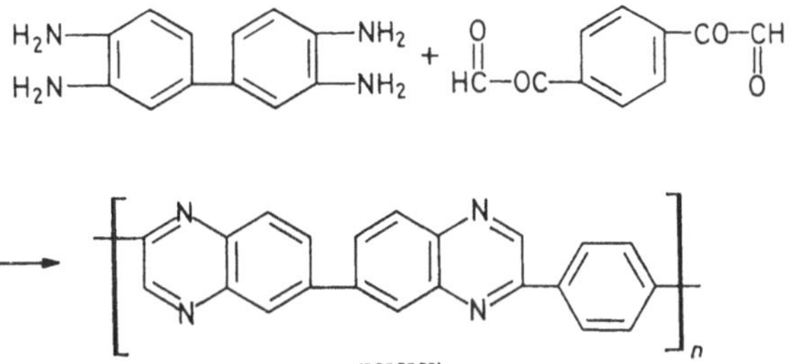

By introducing an oxygen atom between two quinoxaline units[33], polymers soluble in hexamethylphosphoramide were obtained and then could be cast into films stable in air to 500°C. Polyquinoxalines with a sulphur atom and with a sulphone group between the quinoxaline segments[34] have been prepared. These are much more soluble than the others and still show great heat stability.

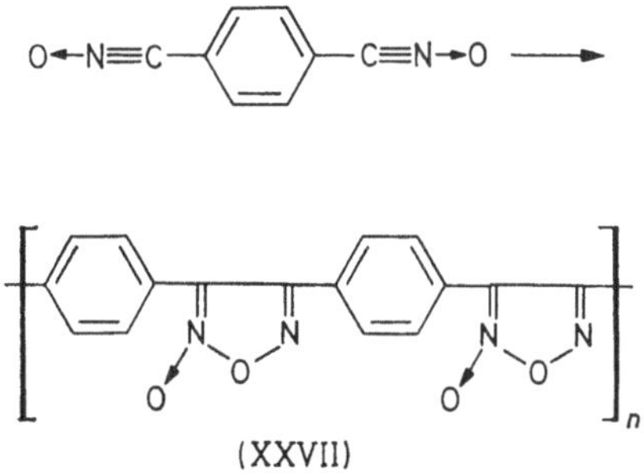

In the last few years a number of 1,3-dipolar addition reactions have been used to yield new heterocyclic polymers. Overberger and Fujimoto[35] have discovered that terephthalonitrile oxide in solution polymerizes to give a product with alternate phenyl and furoxan units (XXVII). When tereph-thalonitrile oxide and 1,4-diethynylbenzene are allowed to react, a polymer with alternate phenyl and isooxazole units results (XXVIII). The bis-nitrile oxide reacts with terephthalonitrile to yield a polymer with oxadiazole units (XXIX).

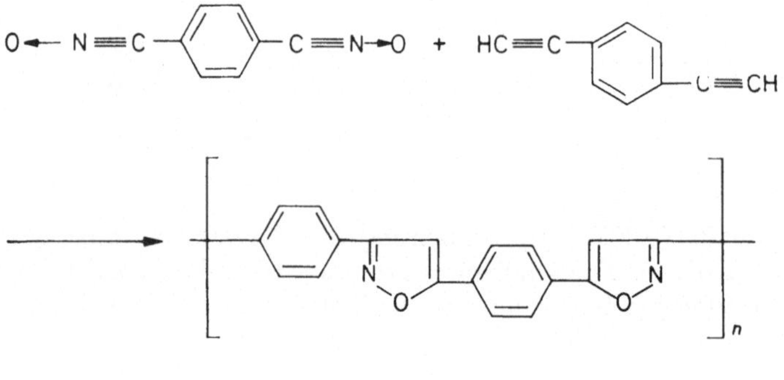

(XXVIII)

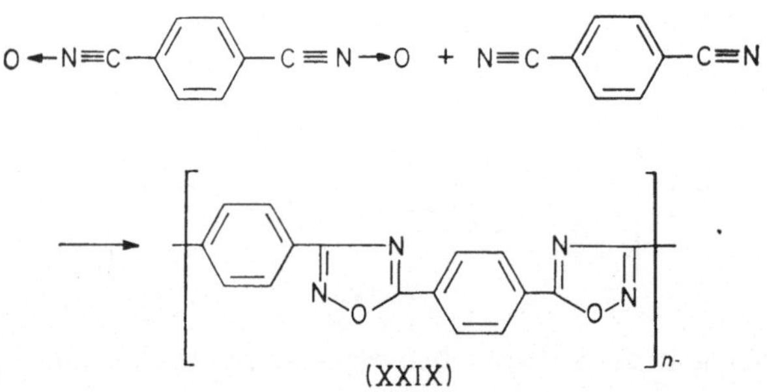

(XXIX)

Akiyama, Iwakura, Shiraishi and Iami[36] have been able to homopoly-merize p-cyanobenzonitrile oxide to obtain a product with alternate phenyl and oxadiazole units (XXX) in a solid state reaction.

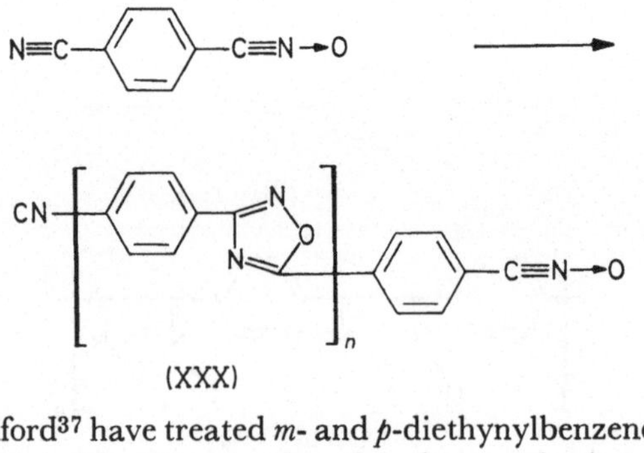

(XXX)

Stille and Bedford[37] have treated m- and p-diethynylbenzene with sydnones to obtain polymers with alternate phenyl and pyrazole rings (XXXI).

Stille and Harris[38] have condensed bis-nitrile imines with diacetylenes to obtain polymers with phenyl and pyrazole rings (XXXII) which are stable to about 420°C.

All these new 1,3-dipolar addition reactions have yielded polymers which are still rather incompletely characterized.

An examination of the results obtained with these various aromatic heterocyclic polymers shows that when the recurring units are connected

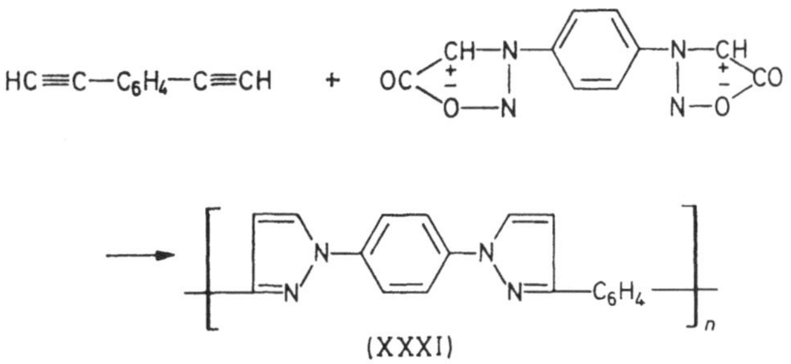

$$HC{\equiv}C-C_6H_4-C{\equiv}CH \quad + \quad OC \overset{CH-N}{\underset{O-N}{<}} \left\langle \right\rangle \overset{N-CH}{\underset{N-O}{>}} CO$$

$$\longrightarrow \left[\begin{array}{c} \\ \end{array} \right]_n \quad -C_6H_4-$$

(XXXI)

with single bonds between the aromatic segments, the polymers are stable in nitrogen to about 500–600°C. There is not much variation between the various heterocyclic recurring units. The stability of these polymers in air does vary somewhat and it seems that the ones with the least hydrogen content are least susceptible to oxidation. There is a marked variation in

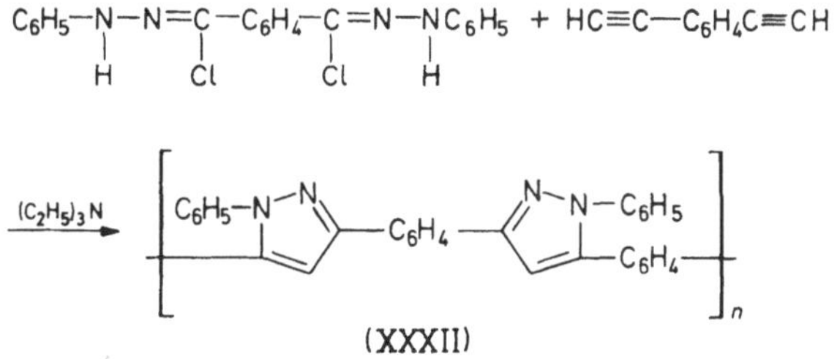

$$C_6H_5-\underset{H}{N}-N{=}\underset{Cl}{C}-C_6H_4-\underset{Cl}{C}{=}N-\underset{H}{N}C_6H_5 \quad + \quad HC{\equiv}C-C_6H_4C{\equiv}CH$$

$$\xrightarrow{(C_2H_5)_3N} \left[C_6H_5-N \quad -C_6H_4- \quad N-C_6H_5 \quad -C_6H_4- \right]_n$$

(XXXII)

solubility behaviour and the polybenzimidazoles seem to be the most tractable polymers in the series.

The remarkable thermal stability of 'black orlon' which is obtained by a regulated pyrolysis of polyacrylonitrile has attracted much interest. This pyrolysed material seems to be a polyquinizarine or a partially hydrogenated polyquinizarine[39] (XXXIII).

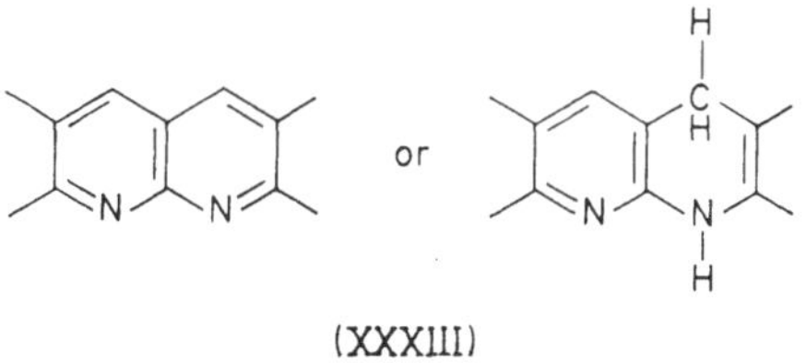

or

(XXXIII)

The stability of this two strand or ladder structure has led to many attempts to synthesize such structures with higher molecular weight. The first successful synthesis of such an aromatic heterocyclic structure came in 1965 in three laboratories at about the same time[40]. The reaction of pyromellitic anhydride with 1,2,4,5-tetraaminobenzene gave a ladder structure (XXXIV). The

363

reaction is rather complex and there are several possible intermediate steps. There are also possibilities of isomerism which are not yet fully investigated. A considerable amount of work has been done with model structures of a non-polymeric nature and the reaction has also been applied to such tetraamines as 3,3'-diaminobenzidine and tetraaminodiphenyl ether to

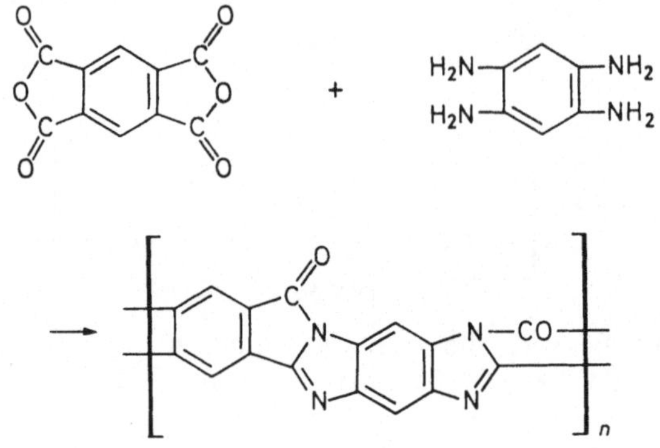

(XXXIV)

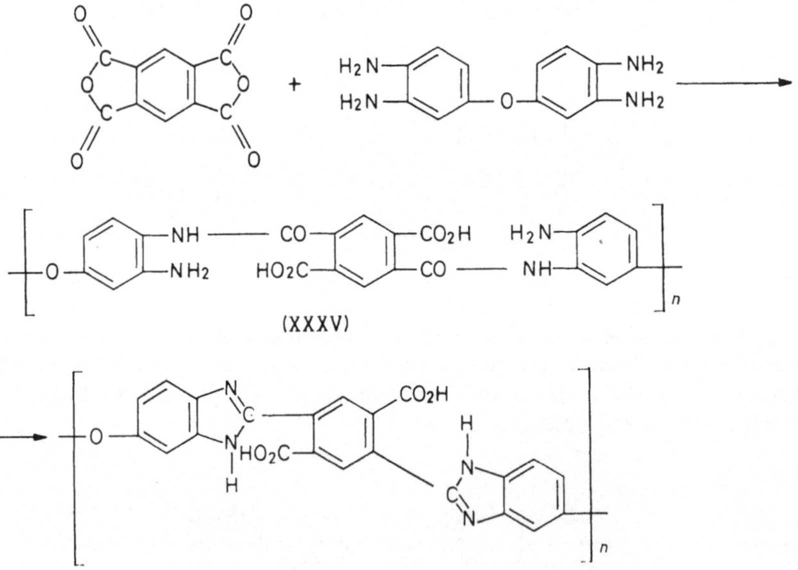

(XXXV)

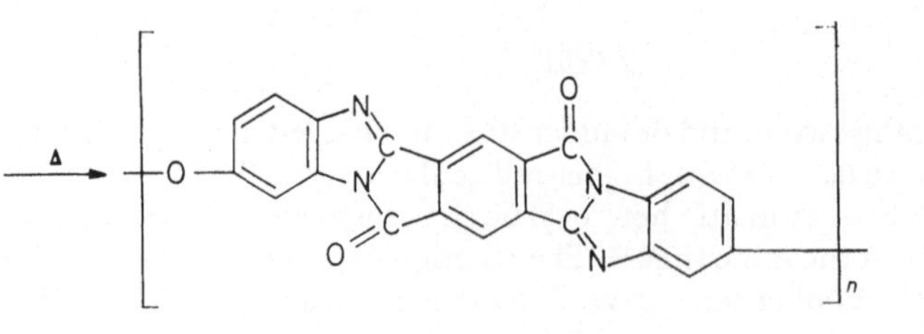

364

obtain non-ladder structures which are somewhat more tractable than the derivative from tetraaminobenzene.

The tetraaminobenzene derivative does not melt and is soluble only in sulphuric acid. Thermogravimetric analysis indicates it is stable in nitrogen to 600°C and the weight loss which occurs up to that stage seems to be caused by water loss due to the fact that the amide link had not been completely formed.

Bell and Pezdirtz have been able to control the reaction of pyromellitic anhydride and tetraaminodiphenyl ether to obtain a soluble intermediate amide-acid type of polymer (XXXV).

Before the last ring closure, the polymer can be dissolved in dimethylacetamide and cast into films. Then by further heating, the final ring closure can be achieved. The polymer thus obtained shows good thermal stability to 500°C. The films so prepared have exceptional stability toward radiation. Their tensile properties were essentially unchanged after exposure to 10 000 megarads.

Colson, Michel and Paufler made films in a similar manner from the polymer derived from 3,3'-diaminobenzidine. Their films were deep red in colour, quite flexible and had a modulus of 700 000 psi; a tenacity of 11 000 psi; and elongation of 2%. They showed no exotherm below 600°C and there was no significant loss in weight in dry air up to a temperature of 550–600°C.

Presumably the polymer from tetraaminobenzene which would have a full ladder structure can be made in a similar step-wise procedure. Hence, there is potentially a way to obtain films or fibres with the complete ladder structure.

Van Deusen[41] has made a similar type of structure from 1,4,5,8-naphthalene tetracarboxylic acid and tetraaminobenzene in polyphosphoric acid (XXXVI). This polymer lost no weight in nitrogen below 600°C and in air there is very little break in the weight loss curve below 500°C. It can be spun

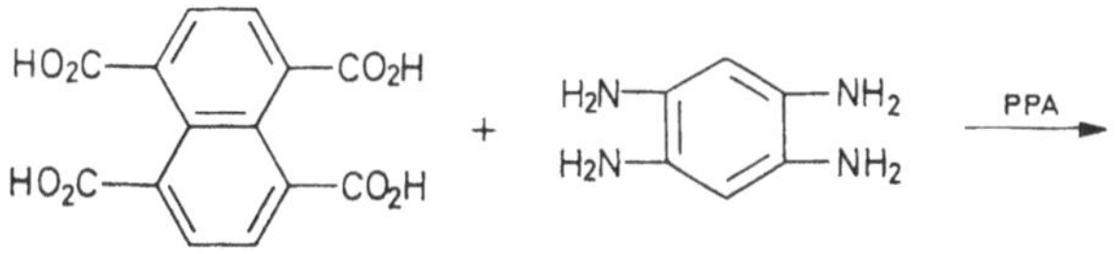

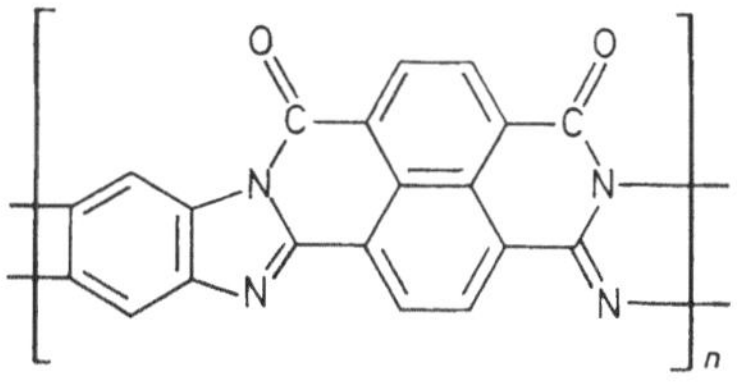

(XXXVI)

from sulphuric acid solution to give a fibre with a tensile strength of 3·4 g/denier.

A third type of ladder polymer (XXXVII) has been obtained[42] by the condensation of 2,3,7,8-tetrahydroxy-1,4,6,9-tetraazaanthracene with 1,2,4,5-tetraaminobenzene in polyphosphoric acid and by the self condensa-

365

tion of 2,3-dihydroxy-6,7-diaminoquinoxaline in the same reagent. This is a highly coloured polymer which does not melt below 350°C. It is slightly soluble in sulphuric and methanesulphonic acids. In the latter solvent, it shows an inherent viscosity of 2·5. In the thermogravimetric test, it loses weight gradually from 200°C to 600°C and this loss of weight seems to be

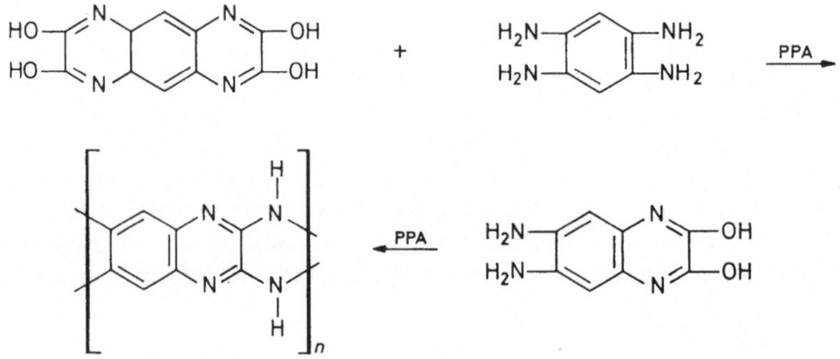

(XXXVII)

due to loss of water showing that the ring closure to give a ladder structure has been incomplete. After 600°C there is a more rapid weight loss until 15% of the weight is lost at 900°C. In air the polymer lost weight rapidly above 400°C. As yet no way has been found to fabricate useful articles from this polymer.

Stille and Mainen[43] have prepared ladder polyquinoxalines from 1,2,4,5-tetraaminobenzene and 2,5-dihydroxy-p-benzoquinone and from the same amine and 1,2,6,7-tetraketopyrene (XXXVIII). The first type showed no

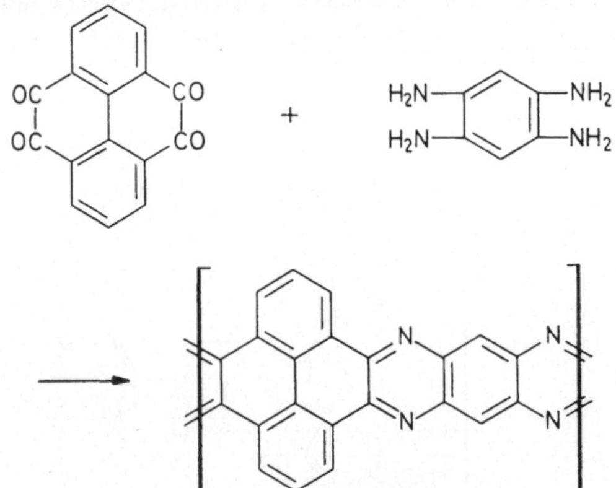

(XXXVIII)

greater thermal stability than do ordinary linear poly-heterocycles and it was thought that ladder formation was incomplete. The second type of polymer showed good solubility in aprotic solvents and excellent thermal stability. In air the major break in the thermogravimetric curve came at 460°C and in nitrogen the break came at 683°C. In both cases the weight loss before the break was less than 4%.

These newer researches on ladder structures seem to show promise that products of this type can be fabricated into films and fibres which will show good thermal and oxidative stability at temperatures at least 100°C above those of non-ladder structures. All of the polymers which have been tested thus far, show evidence that complete ladder structures have not been obtained. Hence, the 600°–650°C limit may not be the final limit which can be obtained for such structures.

Acknowledgement

In conclusion, I want to express my appreciation to the Air Force Materials Laboratory, Research and Technology Division, Air Force System Command, Wright-Patterson Air Force Base, Ohio for their generous financial support of research in my laboratory. I am also deeply grateful to the many research associates who have worked with me on the programme which J have reported here.

References

[1] H. C. Brown. *J. Polymer Sci.* **44**, 9 (1960).
[2] C. E. Sroog, A. L. Endrey, S. V. Abramo, C. E. Berr, W. M. Edwards and K. L. Oliver. *J. Polymer Sci.* **A3**, 1373(1965); R. S. Irwin and W. Sweeny, *ibid.* C, High Temperature Resistant Fibers. **19**, 41 (1967).
[3] J. Preston. *J. Polymer Sci.* A1, **4**, 529 (1966); F. Dobinson, J. Preston. *ibid.* p. 2093; J. Preston, R. W. Smith and C. J. Stehman. Paper presented at Symposium on High Temperature Fibers, Winter Meeting, American Chemical Society, Phoenix, Arizona, January, 1966; J. Preston and W. B. Black. *J. Polymer Sci.* B, **4**, 267 (1966); J. Preston and R. W. Smith. *ibid.* p. 1033; J. Preston, R. W. Smith and C. J. Stehman. *ibid.* Part C, High Temperature Resistant Fibers. **19**, 7 (1967).
[4] C. S. Marvel and G. E. Hartzell. *J. Amer. Chem. Soc.* **81**, 448 (1959); D. A. Frey, M. Hasegawa and C. S. Marvel. *J. Polymer Sci.* A, **1**, 2057 (1963); G. L. LeFebvre and F. Dawans. *ibid.* A, **2**, 3277 (1964); P. E. Cassidy, C. S. Marvel and S. Ray. *J. Polymer Sci.* A, **3**, 1553 (1965).
[5] P. Kovacic and A. Kyriakis. *Tetrahedron Letters* 467 (1962), *J. Amer. Chem. Soc.* **85**, 454 (1963); P. Kovacic and R. M. Lange. *J. Org. Chem.* **28**, 968 (1963); P. Kovacic and F. W. Koch. *ibid.* p. 1864; P. Kovacic and J. Oziomek. *ibid.* **29**, 100 (1964); P. Kovacic, F. W. Koch and C. E. Stephan. *J. Polymer Sci.* A, **2**, 1193 (1964); P. Kovacic, V. J. Marchionna and J. P. Kovacic. *ibid.* A, **3**, 4297 (1965); P. Kovacic, J. T. Uchic and L. C. Hsu. *Polymer Preprints*, Amer. Chem Soc., Div. of Polymer Chem. **8**, 31 (1967).
[6] G. K. Ostrum, D. D. Lawson and J. D. Ingham. A.C.S. *Polymer Preprints* **7**, (2) 895 (1966).
[7] N. Bilow and L. J. Miller. *J. Macromol. Sci.*, A–I, **1**, 183 (1967).
[8] H. Vogel and C. S. Marvel. *J. Polymer Sci.* **50**, 511 (1961).
[9] K. C. Brinker and J. N. Robinson. *U.S. Patent* 2,895,948 (1959).
[10] H. Vogel and C. S. Marvel. *J. Polymer Sci.* A, **1**, 1531 (1963); L. Plummer and C. S. Marvel. *ibid.* A, **2**, 2559 (1964); R. T. Foster and C. S. Marvel. *ibid.* A, **3**, 417 (1965); T. K. Lakshmi Narayan and C. S. Marvel. *ibid.* In the press.
[11] V. V. Korshak. *et al.*, *Bull. Acad. Sci. U.S.S.R. Div. Chem. Sci.* 1677, 1859 (1963); *Vysokomolekul Soedin* **6**, 901, 1251, 1394 (1964); *Doklady U.S.S.R. Chem. Sect.* **149**, 195 (1963).
[12] Y. Iwakura, K. Uno and Y. Imai. *J. Polymer Sci.* A, **2**, 2605 (1964).
[13] We are indebted to Dr. M. Morton of the University of Akron for this determination.
[14] A. B. Conciatori, E. C. Chenvey, T. C. Bohrer and A. E. Prince, Amer. Chem Soc., Symposium on High Temperature Fibers, Phoenix, Arizona, January 1966, *J. Polymer Sci.* Part C. In the press.
[15] H. H. Levine. Private communication.
[16] J. E. Mulvaney, J. J. Bloomfield and C. S. Marvel. *J. Polymer Sci.* **62**, 59 (1962); N. A. Adrova, L. K. Prokhorva. *et al.*, *Doklady U.S.S.R. Chem. Sect.* **158**, 823 (1964).
[17] P. M. Hergenrother, W. Wrasidlo and H. H. Levine. *J. Polymer Sci.* A, **3**, 1665 (1965).

[18] T. Kubota and R. Nakanishi. *J. Polymer Sci.* B, **2**, 655 (1964); Y. I. Braz, I. E. Kardash et al., *Vysokomolekul Soedin.* **8**, 272 (1966).

[19] W. W. Moyer, Jr., C. Cole and T. Anyos. *J. Polymer Sci.* A, **3**, 2107 (1965).

[20] D. T. Longone and H. H. Un. *J. Polymer Sci.* A, **3**, 3117 (1965).

[21] C. J. Abshire and C. S. Marvel. *Macromol. Chem.* **44–46**, 388 (1961).

[22] R. Huisgen, J. Sauer and H. J. Sturm. *Angew. Chem.* **70**, 272 (1958); R. Huisgen. *ibid.* **72**, 359 (1960); R. Huisgen, J. Sauer, H. J. Sturm and J. H. Markgraf. *Ber. dtsch. chem. Ges.* **93**, 2106 (1960).

[23] R. Huisgen, J. Sauer and M. Seidel. *Chem & Ind.* 1114 (1958).

[24] A. H. Frazer and F. T. Wallenberger. *Chem. Engng News*, March 11, 1963, p. 60; *J. Polymer Sci.* A, **2**, 1137, 1147, 1157, 1171, 1181 (1964); A. H. Frazer and T. A. Reed. *ibid.* Part C, High Temperature Resistant Fibers. **19**, 89 (1967).

[25] M. R. Lilyquist and J. R. Holsten. *Polymer Preprints* Amer. Chem. Soc. **4**, (2) 6 (1963); J. R. Holsten and M. R. Lilyquist. *J. Polymer Sci.* A, **3**, 3905 (1965); J. R. Holsten, G. B. Butler and M. R. Lilyquist, *French Pat.* 1,398,146 (1965); M. R. Lilyquist and J. R. Holsten. *J. Polymer Sci.* Part C, High Temperature Resistant Fibers. **19**, 77 (1967).

[26] A. H. Frazer and T. A. Reed. *J. Polymer Sci.* Part C. High Temperature Resistant Fibers. **19**, 89 (1967).

[27] A. H. Frazer and W. P. Fitzgerald Jr. *J. Polymer Sci.* Part C, High Temperature Resistant Fibers. **19**, 95 (1967).

[28] J. Preston and W. B. Black. *J. Polymer Sci* B, **3**, 845 (1965).

[29] J. P. Schaefer and J. L. Bertram. *J. Polymer Sci.* B, **3**, 95 (1965).

[30] V. V. Korshak, E. S. Krongauz, A. M. Berlin and P. N. Gribkova. *Dokl. Acad. Nauk, U.S.S.R.* **149**, 602 (1963); **152**, 1108 (1963); *Vysokomolekul Soedin*, **6**, 1078 (1964).

[31] J. K. Stille and J. R. Williamson. *J. Polymer Sci.* B, **2**, 209 (1964); A, **2**, 3867 (1964).

[32] G. P. deGaudemaris and B. J. Sillion. *J. Polymer Sci.* B, **2**, 203 (1964).

[33] J. K. Stille, J. R. Williamson and F. E. Arnold. *J. Polymer Sci.* A, **3**, 1013 (1965).

[34] J. K. Stille and F. E. Arnold. *J. Polymer Sci.*, A-1, **4**, 551 (1966).

[35] C. G. Overberger and S. Fujimoto. *J. Polymer Sci.* B, **3**, 735 (1965); Y. Imai, M. Akiyama, K. Uno and Y. Iwakura, *Makromolekul. Chem.* **95**, 275 (1966).

[36] M. Akiyama, Y. Iwakura, S. Shiraishi and Y. Imai. *J. Polymer Sci.* B, **4**, 305 (1966).

[37] J. K. Stille and M. A. Bedford. *J. Polymer Sci.* B, **4**, 329 (1966).

[38] J. K. Stille and F. W. Harris. *J. Polymer Sci.* B, **4**, 333 (1966).

[39] R. C. Houtz. *Textile Rec. J.* **20**, 1786 (1950); W. J. Burlant and J. L. Parsons. *J. Polymer Sci.* **22**, 249 (1956); N. Grassie and J. N. Hay. *ibid.* **56**, 189 (1962); A. R. Monahan. *Polymer Preprints*, **2**, (7) 204 (1966).

[40] F. Dawans and C. S. Marvel. *J. Polymer Sci.* A, **3**, 3549 (1965); V. L. Bell and G. F. Pezdirtz. *Polymer Preprints*, **6**, 747 (1965); *J. Polymer Sci.* B, **3**, 977 (1965); J. G. Colson, R. H. Michel and R. M. Paufler. *ibid.* A-1, **4**, 59 (1966); V. L. Bell and R. A. Jewell. *Polymer Preprints* **8**, (1) 235 (1967).

[41] R. L. Van Deusen. *J. Polymer Sci.* B, **4**, 211 (1966); R. L. Van Deusen, O. K. Goins and A. J. Sicree. *Polymer Preprints* **7**, (2) 528 (1966); W. H. Gloor. *ibid.* 819; A. A. Berlin, B. I. Liogon' kii, G. M. Shamroev and G. V. Belova. *Izvest. Akad. Nauk S.S.S.R.*, Ser. Khim **5**, 945 (1966).

[42] H. Jadamus, F. De Schryver, W. De Winter and C. S. Marvel. *J. Polymer Sci.* A-1, **4**, 2831 (1966); F. De Schryver and C. S. Marvel. *ibid.* A–I, **5**, 545 (1967).

[43] J. K. Stille and E. L. Mainen. *J. Polymer Sci.* B, **4**, 39, 665 (1966); J. K. Stille, E. L. Mainen, M. E. Freeburger and F. W. Harris. *Polymer Preprints* **8** (1) 244 (1967).

INFRARED SPECTROSCOPY AND POLYMER STRUCTURE

S. KRIMM

Harrison M. Randall Laboratory of Physics, University of Michigan, Ann Arbor, Michigan, U.S.A.

INTRODUCTION

The power of infrared spectroscopy as a tool in studying the structure of small molecules is well known. This insight into structure results largely from our ability to analyse the spectrum theoretically by means of a normal coordinate analysis. This technique consists of calculating the normal vibration frequencies of a molecule from information on its structure and internal force field. From such a calculation we obtain the values of the frequencies and the forms of the vibrations of the molecule which can be observed in infrared absorption and in Raman scattering. By comparing such calculated frequencies with the observed spectrum it is possible to evaluate our assumptions concerning the structure and force field of the molecule.

A review article published in 1960[1] stressed the importance of extending such theoretical methods of analysis to the study of the infrared spectra of high polymers. Only in this way could the full power of spectroscopy be brought to bear on answering questions related to the structure of, and interactions between, macromolecules. Prior to 1960 the few normal coordinate calculations which had been done (see 1 for references to such studies) were limited in their detailed predictive capability by the lack of a sufficiently complete force field for the molecule and by simplifying approximations made about the structure. The derivation of satisfactory force fields for polymers, which is still the most basic problem associated with a normal coordinate analysis, began in the early 1960s, with the use of extended Urey–Bradley force fields[2, 3] and the development of a detailed valence force field for hydrocarbons[4, 5]. Since then the technique of normal coordinate analysis has been applied to the interpretation of the infrared spectra of many regular polymer structures (see, for example, polyethylene[4, 6-9]; polypropylene, both isotactic[10-14] and syndiotactic[15, 16]; isotactic polybutene-1[17]; syndiotactic 1,2-polybutadiene[18]; polyoxymethylene[19]; polyethylene oxide[20, 21]; polytetrahydrofuran[22]; poly(vinyl chloride)[23, 24]; polyacrylonitrile[25, 26]; poly-3,3-bis(chloromethyl) oxacyclobutane[27]). In some of these cases the chain structure was already known from x-ray diffraction studies; in other cases vibrational analysis shed light on aspects of the structure which were not obtainable from x-ray studies.

In this paper the author wishes to illustrate some of the facets of polymer structure which can be illuminated by means of detailed vibrational analysis. These aspects include not only the structure of the individual regular chain in the crystalline phase, but also features of the arrangement of chains in the crystal. In addition, it is possible to determine to a certain extent the nature of non-crystalline structures present in a polymer system. Examples will be chosen from work in the author's laboratory on poly(vinyl chloride) (PVC) and polyethylene.

STRUCTURE OF CRYSTALLINE POLY(VINYL CHLORIDE)

The ability to distinguish between different regular chain structures of a polymer by means of normal coordinate analyses is well illustrated by the case of poly(vinyl chloride).

Early x-ray diffraction studies[28] had suggested, as a result of the observation of an approximately 5·1Å fibre-axis identity period, that the structure in the crystalline regions of PVC was that of a head-to-tail planar zig-zag syndiotactic chain, as shown in *Figure 1*. A more recent investigation[29] supports this proposal, although variations in x-ray spacings seem to be

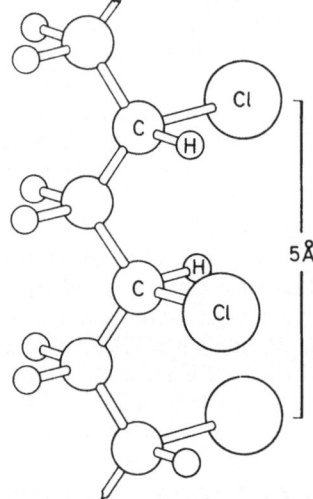

Figure 1. Planar zig-zag structure of syndiotactic PVC. The structure can be described as −(−*T*−)−$_n$, where *T* denotes a *trans* arrangement across a carbon–carbon bond

found between polymers prepared by different methods[30]. The x-ray diffraction pattern of PVC is in general quite poor, probably primarily as a result of poor order[31] and small crystallite size[32], and therefore a complete structure analysis has not been feasible. It is therefore particularly important to have supporting evidence on this structure from spectroscopic studies.

Previous normal coordinate analyses of PVC[23, 24] were based on an assumed planar zig-zag syndiotactic structure. They depended, however, on simplified force fields which had not been checked extensively on small molecules. In addition, calculations were made only for the one structure, so it was not possible to tell if other structures would have given comparably good agreement with the observed frequencies. Thus, while fair agreement with the observed spectrum was obtained, the uncertainties in the force fields left a question concerning the significance of the results.

370

We have recently derived[33, 34] a detailed force field for secondary chlorides which is an extension of the valence force field for hydrocarbons[4, 5]. It gives very good agreement when tested out on small molecules: for example, the average deviation between observed and calculated frequencies in the region of 600 to 1400 cm^{-1} for the most stable conformers of DL- and *meso*-2,4-dichloropentanes is 0·6% and 0·8% respectively. This force field has been applied[33] to the calculation of the normal vibration frequencies of the planar zig-zag structure of syndiotactic PVC, *Figure 1*, and to the two other structures shown in *Figures 2* and *3*. The structure of *Figure 2*, a two-fold helix

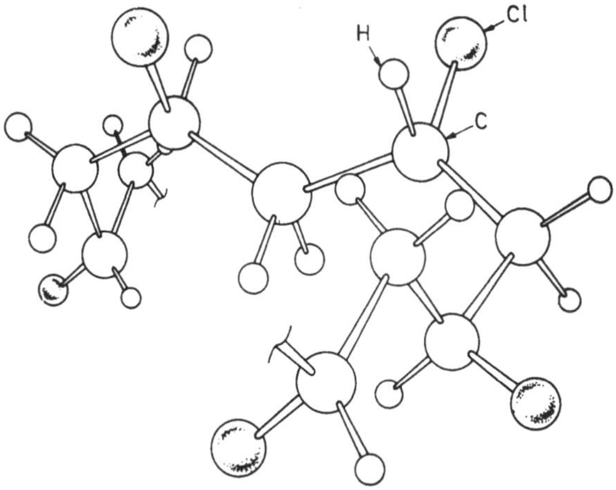

Figure 2. Helical (D_2) structure of syndiotactic PVC. The structure can be described as $-(-TTGG-)-_n$, where G denotes a *gauche* arrangement across a carbon–carbon bond.

with D_2 symmetry, is similar to that found in one form of syndiotactic polypropylene[35], and since the latter polymer also occurs in a planar zig-zag conformation[36] it was felt that the D_2 helical structure might similarly be a possible one for PVC. If PVC were isotactic, its conformation, by analogy with isotactic polypropylene[37], might be a three-fold helix, and therefore the spectrum corresponding to the C_3 helix in *Figure 3* was calculated.

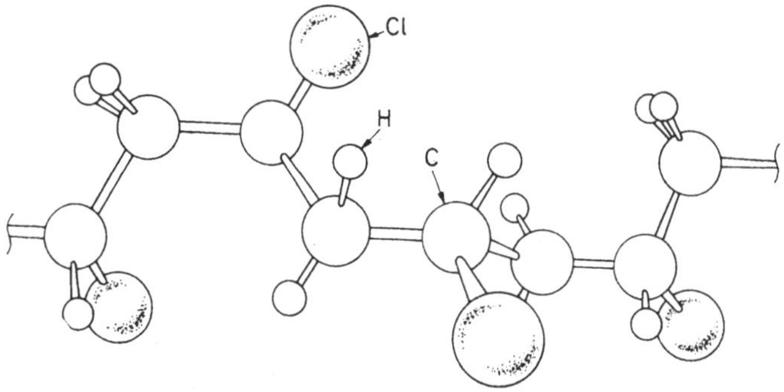

Figure 3. Helical (C_3) structure of isotactic PVC. The structure can be described as $-(-TG-)-_n$.

371

Differences between the calculated frequencies of the above three structures are evident throughout the spectrum, but for purposes of easiest comparison we will concentrate here on the region of the carbon–chlorine stretching vibrations. The calculated frequencies of these modes[33, 38] are given in *Table 1*, together with the values of the two observed bands found

Table 1. Carbon–chlorine stretching frequencies of crystalline PVC

Observed (cm^{-1})	Calculated (cm^{-1})		
	Planar Zig-Zag Syndiotactic	Helical (D$_2$) Syndiotactic	Helical (C$_3$) Isotactic
601 σ*	607 σ		
		622 π	
639 σ	634 σ		
			646 π
		651 σ	
			700 σ
		707 σ	
		[762]†	

* σ = perpendicular, π = parallel; † [ ] = infrared inactive.

in this region in the essentially completely crystalline polymer which can be prepared by polymerization of the monomer in a urea canal complex[39–42]. The dichroic characteristics of the observed bands were obtained from polarized spectra of oriented films of less highly crystalline material[43].

On the basis of the results in *Table 1*, the assignment of a planar zig-zag syndiotactic structure to the chain in the crystalline regions of PVC is highly compelling. The remainder of the calculated spectrum exhibits comparably good agreement with the observed bands[33, 38], as is shown in *Table 2*. (In this table bands labelled σ* show parallel dichroism in the early stages of drawing[44], thus serving to identify B_1 species modes in the spectrum and providing another constraint on possible assignments.) While this result is not particularly surprising in the present case, it does illustrate the potential which a normal coordinate analysis provides for discriminating between various postulated structures of a regular chain.

NON-CRYSTALLINE POLY(VINYL CHLORIDE) STRUCTURES

A highly stereoregular polymer will usually also be highly crystalline in the solid state, and this implies the presence of a single predominant conformation of the chain. This was the case for urea-complex PVC, and this is true of other stereoregular polymers. As we have noted, the normal vibration analysis of the linearly periodic structures associated with such conformations is relatively straightforward, permitting both a ready interpretation of the spectrum as well as a verification of the structure of the chain.

A polymer with poor steric regularity, however, cannot usually take up a regular chain conformation. For example, the presence in a predominantly syndiotactic PVC chain of two neighbouring chlorine atoms having an isotactic arrangement would tend to cause the chain to depart at that point from a planar zig-zag structure, primarily because of the steric repulsion which would exist between two such chlorine atoms if the chain were to remain planar zig-zag. Since the energy as a function of torsion angle about a carbon–carbon bond is locally a minimum near the staggered positions (120° apart), the chain in the above example would be expected to twist away from the planar zig-zag form by any appropriate ~120° rotation which would relieve the steric repulsion. The introduction of such new rotationally isomeric structures would in general be expected to alter the

Table 2. Calculated and observed frequencies for planar zig-zag syndiotactic poly(vinyl chloride)

Av. $|\nu_{obs} - \nu_{calc}|$ (in 600–1500 cm^{-1} region) = 1·3%

Calculated (cm^{-1})		*Observed* (cm^{-1})	*Potential Energy Distribution (contributions $\geqslant$ 10%)*
(125)	A_2		CCCl bend (71) − CCC bend (24)
291	B_1	315 m σ	CCCl bend (45) − CCC(Cl) bend (22) + CC stretch (11)
300	B_2	345 w π	CCCl bend (84)
344	A_1	358 m σ	CCCl bend (63) − CCH(Cl) bend (23)
461	B_1	490 w π	CCC(Cl) bend (44) − CCH(Cl) bend (22) + CH$_2$ rock (14)
(562)	A_2		CCC bend (54) + CCCl bend (18)
607	B_1	601 vs σ*	CCl stretch (75)
634	A_1	639 s σ	CCl stretch (74)
833	B_2	830 m π	CH$_2$ rock (78) + CCH(Cl) bend (14)
982	B_1	962 m σ*	CH$_2$ rock (41) − CC stretch (37)
(1046)	A_2		CC stretch (68) − CCH(Cl) bend (17) − CH$_2$ twist (11)
1083	A_1	1086 w σ	CC stretch (49) − CCC bend (18) − CCC(Cl) bend (18)
1099	B_1	1102 m σ*	CC stretch (51) + HCCl bend (19) + CH$_2$ rock (17)
1102	B_2	1118 vw π	CC stretch (48) + CH$_2$ wag (43)
(1158)	A_2		CH$_2$ twist (66) + CC stretch (18) + CCH(Cl) bend (11)
1173	A_1	1190 w σ(?)	HCCl bend (56) + CH$_2$ twist (26) − CCH(Cl) bend (15)
1185	B_2	1227 m π	CCH(Cl) bend (53) + CH$_2$ wag (26)
1253	B_1	1257 s σ*	HCCl bend (58) − CCH(Cl) bend (31)
1320	A_1	1336 m σ	CH$_2$ twist (58) − HCCl bend (22) + CCH(Cl) bend (14)
(1322)	A_2		CCH(Cl) bend (67) − CH$_2$ twist (18)
1353	B_1	1351 w σ*	CH$_2$ wag (80) − CC stretch (15)
1424	B_2	1380 w π	CC stretch (36) − CH$_2$ wag (33) + CCH(Cl) bend (27)
(1449)	A_2		CH$_2$ bend (77) − CCH bend (20)
1455	A_1	1426 vs σ	CH$_2$ bend (76) − CCH bend (22)

vibrational spectrum from that of the regular chain. Through the under-standing of such differences we hope to learn something about the kinds of non-crystalline structures which are present[45].

Ordinary commercial PVC has long been recognized to be a not very highly stereoregular polymer. Such material exhibits differences in its infrared spectrum when compared with highly syndiotactic PVC, particu-larly in the C–Cl stretching region. In *Figure 4*[42] this region of the spectrum is shown for a PVC polymerized at 50°C; for comparison the spectrum of a urea-complex polymer is also given. New bands appear in the spectrum of the former at 612, 622, 635, 685, and 693 cm^{-1}; furthermore, their relative intensity is found to be a function of the physical state of the specimen. The latter observation was interpreted[46] as indicating a dependence of conforma-tional structure on physical state, the new bands being considered to be associated with new conformations of the polymer chain. It is thus apparent that an understanding of the nature of the non-crystalline structures, and of their variation with the physical state of the polymer, depends upon a

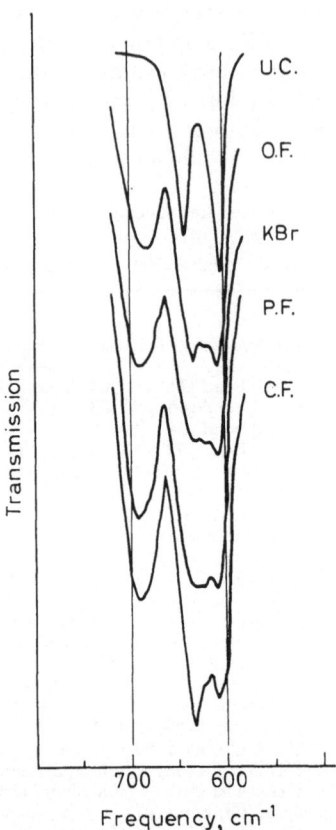

Figure 4. Infrared spectra of a PVC polymerized at 50°C, in various physical states. O.F. = oriented film, KBr = KBr pellet, P.F. = pressed film, C.F. = cast film. The spectrum of a urea-complex PVC (U.C.) is shown for comparison[42]

satisfactory analysis of these new bands. Since their presence is associated with a non-stereoregular chain, it is also clear that the proper interpretation of these bands is basic to any infrared method of determining the tacticity of the polymer.

The analysis of this region of the spectrum of PVC has been the subject of some disagreement. We will therefore briefly review the previous assignments which have been suggested, based mainly on experimental studies of model compounds, and then indicate how our recent calculations help to resolve this question.

Empirical model compound studies

The basis for the experimental association of C–Cl stretching frequency with polymer chain conformation was the preliminary observation of a correlation in small secondary chloride molecules[47]. This was refined by further studies on a large number of model compounds which revealed[48] that the C–Cl stretching frequency depends on the nature of the two substituents *trans* to the Cl atom across neighbouring C—C bonds as well as on

374

the local conformation of the carbon chain. The experimental data on model compounds of unique conformation are summarized in *Figure 5*, which incorporates the analysis of the spectrum of chlorocyclohexane in specifying the $S_{H'H'}$ and S_{CC} frequencies[34].

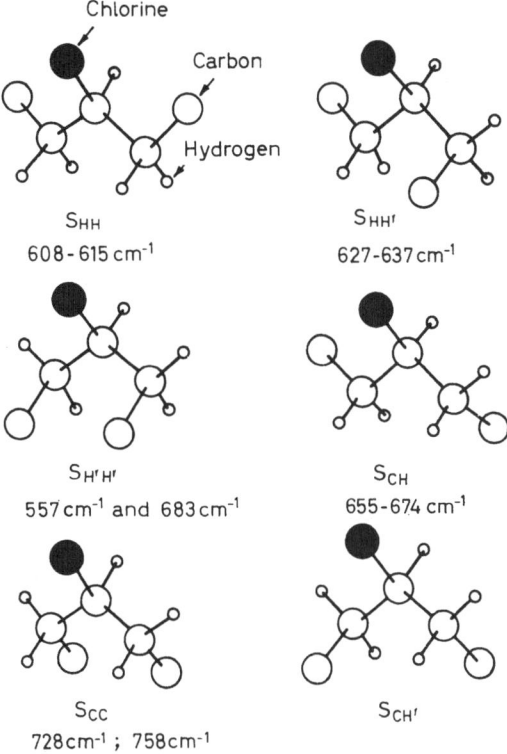

Chlorine

Carbon

Hydrogen

S_{HH}
608 - 615 cm^{-1}

$S_{HH'}$
627 - 637 cm^{-1}

$S_{H'H'}$
557 cm^{-1} and 683 cm^{-1}

S_{CH}
655 - 674 cm^{-1}

S_{CC}
728 cm^{-1} ; 758 cm^{-1}

$S_{CH'}$

Figure 5. Conformations of secondary chlorides and empirically determined C–Cl stretching frequencies[34]

These correlations were used to assign the stable conformers of DL- and *meso*-2,4-dichloropentane[49] and of syndiotactic, heterotactic, and isotactic 2,4,6-trichloroheptane[50, 51]. For the former molecule, for example, the room temperature spectra were interpreted to indicate the presence of only one stable conformer of each isomer, viz., the TT form for the DL compound and the TG' form for the *meso* compound[49, 50] (see *Figure 6*[33, 52]). The TT form gives rise to two S_{HH} frequencies and the TG' form to one S_{HH} and one S_{CH} frequency. From a similar analysis of the spectra of the 2,4,6-trichloroheptanes it was concluded[50, 51] that in the spectrum of PVC the absorption in the S_{HH} region was associated entirely with syndiotactic pairs of Cl atoms while absorption in the S_{CH} region was due to isotactic pairs. This assignment has been used[50, 53–55] to obtain a measure of the tacticity in the polymer sample.

Other studies, however, had shown that the situation was not so simple. Model compound correlations[48] had suggested the presence of the $S_{HH'}$

'DL- Form' Conformation	TT	TG (GT)	TG' (G'T)	GG	GG' (G'G)	G'G'
Structure	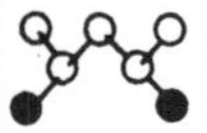	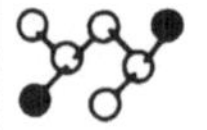		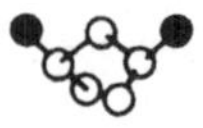	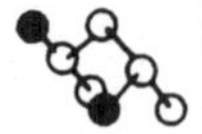	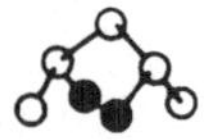
Isomeric structures	S_{HH}, S_{HH}	S_{HH}, S_{CH}	S_{HH}, $S_{HH'}$	S_{CH}, S_{CH}	$S_{HH'}$, S_{CH}	$S_{HH'}$, $S_{HH'}$
Calculated frequencies	624 605 —	675 615 —	632 (608) 572	715 627 —	699 619 531	694 619 595
'Meso Form' Conformation	TT	TG (G'T)	TG' (GT)	GG (G'G')	GG'	G'G
Structure	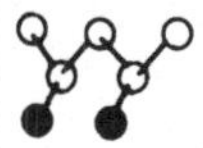		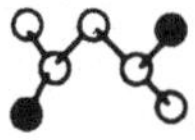	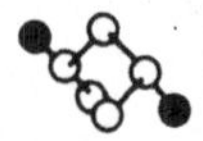	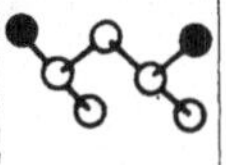	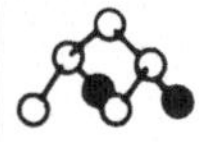
Isomeric structures	S_{HH}, S_{HH}	S_{HH}, $S_{HH'}$	S_{HH}, S_{CH}	$S_{HH'}$, S_{CH}	S_{CH}, S_{CH}	$S_{HH'}$, $S_{HH'}$
Calculated frequencies	615 (614) —	643 618 564	675 614 —	699 617 537	714 627 —	692 (615) 595

Figure 6. Conformers of DL- and meso-2,4-dichloropentane and their calculated C–Cl stretching frequencies[33,52], ( ) denotes expected weak intensity

conformer in 2-chlorobutane at room temperature; this has now been confirmed by normal coordinate analyses[34]. The presence of this as well as other higher energy conformations was therefore believed to be possible in PVC[39, 41, 42, 46]. In fact, it has now been shown[56] that isomeric structures other than S_{HH} can be introduced into the highly stereoregular urea-complex PVC by pressing. This is shown in Figure 7, where it can be seen that the development of absorption near 690 cm^{-1} suggests the introduction of S_{CH} structures while the increase in absorption near 635 cm^{-1} would be consistent with the presence of the $S_{HH'}$ structure. It appears, therefore, that the simplified interpretations suggested for the origins of the additional bands in ordinary PVC are in need of modification.

One part of the problem has been the absence of a detailed analysis, both experimental and theoretical, of the spectra of the model compounds. We discuss such an analysis for the 2,4-dichloropentanes in the next section, and show that even for this model system conformers other than the lowest energy one are present at room temperature. This should make it less difficult to acknowledge their possible existence in the polymer. The other part of the problem has been the reluctance to recognize that polymers in the solid state are not completely equivalent to small molecules in solution, in the liquid state, or in the relatively highly ordered structure of the crystal. Polymer chains in the solid state are subject to constraints imposed by the special conditions of their morphology, such as crystalline–non-crystalline

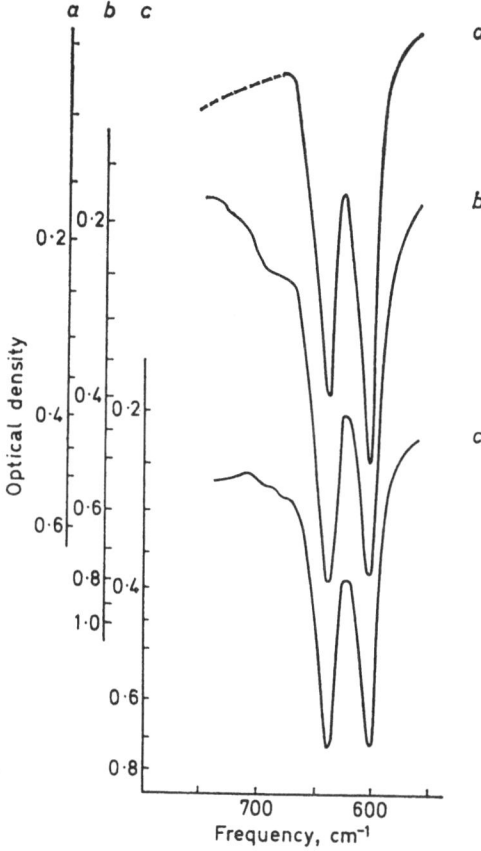

Figure 7. Infrared spectra of urea-complex PVC[56]: (*a*) powder in nujol mull (dotted portion of curve represents nujol band); (*b*) film pressed from powder at about 200°C and cooled rapidly; (*c*) film pressed from powder at about 200°C and cooled slowly (by being left over-night in press)

ratio, single crystal and spherulitic texture, chain entanglements, and so forth. Under these circumstances it is not unlikely that chains will be forced to assume spatial arrangements which occasionally involve higher energy conformations than are normally found in comparable small molecules. The results on pressed urea-complex PVC[56] bear this out. It is therefore not completely valid to transfer the results of model compound studies to polymers without taking such factors into account.

Normal coordinate analysis of model compounds

The availability of a detailed force field for secondary chlorides which reproduces very well the observed spectra of the most stable conformers of both DL- and *meso*-2,4-dichloropentanes[34] has made it feasible to calculate the vibrational spectra of the other conformers of these molecules. This calculation has been done[33, 52], and the predicted C–Cl stretching frequencies are given in *Figure 6*. The observed spectra in this region[52], both at room temperature and at low temperatures, are shown in *Figures 8 and 9*.

The strong bands in the liquid at room temperature are clearly accounted for by one conformer, the *TT* in the case of the DL form and the *TG'* in the case of the *meso* form. It is evident from *Figures 8 and 9*, however, that there

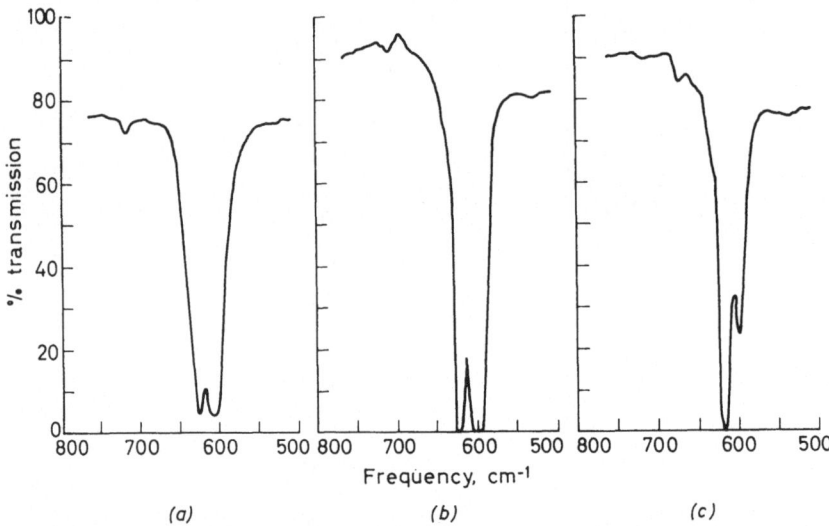

(a)   (b)   (c)

Figure 8. Infrared spectra of DL-2,4-dichloropentane: (*a*) at room temperature; (*b*) at − 110°C; (*c*) at − 133°C (after previous sample had been warmed to room temperature)

are additional weak bands in the C–Cl stretching region. These do not appear to be assignable to impurities or to overtones or combinations[52], but they can be explained satisfactorily as fundamental vibrations of other conformers. For example, in the spectrum of the DL form at room temperature, *Figure 8(a)*, the band at 715 cm⁻¹ can be associated with the *GG* conformer and the band at 682 cm⁻¹ can be correlated with the *TG* conformer (cf. *Figure 6*). Different conditions of cooling (which are not yet fully understood) can result in the retention of different amounts of these two structures, as seen in *Figures 8(b)* and *8(c)*. The other C–Cl stretching frequencies of these conformers, viz., near 627 cm⁻¹ for the *GG* form and near 615 cm⁻¹ for the *TG* form, undoubtedly are overlapped by the two strong bands of the *TT* conformer which are observed at 627 and 606 cm⁻¹. The changes in the relative intensities of these bands at low temperatures are consistent with the above assignment[52]. In the spectrum of the *meso* form at room temperature, *Figure 9(a)*, the band at 642 cm⁻¹ is well correlated with the frequency predicted for the *TG* conformer (cf. *Figure 6*). Its counterpart, predicted at 618 cm⁻¹, would be expected to be overlapped by the strong component associated with the main *TG'* conformer, which is predicted at 614 cm⁻¹ and observed at about 615 cm⁻¹. The weakening at low temperature of the 615 cm⁻¹ band relative to that at 680 cm⁻¹, which is paralleled by the disappearance of the 642 cm⁻¹ band [cf. *Figure 9(b)*], is consistent with this assignment. Additional support for this correlation comes from the presence of a band at 552 cm^{-1} in the spectrum of the liquid which disappears at low temperature along with the band at 642 cm^{-1} (cf. *Figure 9*). This band is predicted at 564 cm^{-1} (see *Figure 6*), in a region in which no other modes are expected except for C–Cl stretching vibrations of three of the six conformers[52]. This region is therefore highly diagnostic for such structures. It is gratifying to note, for example, that the spectrum of the DL

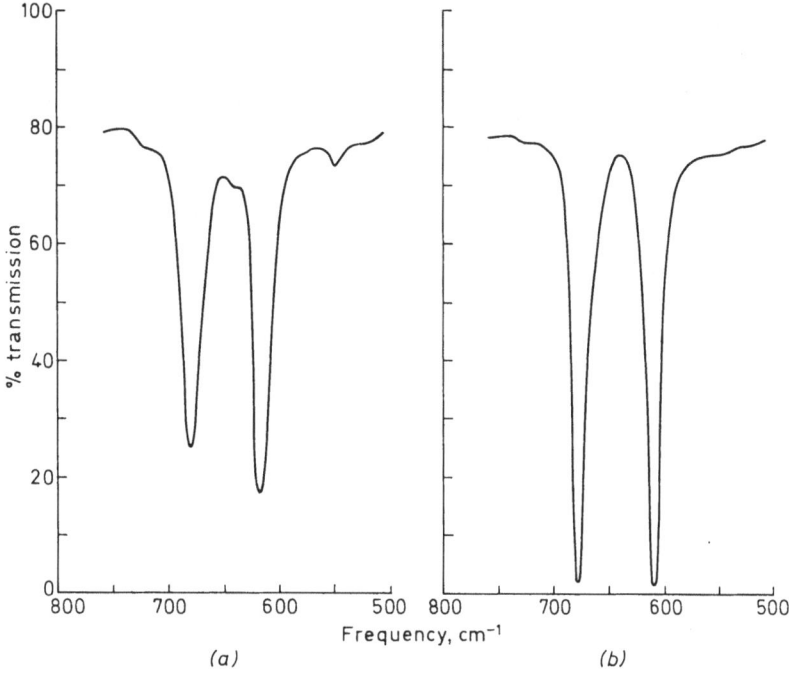

Figure 9. Infrared spectra of *meso*-2,4-dichloropentane: (*a*) at room temperature; (*b*) at − 105°C

form shows no absorption is this region, as is predicted for the conformers found to be present from the analysis of the absorption bands above 600 cm^{-1}. Thus, the observation in the spectrum of the *meso* form of another band at 523 cm^{-1} which disappears on cooling strongly suggests that the *GG* conformer is also present (cf. *Figure 6*, and the predicted frequency of 537 cm^{-1}). In this case the other C–Cl frequencies, viz. 699 and 617 cm^{-1}, could well be overlapped by the strong bands of the main *TG'* conformer.

It can be seen, therefore, that the spectra of the 2,4-dichloropentanes at room temperature indicate the presence of conformers other than the single most stable ones. The changes in spectra with temperature can therefore be understood in terms of changes in the proportions of different conformers, and need not be attributed entirely to environmental effects[57, 50]. If the additional conformers suggested by the above analysis are present, then we see that associated with the syndiotactic (DL) structure are S_{CH} modes (from the *TG* and *GG* conformers) as well as S_{HH} modes (from the dominant *TT* conformer), while the isotactic (*meso*) structure can give rise to $S_{HH'}$ modes (from the TG and GG conformers) as well as S_{CH} modes (from the dominant TG' conformer). It should therefore not be surprising to find weak bands in the spectrum of ordinary PVC which can be associated with such structures. The assignments of such additional bands to conformations other than the most stable ones[42] therefore is validated by the above analysis of the spectra of the 2,4-dichloropentanes.

Incidentally, it is of interest to note that in the conformers which might have been thought unlikely to occur because of steric repulsions, viz., the TG form of the DL-isomer and the TG and GG forms of the *meso*-isomer, a Cl atom is adjacent to a CH_3 group. While it is possible to relieve this close contact by small rotations about C—C bonds, it may be that these structures are stabilized to some extent by an attractive interaction between CH_3 and Cl, what might be called a C—H $\cdots$ Cl hydrogen bond. Although such a bond may seem unusual, it should be noted that evidence is accumulating for the participation of C—H groups in such weak types of hydrogen bonding, both in small molecules[58, 59] as well as in macromolecules[60].

Normal coordinate analyses of 2,4,6-trichloroheptane[33, 61] indicate similarly that conformers are present at room temperature other than the single ones proposed earlier for each isomer[50]. This is consistent with conclusions reached from NMR studies[51]. As a result we can expect isotactic sequences in a PVC chain to give rise to S_{HH} modes (from the $GTTG'$ conformer) as well as to the S_{CH} and $S_{HH'}$ modes which we have seen are possible from the analysis of the 2,4-dichloropentanes. This is consistent with our earlier suggestion[42], and points to the danger in determining tacticity on the basis of an oversimplified analysis of the infrared spectrum[50, 53–55].

Normal coordinate analyses of PVC fragments

In order to extend the normal coordinate analyses to the interpretation of the spectra of non-crystalline PVC structures, the C–Cl stretching frequencies of a fragment of a PVC chain were calculated[33, 38]. It was hoped in this way to overcome any possible errors associated with a comparison using frequencies calculated for a small model molecule. The fragment chosen was $C-(-CH_2CHCl-)_3-CH_2-C$, and calculations were done for a structure with two syndiotactic pairs of Cl atoms and for one with one isotactic pair and one syndiotactic pair. The C–Cl stretching frequencies of some conformations of these fragments are given in *Table 3*. While these results contain many implications[38], we will concentrate here only on the interpretations they suggest of the spectra of the pressed urea-complex PVCs, which are shown in *Figure 7*.

We have seen that the pressing of urea-complex PVC results in the introduction of new absorption bands near 690 cm^{-1} and near 635 cm^{-1}. The former can be associated with the S_{CH} structure of the $TTTG'TT$ conformation, which is analogous to the TG conformation of DL-2,4-dichloropentane. (The $S_{CH'}$ structure of the $TTG'G'TT$ conformation could also qualify on the basis of frequency agreement, but it is a less likely possibility on the basis of steric and electrostatic repulsions between adjacent Cl atoms[38].) The increased absorption near 635 cm^{-1} had been assigned to the $S_{HH'}$ structure[56], and this is still consistent with the results in *Table 3*. The calculations show, however, that another assignment is possible, namely to one component of the S_{CH} S_{CH}, structure of the $TTTG'G'T$ conformation (which is analogous to the GG conformation of DL-2,4-dichloropentane). We expect in this

Table 3. Carbon–chlorine stretching frequencies of C–(–CH_2CHCl–)$_3$–CH_2—C

	Conformation	Isomeric Structures	C–Cl Frequencies
Syndiotactic fragment (two s pairs)	$TTTGTT$	S_{HH} $S_{HH'}$ S_{HH}	606 615 618 638
	$TTTG'TT$	S_{HH} S_{CH} S_{HH}	616 619 692
	$TTG'G'TT$	S_{HH} $S_{CH'}$ S_{HH}	617 623 687
	$TTTG'G'T$	S_{HH} S_{CH} S_{CH}	615 637 730
Isotactic fragment (one i, one s pair)	$TTGTTT$	S_{HH} $S_{HH'}$ S_{HH}	604 611 623 648
	$TTG'TTT$	S_{HH} S_{CH} S_{HH}	614 619 693

case to find another component near 730 cm^{-1}; in fact, the spectrum of the annealed pressed film [*Figure 7(c)*] shows weak absorption near 750 cm^{-1}. Since this mode might be expected to be weaker than that at 637 cm^{-1} (cf. the results for the GG conformer of DL-2,4-dichloropentane[52]), such an assignment cannot be excluded. It is interesting to note that if this assignment for the 635 cm^{-1} band is correct, it implies that annealing has resulted in the enhancement of the (presumably) more stable GG conformation while fast cooling has frozen in the more accessible but (presumably) higher energy TG conformation. Since the physical properties of the polymer must depend on such conformational structural characteristics, it is obvious that a detailed understanding of the structure in non-crystalline portions of a polymer chain is fundamental to our ability to correlate physical properties with polymer structure.

A cautionary comment is pertinent at this point, and illustrates the subtleties often underlying the observed spectrum. The dispersion curve in the C–Cl stretching region of planar zig-zag syndiotactic PVC is shown in *Figure 10*[33, 62]. While for the infinite chain only those modes with a phase

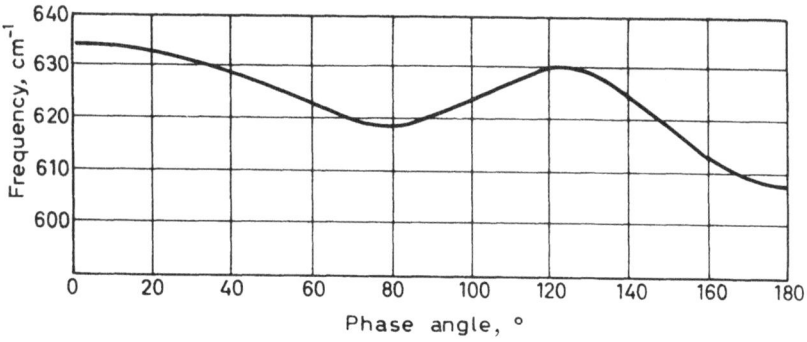

Figure 10. Calculated dispersion curve for C–Cl stretching vibrations in planar zig-zag syndiotactic PVC[33,62]. The phase angle is between adjacent CH_2CHCl groups

angle of 0° or 180° can be infrared active, when defects are introduced into the chain it becomes possible for modes with other phase angles to gain activity[33, 62]. If these were distributed uniformly with respect to phase angle, then it can be seen from *Figure 10* that absorption near the high frequency cut-off would be enhanced with respect to that near the low frequency cut-off. Such defects could be the $S_{HH'}$, S_{CH}, and S_{CH} S_{CH} structures which we have discussed above. This mechanism for intensity enhancement complicates the picture, and makes it more difficult at present to be certain of the origin of the increased absorption near 635 cm^{-1}.

FOLD PLANE STRUCTURE IN POLYETHYLENE CRYSTALS

The preceding discussion has been concerned with the vibrations of a single polymer chain, either in its regular conformation within a crystal or the conformations it assumes in non-crystalline regions. In some cases the interactions between chains manifest themselves in the vibrational spectrum, and it then becomes possible to study certain aspects of the crystal structure itself by analysis of the infrared spectrum. Our recent studies on the polyethylene system illustrate the kinds of questions which can be posed and answered as a result of the extension of normal coordinate analyses to the entire crystal structure of the polymer.

The early observation[63] that the CH_2 rocking mode in polyethylene is a doublet, with components near 720 and 731 cm^{-1}, led to the surmise[64, 65] that such splitting was due to intermolecular interactions between the two chains in the orthorhombic unit cell[66] (see *Figure 11*). Subsequent experimental work[67, 68] supported this interpretation, as did theoretical

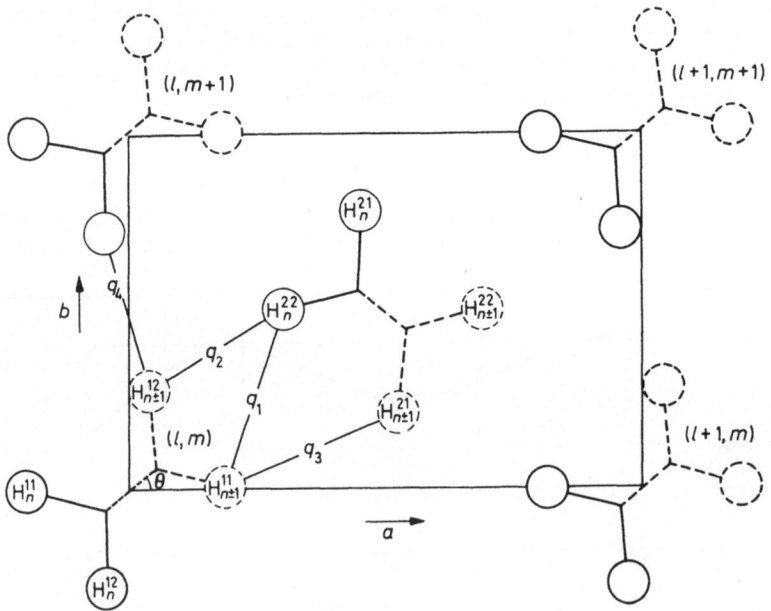

Figure 11. Cross-section of the unit cell of crystalline polyethylene. The qs indicate the dominant hydrogen–hydrogen interactions

studies[69, 70]. Recent normal coordinate analyses[71, 72] provide a detailed accounting not only of such splittings in the high frequency region of the spectrum, but also of the frequency of the observed[73, 74] lattice translational mode near 71 cm^{-1} and its variation with temperature[73, 72]. As a result of these studies we now have a reasonably good understanding of the inter-molecular force field in crystalline polyethylene.

Polyethylene is widely known to form single crystals consisting of folded chains[75]. Crystals grown from solution generally show a well-developed morphology which indicates that the chains are folded in the (110) planes, although under certain conditions crystal shapes are obtained which indicate that folding can also take place parallel to the (100) planes[75]. The situation in melt-crystallized polyethylene is less clear. Some insight comes from studies of crystals grown from solution under conditions[76-79] such that they display morphologies analogous to the spherulitic structures which occur in the melt crystallized polymer. In these circumstances the crystals are often found to be elongated, with the crystallographic b-axis parallel to the long direction of the crystal. No direct evidence, however, is available on the fold plane in such crystals or in the melt crystallized polymer, although it presumably is parallel to the (100) planes. X-ray diffraction is incapable of giving information on this point since it 'sees' only the interior of the crystal, whose structure[66] is essentially independent of the geometry of the folding.

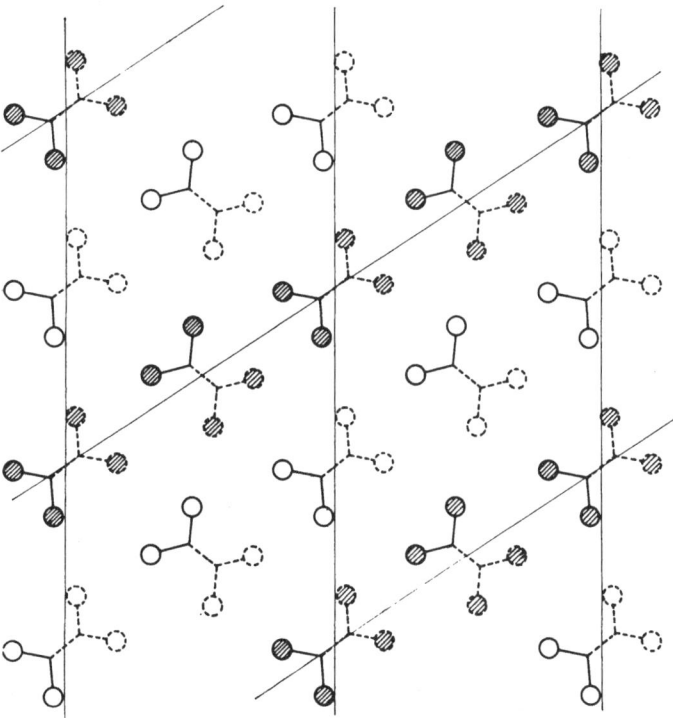

Figure 12. Cross-section of a uniform 1:1 folded chain co-crystal of normal (open circles) and fully deuterated (shaded circles) polyethylene in which the chains fold in the (110) planes. The folds on the top and bottom surfaces of the crystal are not shown

A vibrational analysis as a function of these features of crystallization, however, has been recently shown[80] to be capable of providing a method for studying such aspects of crystalline morphology.

The basis for a spectroscopic study of the geometry of the fold-plane in polyethylene arises from the fact that, as can be seen from *Figure 11*, a chain at the corner of the unit cell interacts with one at the centre of the cell as well as with neighbouring chains along the *b*-axis direction. Therefore, if we co-crystallize normal polyethylene with fully deuterated polyethylene we would anticipate a different pattern of interactions if the chain continuity is along the (110) plane than if it is parallel to the (100) plane. This is illustrated in *Figures 12* and *13* for a uniform 1:1 co-crystal of the two polymers. In the structure of *Figure 12* the interactions between similar

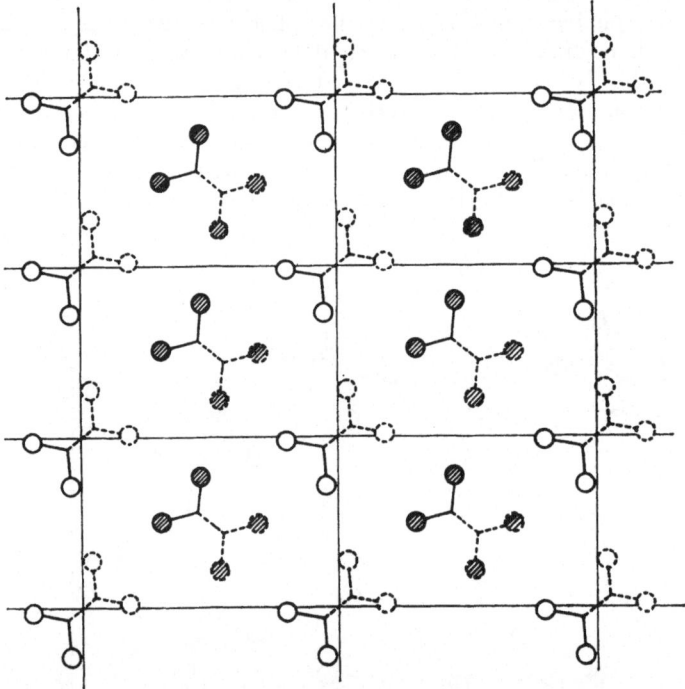

Figure 13. Cross-section of a uniform 1:1 folded chain co-crystal of normal (open circles) and fully deuterated (shaded circles) polyethylene in which the chains fold parallel to the (100) planes

oscillators are preserved between chains along the (110) plane but eliminated in the *b*-axis direction. The new enlarged unit cell still contains two chains of a given kind, however, so we expect splittings in the fundamentals, but of different magnitude than in the pure crystal. In the structure of *Figure 13* the unit cell contains only one chain of a given kind and such splittings should disappear. This is borne out by the calculations[80]. For example, the CD_2 bending mode is predicted to be a doublet whose components are split by 8·1 cm^{-1} in pure $(CD_2)_n$ and by 4·1 cm^{-1} in the 1:1 structure of *Figure 12*, whereas it is predicted to be a singlet in the structure of *Figure 13*. Thus, it is possible to use the predictions concerning the splittings in such a mixed crystal system to determine characteristics of the fold structure in poly-ethylene.

Experiments to test these ideas have been carried out[81, 82]. Preliminary results are reported here. In *Figure 14(a)* is shown the CD_2 bending region of the spectrum of a film of fully deuterated polyethylene cast from *p*-xylene.

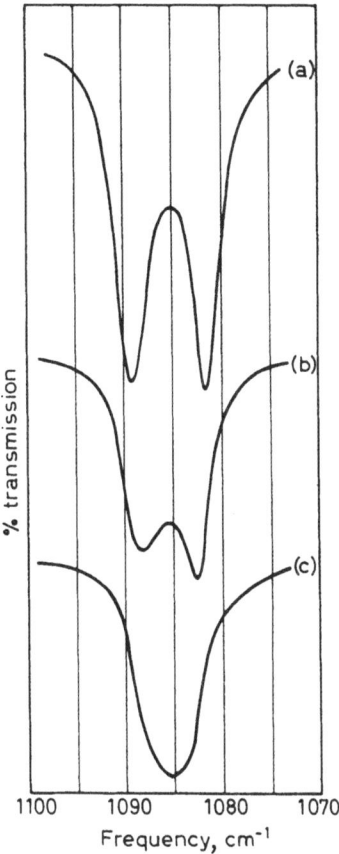

Figure 14. Infrared spectra of CD_2 bending mode: (*a*) in a cast film of fully deuterated polyethylene; (*b*) in a cast film of $1(CD_2)_n : 2(CH_2)_n$ composition; (*c*) in a film obtained by melt-crystallizing sample (*b*)

(The same results are obtained from cast films as from single crystals grown from dilute solution.) The observed splitting of 7.5 cm^{-1} compares well with the predicted value of 8.1 cm^{-1}. Such a film, when melt crystal-lized at 190°C and cooled to room temperature, shows the same 7.5 cm^{-1} splitting in the CD_2 bending doublet. The comparable spectrum of a cast film whose composition was $1(CD_2)_n$ to $2(CH_2)_n$ is shown in *Figure 14(b)*. The splitting of the CD_2 bending mode varies somewhat with the conditions under which the film is prepared (probably due to aggregation of the higher melting deuterated polyethylene); the smallest splitting that we have observed is the $\sim$5·7 cm^{-1} shown in *Figure 14(b)*. The predicted splitting for the CD_2 bending mode of the structure of *Figure 12*, in which the $(CD_2)_n$ chain is similarly surrounded by $(CH_2)_n$ chains as in our 1 : 2 sample, is, as we have noted, 4·1 cm^{-1}. The agreement seems good enough to permit us to conclude that in the cast film the chains are indeed folded along (110) planes. When this 1 : 2 cast film is melt-crystallized at 155°C, the spectrum of *Figure 14(c)* is obtained. The doublet is now replaced by a singlet, indicating that a fold structure similar to that of *Figure 13*, viz., with folds parallel to the (100) planes, has developed. It is important to note that re-casting of a film from this melt-crystallized sample brings about a return of the spectrum to that of *Figure 14(b)*, indicating that there has been no artifact accompanying the melting and that the fold geometry is indeed a property of the preparative technique. Other aspects of the fold geometry can be studied by analogous methods[80], thus demonstrating that inter-chain interactions in polymers can provide an important spectroscopic probe of crystalline morphology.

Acknowledgement

The author is indebted to the National Science Foundation for its support of much of this work. A grant from the Michigan Memorial–Phoenix Project supported the purchase of the perdeuteropolyethylene.

References

1 S. Krimm. *Fortschr. Hochpolym. Forsch.* **2**, 51 (1960).
2 H. Tadokoro. *J. Chem. Phys.* **33**, 1558 (1960).
3 T. Miyazawa. *J. Chem Phys.* **35**, 693 (1961).
4 J. H. Schachtschneider and R. G. Snyder. *Spectrochim. Acta* **19**, 117 (1963).
5 R. G. Snyder and J. H. Schachtschneider. *Spectrochim. Acta* **21**, 169 (1965).
6 M. Tasumi, T. Shimanouchi and T. Miyazawa. *J. Mol. Spectros.* **9**, 261 (1962).
7 Yu. Ya. Gotlib and L. V. Kudinskaya. *Optics and Spectroscopy* **10**, 168 (1961).
8 T. P. Lin and J. L. Koenig. *J. Mol. Spectros.* **9**, 228 (1962).
9 M. Tasumi, T. Shimanouchi and T. Miyazawa. *J. Mol. Spectros.* **11**, 422 (1963).
10 T. Miyazawa, Y. Ideguchi and K. Fukushima. *J. Chem. Phys.* **38**, 2709 (1963).
11 R. G. Snyder and J. H. Schachtschneider. *Spectrochim. Acta* **20**, 853 (1964).
12 T. Miyazawa. *J. Polymer Sci.* **C7**, 59 (1964).
13 J. H. Schachtschneider and R. G. Snyder. *J. Polymer Sci.* **C7**, 99 (1964).
14 H. Tadokoro, M. Kobayashi, M. Ukita, K. Yasufuku, S. Murahashi and T. Torii. *J. Chem. Phys.* **42**, 1432 (1965).
15 J. H. Schachtschneider and R. G. Snyder. *Spectrochim. Acta* **21**, 1527 (1965).
16 M. Miyazawa. *J. Chem. Phys.* **43**, 4030 (1965).
17 M. Ukita. *Bull. Chem. Soc. Japan* **39**, 742 (1966).
18 G. Zerbi and M. Gussoni. *Spectrochim. Acta* **22**, 2111 (1966).
19 H. Tadokoro, M. Kobayashi, Y. Kawaguchi, A. Kobayashi and S. Murahashi. *J. Chem. Phys.* **38**, 703 (1963).
20 T. Miyazawa, K. Fukushima, and Y. Ideguchi. *J. Chem. Phys.* **37**, 2764 (1962).
21 T. Yoshihara, H. Tadokoro and S. Murahashi. *J. Chem. Phys.* **41**, 2902 (1964).
22 K. Imada, H. Tadokoro, A. Umehara and S. Murahashi. *J. Chem. Phys.* **42**, 2807 (1965).
23 T. Shimanouchi and M. Tasumi. *Bull. Chem. Soc. Japan* **34**, 359 (1961).
24 V. G. Boitsov and Yu. Ya. Gotlib. *Optics and Spectroscopy*, Suppl. 2, 65 (1966).
25 R. Yamadera, H. Tadokoro and S. Murahashi. *J. Chem. Phys.* **41**, 1223 (1964).
26 R. Yamadera and S. Krimm. *I.U.P.A.C. Symposium on Macromolecular Chemistry* Tokyo, 1966. To be published.
27 S. Enomoto, C. G. Opaskar and S. Krimm. *J. Polymer Sci.* **C16**, 2263 (1967).
28 C. S. Fuller. *Chem Review* **26**, 143 (1940).
29 G. Natta and P. Corradini. *J. Polymer Sci.* **20**, 251 (1956).
30 P. H. Burleigh. *J. Amer. Chem. Soc.* **82**, 749 (1960).
31 M. Mammi and V. Nardi. *Nature, Lond.* **199**, 247 (1963).
32 V. P. Lebedev, N. A. Okladnov, K. S. Minsker and B. P. Shtarkman. *Vysokomol. soyedin* **7**, 655 (1965); *Polymer Science U.S.S.R.* **7**, 724 (1965).

[33] C. G. Opaskar, Ph.D. Thesis, University of Michigan, 1966.

[34] C. G. Opaskar and S. Krimm. *Spectrochim. Acta*, **23A**, 2261 (1967).

[35] G. Natta, I. Pasquon, P. Corradini, M. Peraldo, M. Pegoraro and A. Zambelli *Atti. R. Accad. Lincei* **28**, 539 (1960).

[36] G. Natta, M. Peraldo and G. Allegra. *Makromol. Chem.* **75**, 215 (1964).

[37] G. Natta, P. Corradini and M. Cesari. *Atti. R. Accad. Lincei* **21**, 365 (1956).

[38] C. G. Opaskar, S. Krimm, V. L. Folt and J. J. Shipman. To be published.

[39] S. Krimm, A. R. Berens, V. L. Folt and J. J. Shipman. *Chem. & Ind.* 1512 (1958); *ibid.* 433 (1959).

[40] T. Shimanouchi, S. Tsuchiya and S. Mizushima. *J. Chem. Phys.* **30**, 1365 (1959).

[41] S. Krimm, V. L. Folt, J. J. Shipman and A. R. Berens. *J. Polymer Sci.* **A1**, 2621 (1963).

[42] S. Krimm, V. L. Folt, J. J. Shipman and A. R. Berens. *J. Polymer Sci.* **B2**, 1009 (1964).

[43] S. Krimm and C. Y. Liang. *J. Polymer Sci.* **22**, 95 (1956).

[44] M. Tasumi and T. Shimanouchi. *Spectrochim. Acta* **17**, 731 (1961).

[45] S. Krimm. *J. Polymer Sci.* **C7**, 3 (1964).

[46] S. Krimm and S. Enomoto. *J. Polymer Sci.* **A2**, 669 (1964).

[47] S. Mizushima, T. Shimanouchi, K. Nakamura, M. Hayashi and S. Tsuchiya *J. Chem. Phys.* **26**, 970 (1957).

[48] J. J. Shipman, V. L. Folt and S. Krimm. *Spectrochim. Acta* **18**, 1603 (1962).

[49] T. Shimanouchi and M. Tasumi. *Spectrochim. Acta* **17**, 755 (1961).

[50] T. Shimanouchi, M. Tasumi and Y. Abe. *Makromol. Chem.* **86**, 43 (1965).

[51] D. Doskočilová, J. Štokr, B. Schneider, H. Pivcová, M. Kolinský, J. Petránek and D. Lím. Preprint P5, *I.U.P.A.C. Symposium on Macromolecular Chemistry*, Prague, 1965.

[52] C. G. Opaskar, S. Krimm, J. Burr, V. L. Folt and J. J. Shipman. To be published.

[53] M. Takeda and K. Iimura. *J. Polymer Sci.* **57**, 383 (1962).

[54] H. Germar. *Kolloid Z.* **193**, 25 (1963).

[55] B. Schneider, J. Štokr, D. Doskočilová, M. Kolinský, S. Sýkora and D. Lím. Preprint P599, *I.U.P.A.C. Symposium on Macromolecular Chemistry*, Prague, 1965.

[56] S. Krimm, J. J. Shipman, V. L. Folt and A. R. Berens. *J. Polymer Sci.* **B3**, 275 (1965).

[57] K. Iimura, T. Hama, T. Shibuya and M. Takeda. *Bull. Chem. Soc. Japan* **37**, 1758 (1964).

[58] D. J. Sutor. *J. Chem. Soc.* 1105 (1963).

[59] A. Allerhand and P. von R. Schleyer. *J. Amer. Chem. Soc.* **85**, 1715 (1963).

[60] S. Krimm, K. Kuroiwa and T. Rebane. *Conformation of Biopolymers*, Ed. G. N. Ramachandran, Academic Press, London and New York, Vol. 2, 1967, p. 439.

[61] C. G. Opaskar and S. Krimm. To be published.

[62] C. G. Opaskar and S. Krimm. Preprints of American Chemical Society, Chicago, September 1967. To be published.

[63] H. W. Thompson and P. Torkington. *Proc. Roy. Soc.* **A184**, 3 (1945).

[64] R. S. Stein and G. B. B. M. Sutherland. *J. Chem. Phys.* **21**, 370 (1953).

[65] R. S. Stein and G. B. B. M. Sutherland. *J. Chem. Phys.* **22**, 1993 (1954).

[66] C. W. Bunn. *Trans. Faraday Soc.* **35**, 482 (1939).

[67] S. Krimm. *J. Chem. Phys.* **22**, 567 (1954).

[68] S. Krimm, C. Y. Liang and G. B. B. M. Sutherland. *J. Chem. Phys.* **25**, 549 (1956).

[69] R. S. Stein. *J. Chem. Phys.* **23**, 734 (1955).

[70] R. G. Snyder. *J. Mol. Spectros.* **7**, 116 (1961).

[71] M. Tasumi and T. Shimanouchi. *J. Chem. Phys.* **43**, 1245 (1965).

[72] M. Tasumi and S. Krimm. *J. Chem. Phys.* **46**, 755 (1967).

[73] J. E. Bertie and E. Whalley. *J. Chem. Phys.* **41**, 575 (1964).

[74] S. Krimm and M. I. Bank. *J. Chem. Phys.* **42**, 4059 (1965).

[75] P. H. Geil. *Polymer Single Crystals* Interscience, New York, 1963.

[76] D. C. Bassett, A. Keller and S. Mitsuhashi. *J. Polymer Sci.* **A1**, 763 (1963).

[77] I. Heber. *Kolloid Z.* **189**, 112 (1963).

[78] H. D. Keith. *J. Polymer Sci.* **A2**, 4339 (1964).

[79] H. D. Keith. *J. Appl. Phys.* **35**, 3115 (1964).

[80] M. Tasumi and S. Krimm. *J. Polymer Sci.* **A2**, in press.

[81] M. I. Bank, Ph.D. Thesis, University of Michigan, 1968.

[82] M. I. Bank and S. Krimm. To be published.

387

POLYMER DEGRADATION AND ELECTRON SPIN RESONANCE SPECTROSCOPY

N. Grassie

Chemistry Department, University of Glasgow, Scotland

Two years ago, at the symposium in Prague, I chose as the theme of a lecture similar to this one, the application of modern analytical techniques to the solution of polymer degradation problems[1]. I tried to demonstrate some of the contributions which had been made by the use of, for example, infrared, mass and nuclear magnetic resonance spectroscopy as well as gas chromatography and radiochemical techniques. I looked forward to rapid developments in the utilization of such methods and especially to the simultaneous application of a number of them in order to try to solve some of the more difficult problems of polymer degradation.

One of the techniques to which I made brief reference at that time was electron spin resonance spectroscopy. Of all the modern analytical techniques, e.s.r. is probably most limited in its application in the sense that it is concerned only with free radicals. On the other hand information about free radicals cannot be obtained in any other comparably direct way and its potential value in the study of polymer degradation processes is clear since so many of them proceed by free radical mechanisms. During recent years, limited but definite progress has been made in the application of e.s.r. to polymer degradation problems and its was against this background that the organizers of this conference invited me to talk on the subject of "Electron Spin Resonance and Polymer Degradation".

Since I am certainly not competent to make authoritative judgements upon the interpretation of e.s.r. data, I had to assume that I was invited to this task because of my interest in polymer degradation. In order to emphasize this fact, I have reversed the title originally suggested so that it now reads "Polymer Degradation and Electron Spin Resonance". It is my intention to review progress in the application of e.s.r. data to polymer degradation, aiming my lecture especially at the majority of you who, I hope, like myself, would like to learn something of the potentiality of the technique in this direction. My intention is to try to stimulate interest rather than give an authoritative review. Firstly, I shall discuss briefly and quite generally the kind of information which e.s.r. measurements can give. Secondly, I shall describe some of the results from the literature which have interested me. Finally, I shall discuss polymer degradation problems in which I am currently interested and to which e.s.r. might be relevant.

Electron spin resonance spectra are associated with the unpaired electrons which exist in free radicals[2]. An electron may only have a spin, m_s, of either $+\frac{1}{2}$ or $-\frac{1}{2}$ so that when a magnetic field, H, is applied to a system containing free electrons they will become aligned with their spins parallel or antiparallel to the magnetic field. The energies of the two kinds of electrons are $\frac{1}{2}g\mu H$ above and below their energies in absence of an applied field, μ being a constant and g a variable factor depending upon the interaction of the unpaired electron with the rest of the system. Thus, an energy transition of magnitude $h\nu = g\mu H$ is possible within the system, as illustrated in the second step of *Figure 1*. Clearly, absorption of radiation can take place at any frequency depending upon field strengths but high sensitivity is obtained at high field strengths and in practice the field strength and the

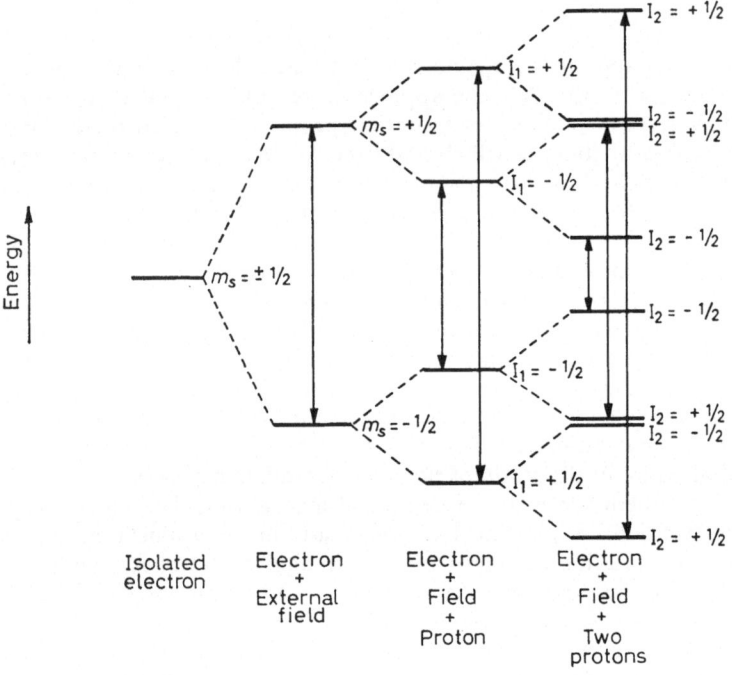

Figure 1. Hyperfine splitting of energy levels by protons

frequency of the applied radiation are only limited by the problems of instrument design. Resonance absorption could obviously be detected by maintaining the field constant and varying frequency or by maintaining constant frequency and scanning the field. For various reasons, the latter is preferred and all commercial instruments are based on this principle. Frequencies of the order of 10 000 Mc/s and magnetic fields up to 5000 gauss are used. The area under the absorption curve is directly proportional to the concentration of free radicals. It turns out that the value of g does not vary very much with the environment of the free radical so it is difficult to derive information about the nature of the radicals involved from the position of absorption in the e.s.r. spectrum as shifts in nuclear magnetic resonance spectra can give information about the environment of protons. The real source of information is the well-defined and characteristic hyperfine splitting which can occur.

If the unpaired electron finds itself in the environment of one atomic nucleus possessing spin, there will be interaction between their magnetic fields. In the case of a proton, which has a spin (I) of $\pm\frac{1}{2}$, the spins will be aligned, in the applied magnetic field, parallel or antiparallel to each other and the various energy levels may be represented as in the third step in *Figure 1*. Since the nuclear spins remain unchanged during the electron spin resonance transition, only the energy transitions indicated are possible and two peaks will appear in the e.s.r. spectrum. This is described as *hyperfine splitting*. The magnitude of the splitting is obviously a direct measure of the interaction between electron and nucleus and can thus often give a great deal of information about the location of a free radical centre.

When an unpaired electron is equally coupled with a second proton, the energy level diagram may be extended as illustrated in the fourth step in *Figure 1*. Three transitions are now clearly possible, which are represented by three equally spaced spectral lines with intensities in the ratio $1:2:1$. In general, for n equivalent protons the number of lines is $2nI + 1$ with intensities proportional to the coefficients of the binomial expansion. Very much more complicated spectra arise when, for example, the nuclei are not all magnetically equivalent with respect to the unpaired electron or when different kinds of nuclei are involved. A typical e.s.r. absorption spectrum is illustrated in *Figure 2a*. It refers to the $\cdot CH_2OH$ radical obtained by irradiating methanol. Spectra are usually represented, however, in the form of the first derivative as in *Figure 2b*.

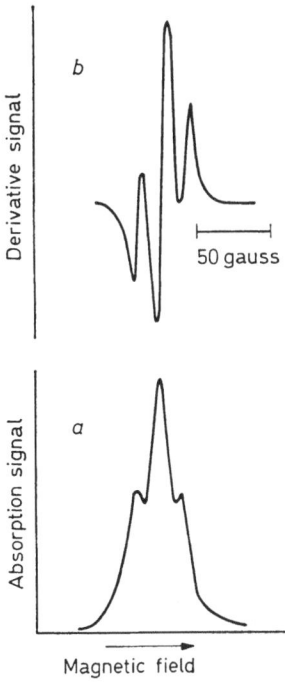

Figure 2. Electron spin resonance spectrum of irradiated methanol, after Fijimoto and Ingram[24] and Willard, Merritt and Dean[2]

391

The problems of detecting and identifying radicals in chemical reactions are intensified by reason of their low concentration, which is a direct result of their high reactivity. Concentrations can be increased by increasing the lifetime of the radicals and it is for this reason that a great many e.s.r. investigations have been carried out on radicals immobilized or trapped in the solid phase. Radicals are most readily formed in such an environment by high energy irradiation and indeed we can discern the birth of the modern interest in e.s.r. as being closely associated with the interest in the chemical effects of high energy radiation which developed rapidly after powerful sources of high energy radiation became available about 20 years ago. Part of this interest was strongly focused upon polymers and its relevance to the wider aspects of polymer degradation was obvious since at that time a number of degradation reactions had been shown to proceed by free-radical mechanisms. Since that time to the present, e.s.r. spectroscopy has demonstrated clearly the kinds of radicals which are to be expected in degrading polymers.

Evidence from e.s.r. spectra of γ-irradiated polystyrene, for example, gives support to the assumption that radicals with the structure A are

$$\sim\sim CH_2-\overset{\bullet}{C}-CH_2\sim\sim$$

A

involved in the photolysis and photo-oxidation of polystyrene[3]. Polystyrene irradiated at 77°K *in vacuo* gives a three-line spectrum, as in *Figure 3*[4].

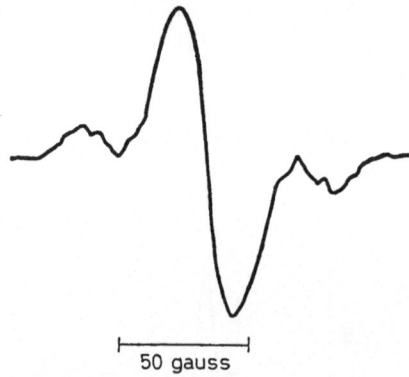

|————————————|
50 gauss

Figure 3. Electron spin resonance spectrum of polystyrene irradiated at 77°K and observed at 300°K, after Florin, Wall and Brown[4]

Abraham and Whiffen[5] believed it to be due to interaction of the unpaired electron in A with two of the four adjacent methylene protons. This was disputed by Tsvetkov, Molin and Voyevodskii[6] who believed that the radical centre is on the benzene ring. The problem was resolved by Florin, Wall and Brown[4], however, who demonstrated that neither deuterium substitution in the main chain nor substitution in the *meta-* and *para-*positions in the ring affect the spectrum but that *ortho-*substitution changes it completely (*Figure 4*). The only possible explanation of these observations is that the radical has the structure *A* and that the three-line spectrum is the result of interaction of the unpaired electron with the two *ortho-*protons. This spectrum is maintained on warming the polystyrene to room temperature although over a prolonged period changes occur which suggest the development of interaction of the unpaired electrons with the chain protons[7]. The nature of these changes is not precisely understood but seems worthy of re-examination in the light of what is now known about the mechanisms of thermal and photo-degradation of polystyrene.

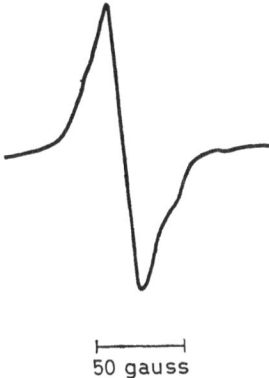

50 gauss

Figure 4. Electron spin resonance spectrum of poly(2,5-dichlorostyrene) irradiated at 300°K and observed at 300°K, after Florin, Wall and Brown[4]

The high-energy irradiation of polyethylene has been shown to result in three different radicals[8]. A sextet spectrum, associated with a relatively stable radical, has been attributed to the alkyl radical,

$$\sim\sim CH_2\text{---}\overset{\bullet}{C}H\text{---}CH_2\sim\sim$$

A more complicated spectrum, probably a septet of doublets, has been associated with the more stable allyl radical,

$$\sim\sim CH_2\text{---}\overset{\bullet}{C}H\text{---}CH\text{==}CH\sim\sim$$

Finally, a very stable radical with a singlet spectrum is believed to be polyenic

$$\sim\sim CH_2\text{---}\overset{\bullet}{C}H\text{---}(CH\text{==}CH)_n\text{---}CH_2\sim\sim$$

The alkyl radical predominates in polymer lightly irradiated (<1 Mrad) at low temperatures. With intermediate doses (10–100 Mrad) the allyl radical predominates while even higher doses give the polyene which is quite stable for long periods at room temperature. Grishina and Bakh[9] claim that, at dosages greater than 6000 Mrad, cyclic polyenic structures become involved.

These observations are relevant to more conventional polymer degradation studies because they clearly indicate that under the influence of high-energy irradiation, polyethylene can be made to eliminate hydrogen from series of adjacent ethylene units to form a polyene structure in very much the same way as poly(vinyl chloride) eliminates hydrogen chloride. Indeed, the mechanism in polyethylene must be strictly analogous to that proposed for poly(vinyl chloride)[10], the propagation steps of the chain process being,

for poly(vinyl chloride)

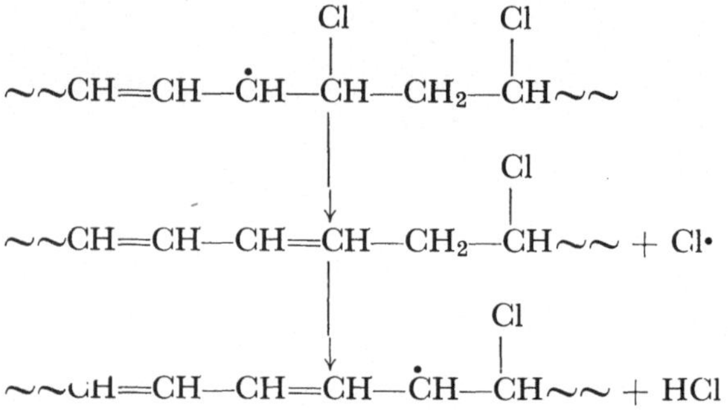

$$\sim\sim CH{=}CH{-}\overset{\bullet}{C}H{-}\underset{\underset{\displaystyle Cl}{|}}{C}H{-}CH_2{-}\underset{\underset{\displaystyle Cl}{|}}{C}H\sim\sim$$

$$\sim\sim CH{=}CH{-}CH{=}CH{-}CH_2{-}\underset{\underset{\displaystyle Cl}{|}}{C}H\sim\sim + Cl\bullet$$

$$\sim\sim CH{=}CH{-}CH{=}CH{-}\overset{\bullet}{C}H{-}\underset{\underset{\displaystyle Cl}{|}}{C}H\sim\sim + HCl$$

for polyethylene

$$\sim\sim CH{=}CH{-}\overset{\bullet}{C}H{-}CH_2{-}CH_2{-}CH_2\sim\sim$$

$$\sim\sim CH{=}CH{-}CH{=}CH{-}CH_2{-}CH_2\sim\sim + H\bullet$$

$$\sim\sim CH{=}CH{-}CH{=}CH{-}\overset{\bullet}{C}H{-}CH_2\sim\sim + H_2$$

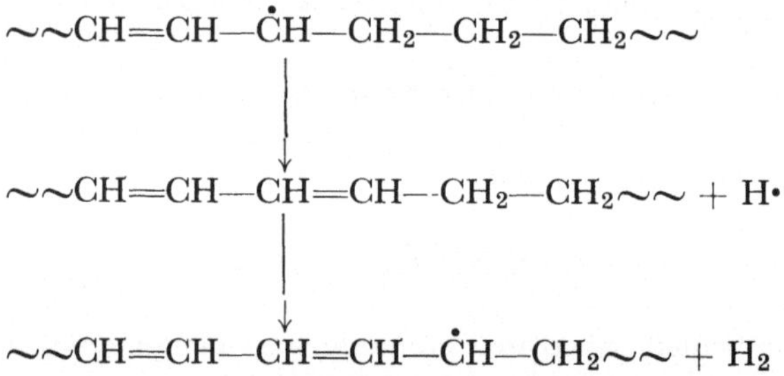

It is interesting that hydrogen is believed to be similarly eliminated from the chain backbone of polystyrene by ultraviolet irradiation[11],

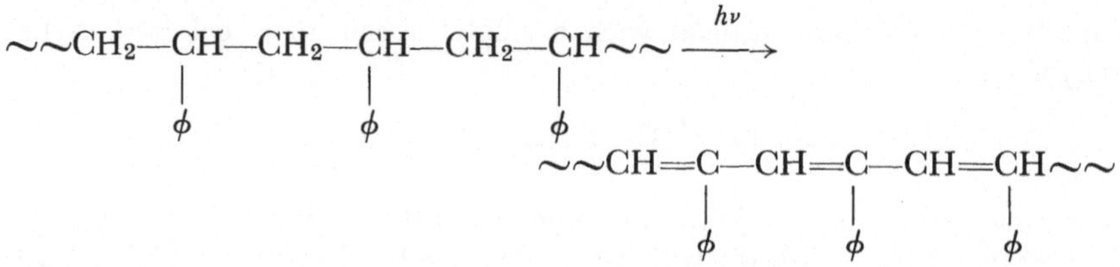

$$\sim\sim CH_2{-}\underset{\underset{\displaystyle \phi}{|}}{C}H{-}CH_2{-}\underset{\underset{\displaystyle \phi}{|}}{C}H{-}CH_2{-}\underset{\underset{\displaystyle \phi}{|}}{C}H\sim\sim \xrightarrow{h\nu}$$

$$\sim\sim CH{=}\underset{\underset{\displaystyle \phi}{|}}{C}{-}CH{=}\underset{\underset{\displaystyle \phi}{|}}{C}{-}CH{=}\underset{\underset{\displaystyle \phi}{|}}{C}H\sim\sim$$

This kind of reaction, which was regarded as an almost exclusive property of poly(vinyl chloride) and related chlorine-containing and vinyl-ester polymers, is now seen to be typical of a wider range of vinyl polymers. This series, poly(vinyl chloride), polystyrene, polyethylene is interesting to speculate upon from the general point of view of polymer degradation.

A great deal of attention has been given to radicals derived from poly-(methyl methacrylate). Nevertheless, profound differences of opinion exist about the interpretation of their e.s.r. spectra and this system serves to illustrate the care which will have to be taken in the application of this technique to the clarification of polymer degradation mechanisms. Radicals produced at room temperature by a variety of methods, including γ-ray, x-ray and u.v. irradiation and occluded during polymerization in gelled or precipitated polymers, give an identical nine-line spectrum [12-14] which is

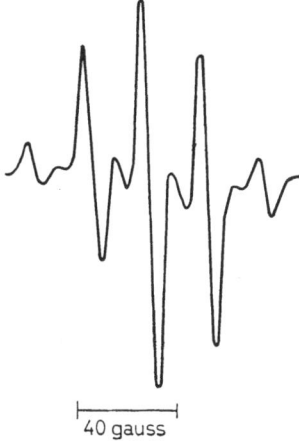

40 gauss

Figure 5. Electron spin resonance spectrum of poly(methyl methacrylate) irradiated and observed at room temperature, after Kourim and Vacek[15]

illustrated in *Figure 5*[15]. There has been general agreement that this is composed of overlapping five- and four-line spectra. The point in question is whether they arise from a single radical or from two different radicals. By comparison with the corresponding spectra of other methacrylates[13, 16] and of various deuterated poly(methyl methacrylates)[15] it has been deduced that the two overlapping spectra are derived from the single radical, *B*.

$$CH_3$$
$$|$$
$$\sim\sim CH_2-C\cdot$$
$$|$$
$$COOCH_3 \qquad B$$

The ester methyl protons are not involved but interaction of the five β protons with the unpaired electron is limited by rotational restrictions so that interaction can occur with only either three or four thus leading to the composite nine-line spectrum. The balance of present opinion favours this theory rather than Bullock and Sutcliffe's suggestion[17] that while the five-

395

line spectrum arises from the above alkyl radical, the four-line spectrum is due to the allyl radical

$$
\begin{array}{ccc}
CH_3 & & CH_3 \\
| & \cdot & | \\
CH_2{=}C{-}CH{-}C{-}CH_2{\sim}{\sim} & & \\
& | & \\
& COOCH_3 &
\end{array}
$$

Ambiguities which might arise in the application of e.s.r. data of this kind to degradation mechanism problems become evident when the spectra obtained by irradiation at different temperatures are compared. The evidence is that irradiation at $77°K$ causes scission of the ester group so that the rather complicated spectrum is due to the radical $\cdot COOCH_3$ and perhaps $\cdot CH_3$ and $\cdot CHO$[18]. As the temperature of the sample is raised the spectrum gradually becomes converted to the nine-line spectrum obtained by irradiation at a higher temperature. It is also interesting that, even at the higher temperatures, monomer is required to be present for the proper resolution of the nine-line spectrum[18, 19]. It seems, therefore, that while the nine-line spectrum of radical B will normally be detectable in degrading poly(methyl methacrylate) it must not be assumed that the initiation process is main-chain scission. Elimination of side chains is more likely, to be followed by a secondary decomposition process which results in the radical B.

With the development of the e.s.r. technique, and especially if its sensitivity can be increased, work of this kind, carried out on polymers in the act of degrading under the influence of heat, visible and ultraviolet radiation and during oxidation, clearly has great potential in the clarification of degradation mechanisms. However, as I said in the conclusion of my lecture in Prague two years ago, these modern analytical techniques will be most effective when they are so widely available that they can be used in combination with each other. Some recent studies on poly(vinyl chloride), in which e.s.r. and u.v. and visible spectroscopy have been applied in combination may be a signpost for the future.

A number of laboratories have investigated this system. The fact that their conditions of irradiation and subsequent observation of e.s.r. spectra vary from one to another, places a limitation on the close comparison of their results. It is nevertheless possible to see how e.s.r. spectral evidence can contribute to a more complete understanding of the degradation of poly-(vinyl chloride). The changes which occur in the e.s.r. spectrum as the temperature of previously irradiated polymer is raised are highly significant. Ohnishi and his coworkers[20] have obtained the series of spectra illustrated in *Figure 6*. The rather complicated spectrum (A) recorded at $-196°C$ after irradiation at $-196°C$ changes even on warming to $-78°C$ (B). By subtracting (B) from (A) they have obtained a well-defined six-line spectrum and therefore suggest that the changes which occur in this temperature range are due to the disappearance of the radical, $\sim\sim CH_2{-}\overset{\cdot}{C}H{-}CH_2\sim\sim$. During storage at $20°C$ the spectrum changes progressively to a well-defined singlet while colour develops. Loy[21] and Atchison[22], on the other hand, have been more interested in the changes in radical concentration which

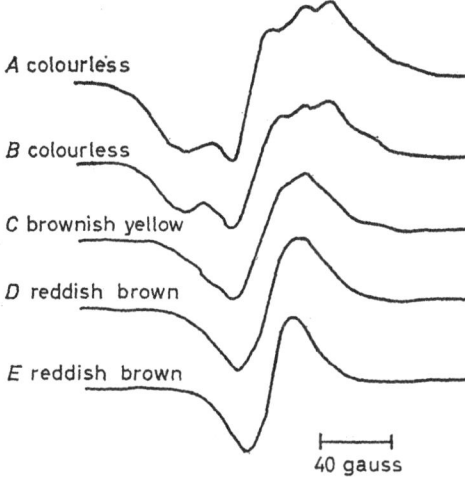

Figure 6. Change of electron spin resonance spectrum of poly(vinyl chloride), irradiated *in vacuo* at −196°C, with rising temperature, after Ohnishi, Nakajima and Nitta[20]: *A*, immediately after irradiation; *B*, after 15 hours at −78°C; *C*, after 5 minutes at 20°C; *D*, after 23 hours at 20°C; *E*, after 12 days at 20°C. Spectra recorded at −196°C

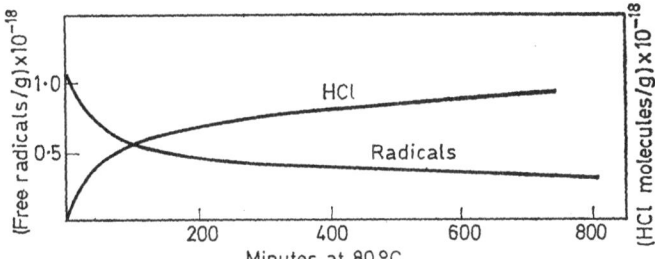

Figure 7. Change in free-radical concentration and HCl produced with time at 80°C of poly(vinyl chloride) previously irradiated at 77°K, after Loy[21]

occur at higher temperatures up to 80°C. They have shown firstly, as in *Figure 7*, that the decay in the concentration of free radicals at 80°C runs closely parallel with the evolution of HCl and secondly, as in *Figure 8*, that there are three distinct phases in the radical decay which can be attributed to three distinct types of radical. It is also clear from *Figure 8* that the amount of HCl produced per radical disappearing is characteristic of the radical. From this and other evidence, Loy has concluded that the rate of radical decay is a measure of the rate of initiation of the dehydrochlorination process and that the rate of evolution of HCl is a direct measure of the chain length in each case. From the data in *Figure 8* he derived chain lengths of 26, 51 and 126 for the short, medium and long lived radicals, respectively.

The development of colour, as well as the evolution of HCl, runs parallel with the change in radical concentration. Atchison[22] has been able to show

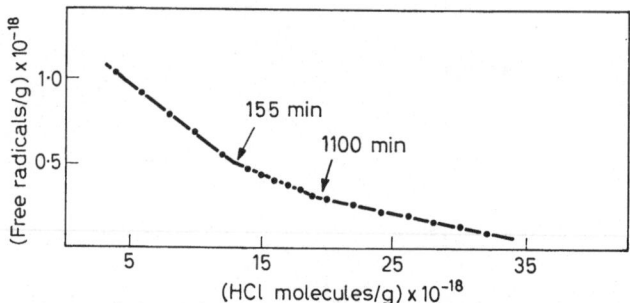

Figure 8. Relationship between free-radical concentration and HCl produced from poly(vinyl chloride) previously irradiated at 77°K, after Loy[21]

that the changes in the optical spectrum, illustrated in *Figure 9*, can be interpreted as a series of sharp peaks superimposed on a broad absorption band. The sharp peaks are clearly due to conjugated polyene sequences of various lengths and the magnitude of the broad band is closely related to the concentration of long lived radicals. While no firm conclusions about the

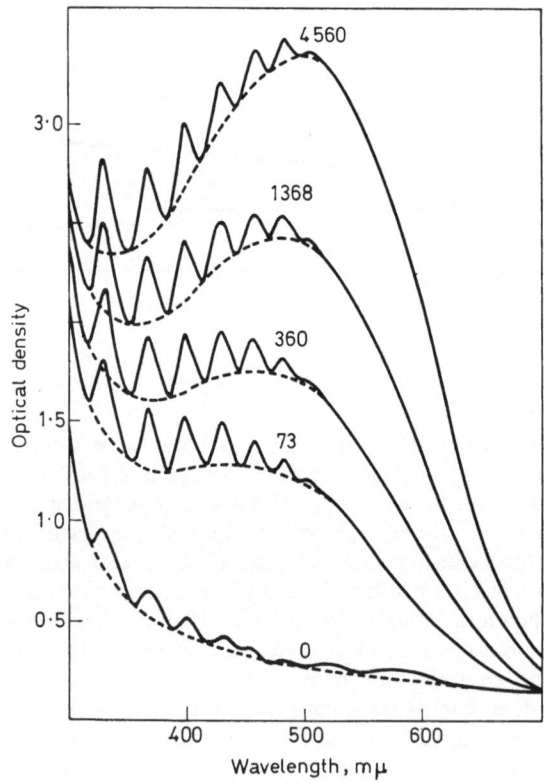

Figure 9. Development of colour in poly(vinyl chloride) during storage at 30°C after irradiation at room temperature, after Atchison[22]. Numbers on curves denote hours at 30°C

nature of the radicals involved or the precise mechanisms of these reactions are yet possible, these kinds of observations clearly illustrate the potential of e.s.r. in the polymer degradation field.

In most work which has been reported in which e.s.r. spectral data have been correlated with reactions in polymers the radicals have been produced at or below normal temperatures by means of high energy radiation and, as in the case of poly(vinyl chloride), it is the subsequent reactions of these radicals which have been studied. Although in many cases, and certainly in the case of poly(vinyl chloride), the overall characteristics of these reactions are similar to those which occur in thermal and photo-initiated degradations it is not at all certain that this is invariably so. Even in an apparently favourable case like poly(vinyl chloride) it is clear that important differences may exist between the mechanisms of the reactions brought about in these different ways. An immediate aim of e.s.r. spectroscopy in this field of study must, therefore, be to observe radicals during the course of thermally and photochemically induced degradation processes. There will be considerable experimental difficulty in carrying out degradation reactions at high temperature or under high light intensities within the spectrometer itself and simultaneously detecting relatively low radical concentrations, but if these problems can be overcome there is no doubt that vitally interesting information would be forthcoming. A great many of the conclusions which have been drawn about degradation processes from kinetic observations could be verified by direct identification of the radicals involved. I can probably best illustrate the kind of application I have in mind by devoting the remainder of this lecture to describing some recent work, the conclusions from which would be very much more convincing if they could be confirmed by e.s.r. data. This work concerns a comparison of the mechanisms of the thermal and photo–degradation of copolymers of methyl methacrylate and acrylonitrile[23].

The thermal reaction was studied at 280°C and the photo reaction at 160°C. As in pure poly(methyl methacrylate) the reactions consist basically of the depolymerization or "unzipping" of the polymer molecule to give monomer as the main product. Unlike the reaction in poly(methyl methacrylate) however, there is a rapid decrease in the molecular weight of the residual polymer. The presence of the acrylonitrile modifies the reaction in a different way in each case.

At the higher temperature of the thermal reaction the volatile product consists of a mixture of the two monomers although the proportion of acrylonitrile is less than in the copolymer. This demonstrates that the depolymerization process can pass through acrylonitrile units. Contrary to expectations the rapid decrease in molecular weight is not the result of scission at the acrylonitrile units. Indeed, *Figure 10* demonstrates that the rate of decrease of molecular weight is greater the lower the acrylonitrile content. The reaction in fact consists of random chain scission in the methyl methacrylate segments of the copolymer molecules to give radicals and this is followed by depolymerization which may pass through several acrylonitrile units before termination occurs, probably by interaction of pairs of radicals.

In the photo-reaction, on the other hand, the rate of bond scission,

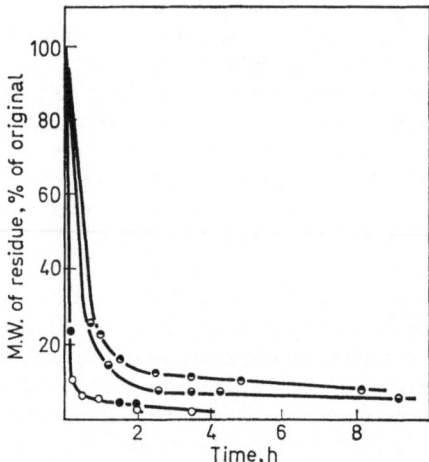

Figure 10. Changes in molecular weight with time, at 280°C, of methyl methacrylate/
acrylonitrile copolymers (○, 410/1; ●, 40/1; ◔, 16/1; ◑, 8/1)

calculated from the decrease in molecular weight, increases with acrylonitrile
content as illustrated in *Figure 11*. By contrast with the thermal reaction
acrylonitrile is not liberated in detectable amounts. Thus, the photo-
reaction consists of chain scission at or near the acrylonitrile units, followed
by depolymerization to the next acrylonitrile unit. It is clear from *Figure 11*
that while the rate of bond scission increases with acrylonitrile content it
does not increase nearly in proportion. In addition, the numbers of monomer

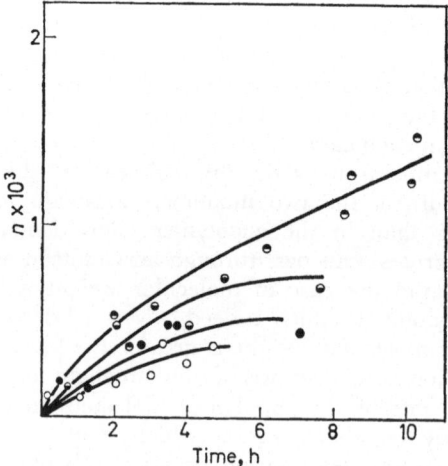

Figure 11. Chain scissions in methyl methacrylate/acrylonitrile copolymers as a function of
time of photo-degradation at 160°C (○, 410/1; ●, 40/1; ◔, 16/1; ◑, 8/1)

molecules produced per scission of the 410/1, 40/1, 16/1 and 8/1 copolymers are 1710, 728, 512 and 256 respectively, whereas one would expect maximum values of 820, 80, 32 and 16 if the reaction consisted only of depolymerization to the next acrylonitrile unit after each chain scission. Bearing in mind that the polymer at 160° is in the form of a highly viscous liquid, these observations have been accounted for in the following way. Once the broken ends have depolymerized to acrylonitrile units, the resulting acrylonitrile terminated radicals are virtually trapped and the closer they are together, that is the richer the copolymer is in acrylontrile, the greater will be the probability that recombination of the broken ends will occur. Thus, in the 8/1 copolymer in particular the effect will be that few of the original scissions are permanent. In copolymers containing progressively less acrylonitrile, the effect of depolymerization will be to bring the radicals further from each other and thus to make the probability of recombination progressively less. These are the essential facts although there is a great deal of additional information in support of these mechanisms for the thermal and photoreactions.

It is clear that if e.s.r. measurements could be made during these two degradation processes it is possible that they might supply the ultimate proof that these deductions, made principally on the basis of rate and molecular-weight measurements, are correct or, alternatively, that they must be modified. In the thermal reaction, for example, one would expect to obtain an e.s.r. spectrum typical of the methyl methacrylate radical. It might be the nine-line spectrum already discussed and which is composed of overlapping four- and five-line spectra or possibly a simpler six-line spectrum due to the five β protons becoming equivalent as a result of the greater mobility of the methyl and methylene groups in the relatively much less viscous polymer at the higher temperature of the thermal reaction. In either case, the spectrum would be expected to be fundamentally different from that obtained during the photo-initiated reaction which should reflect a higher concentration of acrylonitrile type radicals,

$$
\sim\!\!\sim\!CH_2\!-\!\!\!\begin{array}{c} CN \\ | \\ \overset{\displaystyle |}{C}\cdot \\ | \\ H \end{array}
$$

This spectrum will probably be rather complicated. Ideally, interaction with the nitrogen atom should be expected to give three lines of equal intensity. These might be split into triplets by interactions with the methylene protons, each line of which might be further split into a doublet by interaction with the α proton, giving 18 lines in all.

If the experimental difficulties can be overcome such that e.s.r. spectroscopy can be applied to the solution of problems of this kind, then within the next few years, this technique could become one of the most vitally important in allowing rapid new advances to be made in our understanding of polymer degradation processes.

References

[1] N. Grassie. *Pure Appl. Chem.* **12**, 237 (1966).

[2] H. H. Willard, L. L. Merritt, and J. A. Dean. *Instrumental Methods of Analyses*, Chap. 7, Van Nostrand, Princeton, 1965.

[3] N. Grassie and N. A. Weir. *J. Appl. Polymer Sci.* **9**, 975, 987 (1965).

[4] R. E. Florin, L. A. Wall and D. W. Brown. *Trans. Faraday Soc.* **56**, 1304 (1960).

[5] R. T. Abraham and D. H. Whiffen. *Trans. Faraday Soc.* **54**, 1291 (1958).

[6] I. D. Tsvetkov, I. N. Molin, and V. V. Voyevodskii. *Vysokomol. soyed* **1**, 1805 (1959). [*Pol. Sci. U.S.S.R.* **2**, 165 (1961)].

[7] R. E. Florin, L. A. Wall and D. W. Brown. *J. Polymer Sci.* A **1**, 1521 (1963).

[8] S. Ohnishi, S. Sugimoto, and I. Nitta. *J. Polymer Sci.* A. **1**, 605 (1963).

[9] A. D. Grishina and N. A. Bakh. *Vysokomol soyed* **7**, 1698 (1965). [*Pol. Sci. U.S.S.R.* **7**, 1871 (1965).]

[10] R. R. Stromberg, S. Straus, and B. G. Achhammer. *J. Polymer Sci.* **35**, 355 (1959).

[11] N. Grassie and N. A. Weir. *J. Appl. Polymer Sci.* **9**, 999 (1965).

[12] D. J. E. Ingram, M. C. R. Symons, and M. G. Townsend. *Trans. Faraday Soc.* **54**, 409 (1958).

[13] R. J. Abraham, H. W. Melville, D. W. Ovenall and D. H. Whiffen. *Trans. Faraday Soc.* **54**, 1133 (1958).

[14] N. M. Atherton, H. W. Melville, and D. H. Whiffen. *Trans. Faraday Soc.* **54**, 1300 (1958).

[15] P. Kourim and K. Vacek. *Trans. Faraday Soc.* **61**, 415 (1965).

[16] D. W. Ovenall. *J. Polymer Sci.* **41**, 199 (1959).

[17] A. T. Bullock and L. H. Sutcliffe. *Trans. Faraday Soc.* **60**, 625 (1964).

[18] Y. Hajimoto, N. Tamura, and S. Okamoto. *J. Polymer Sci.* A **3**, 255 (1965).

[19] I. S. Ungar, W. B. Gager, and R. I. Leininger. *J. Polymer Sci.* **44**, 295 (1960).

[20] S. Ohnishi, Y. Nakajima, and I. Nitta. *J. Appl. Polymer Sci.* **6**, 629 (1962).

[21] B. R. Loy. *J. Polymer Sci.* **50**, 245 (1961).

[22] G. J. Atchison. *J. Appl. Polymer Sci.* **7**, 1471 (1963).

[23] N. Grassie and E. Farish. *European Polymer J.* **3**, 619, 627 (1967).

[24] M. Fijimoto and D. J. E. Ingram. *Trans. Faraday Soc.* **54**, 1305 (1958).

SYNTHESES ET STRUCTURES NOUVELLES DE POLYMERES

PAUL REMPP

Centre de Recherches sur les Macromolécules, Strasbourg, France

I. INTRODUCTION

Les travaux de synthèse en chimie macromoléculaire ont eu longtemps pour objectif principal la préparation de composés homopolymères à chaîne linéaire, de structure bien établie et de faible polydispersité. Ces polymères, chimiquement bien définis et d'homogénéité satisfaisante ont servi à la mise au point de techniques de caractérisation moléculaire et structurale, et l'étude de leurs propriétés tant en solution diluée qu'à l'état solide a conduit à de nombreux résultats fondamentaux.

L'idée d'associer deux monomères dans une même macromolécule est cependant déjà ancienne. La copolymérisation, par voie radicalaire, de deux composés insaturés conduit à la formation d'un copolymère statistique; la composition instantanée de celui-ci dépend des proportions de chacun des constituants dans le mélange de monomères et des rapports de réactivité radicalaires bien connus[1]. Lorsqu'elles sont menées à des taux de conversion élevés, ces copolymérisations conduisent en général à des produits hétérogènes en composition[2] et polydispersés en masse, mais dont les propriétés ne diffèrent pas de façon déterminante de celles des homopolymères correspondants. Il faut souligner que dans un tel coplymère la probabilité de trouver des séquences homopolymères de quelque longueur est quasiment nulle.

Bien différent est le cas des copolymères constitués à dessein de séquences homopolymères relativement longues, reliées les unes aux autres par des liaisons homopolaires. Les propriétés de ces composés diffèrent sensiblement de celles des homopolymères parents, et c'est ce qui explique l'intérêt que manifestent en leur faveur les chercheurs de nombreux laboratories universitaires et industriels. Notre propos sera de passer en revue les méthodes mises au point au cours de ces dernières années en vue la préparation de copolymères greffés et séquencés. Ces deux types de produits ne différent que par la position relative des séquences homopolymères, les copolymères séquencés étant essentiellement linéaires, alors que les copolymères greffés sont ramifiés.

Les homopolymères à chaîne ramifiée suscitent depuis peu un regain d'intérêt, principalement comme terme de comparaison ou comme produits de référence; nous passerons également en revue les méthodes de préparation d'homopolymères ramifiés à structure définie, en "peigne" ou en "étoile". Remarquons dès à présent qu'un homopolymère en peigne n'est autre qu'un copolymère greffé dans lequel squelette et greffons sont de même nature chimique.

403

O

La préparation de copolymères contenant des séquences homopolymères ne peut s'effectuer que de deux manières simples :

(1) On peut créer, sur une chaîne polymérique des sites promoteurs, susceptibles d'amorcer la polymérisation d'un deuxième monomère. Si ces sites sont situés aux extrémités de la chaîne, il y a formation de copolymères séquencés. S'ils sont formés en d'autres positions, le long de la chaîne, la réaction donnera lieu à la formation d'un copolymère greffé.

(2) On peut également avoir recours à une réaction chimique entre fonctions portées par deux chaînes macromoléculaires différentes. Ici encore, si les fonctions réactives sont situées exclusivement en bout de chaîne, on obtient un copolymère séquencé et dans le cas contraire on assiste à un processus de greffage. Mais il convient de se rappeler que si des réactants macromoléculaires possèdent l'un et l'autre plus de deux sites réactifs, la réaction peut conduire à une gélification par formation d'un réseau tridimensionnel.

La polymérisation successive de deux monomères sous l'action d'un promoteur radicalaire conduit en général à un mélange d'hompolymères car la durée de vie d'un site radicalaire est extrêment réduite[3]. Il en va tout autrement des polymérisations anioniques en phase homogène, selon la technique des "polymères vivants", qui sont caractérisées par l'absence de toute réaction de transfert et de terminaison spontanées[4,5]. La mise au point de ces méthodes de polymérisation a de ce fait constitué un progrès décisif pour la synthèse de copolymères séquencés et greffés. Cependant les polymérisations anioniques ne s'appliquent qu'à un nombre limité de monomères et de ce fait les autres méthodes demeurent utiles, et elles feront l'objet de la première partie de cet exposé. Nous indiquerons ensuite comment et à quels systèmes les techniques de polymérisation anioniques peuvent s'appliquer.

II. METHODES DE SYNTHESE PROCEDANT PAR VOIE RADICALAIRE

A. Création de sites promoteurs sur une chaîne

Pour créer sur une chaîne polymérique des sites réactifs, susceptibles de servir par la suite de promoteurs radicalaires pour la polymérisation d'un deuxième monomère, on peut procéder de diverses manières. Une revue détaillée de ces méthodes de préparation a été publiée en 1960 par Smets et Hart[6], et nous nous bornerons donc ici à les résumer, en donnant quelques exemples récents.

1. Promoteur bifonctionnel

On peut avoir recours à promoteur bifonctionnel, susceptible d'amorcer la polymérisation d'un premier monomère dans des conditions bien déterminées. La chaîne macromoléculaire porte alors à son extrémité un site promoteur, qui peut servir ultérieurement à la polymérisation d'un 2e. monomère et conduire ainsi à la formation d'un copolymère séquencé. Cette méthode a été employée avec succès notamment par Molyneux[7] et par Burnett[8] (promoteur: dihydroperoxyde de *m*-di-*iso*propylbenzène) et par Van Beylen et Smets[9] (promoteur: azo-bis-cyanopervaleriate de *t*-butyle).

$$HO-O-\underset{\underset{CH_3}{|}}{\overset{\overset{CH_3}{|}}{C}}-\!\!\!\!\!\!\bigcirc\!\!\!\!\!\!-\underset{\underset{CH_3}{|}}{\overset{\overset{CH_3}{|}}{C}}-OOH$$

$$[CH_3]_3C-O-O-\overset{\overset{O}{\|}}{C}-[CH_2]_2-\underset{\underset{CN}{|}}{\overset{\overset{CH_3}{|}}{C}}-N\!=\!N-\underset{\underset{CN}{|}}{\overset{\overset{CH_3}{|}}{C}}-[CH_2]_2-\overset{\overset{O}{\|}}{C}-O-O-C[CH_3]_3$$

2. Méthodes chimiques

Divers auteurs ont mis au point des méthodes chimiques permettant de doter des chaînes macromoléculaires de fonctions perester[10,11] ou peranhydride[12], et ont utilisé ces fonctions, dans une 2e. étape, comme promoteurs en vue de la préparation de copolymères greffés.

3. Colorants et photosensibilité

La photopolymérisation de monomères en présence d'un colorant agissant comme photosensibilisateur, étudiée par Oster[13] a également été appliquée par Smets[14] à la préparation de copolyméres greffés: le colorant— l'éosine—est préalablement fixé sur une chaîne de polyvinylamine.

4. Fonctions peroxydiques ou hydroperoxydiques

L'utilisation de sites peroxydiques ou hydroperoxydiques est fort répandue. Nous avons déjà cité l'utilisation comme promoteur du dihydroperoxyde de *m*-di-*iso*propylbenzène[7,8] qui conduit à des chaînes dotées de fonctions hydroperoxyde terminales, utilisables par la suite pour la polymérisation redox d'un 2e. monomère. Des fonctions hydroperoxyde terminales ont été également obtenues par Riess et Banderet par dégradation peroxydante de polystyrène ou de poly*iso*butène et utilisées pour la préparation de copolymères séquencés[15]. La peroxydation de polypropylène atactique en solution conduit à des fonctions hydroperoxyde utilisables selon Wallace et Hadley[16] pour la préparation de copolymères greffés. Hulahan, Stivala et Levi[17] ont opéré de manière semblable pour réaliser le greffage de méthacrylate d'éthyle sur du poly-*p*-*iso*propylstyrène hydroperoxydé.

L'irradiation γ, en présence d'oxygène, conduit également à la formation de sites hydroperoxydes, répartis au hasard le long de la chaîne macromoléculaire. Cette méthode a été appliquée avec succès au greffage de polyacrylontrile sur du polyéthylène et sur du chlorure de polyvinyle par Chapiro et ses collaborateurs[18,19], et par Turska[20] au greffage de polyacétate de vinyle sur du polyméthacrylate de méthyle.

5. Irradiation γ

L'irradiation γ de polymères gonflés par un 2e. monomère peut également être réalisés en atmosphère inerte. L'irradiation a pour effet la création de sites radicalaires sur le polymère et le greffage peut s'effectuer par amorçage, par transfert ou par recombinaison de radicaux. Le greffage de polystyrène sur du polytétrafluoroéthylène a été réalisé ainsi[21].

Tous les exemples cités jusqu'ici comportent la formation de sites réactifs sur une chaîne macromoléculaire, sites qui amorcent ensuite la polymérisation d'un 2e. monomère.

B. Réactions entre macromolécules

On peut également citer quelques cas de réaction entre macromolécules, conduisant à la formation de copolymères greffés ou séquencés.

1. Transfert

Le cas le plus évident est celui où le greffage résulte d'une réaction de transfert au polymère. Tel est le cas de la polymérisation radicalaire de l'acétate de vinyle, effectuée en présence de polyoxyéthylène, qui conduit selon Kahrs et Zimmermann[22] à un greffage partiel du polyacétate de vinyle sur la chaîne de polyoxyéthylène. D'autres systèmes ont été étudiés, notamment par Kobryner et Banderet[23].

2. Transestérification

La synthèse de copolymères greffés par transestérification de polyméthacrylate de méthyle avec des polyesters, qui a été réalisée par Kolesnikov et Tsen Khan Ming[24] comporte également une réaction entre deux macromolécules.

Toutes ces méthodes de synthèse de copolymères greffés ou séquencés procédant par voie radicalaire conduisent à des produits polydispersés en masse et en composition et contenant en général une proportion importante d'homopolymères. Elles exigent donc des techniques de séparation, de purification, voire de fractionnement parfaitement au point, et un contrôle rigoureux de l'homogénéité du produit purifié. Ce sont par ailleurs des méthodes de laboratoire, et sauf rares exceptions, ces méthodes ne sont pas utilisables industriellement pour la préparation de copolymères à séquences.

III. METHODES PROCEDANT PAR VOIE IONIQUE

L'avènement en 1956 des techniques de polymérisation anionique en milieu polaire et aprotique[4] a constitué un progrès décisif pour la préparation de copolymères séquencés, et ce pour deux raisons essentielles:

—l'absence de toute réaction de désactivation spontanée: les chaînes polymériques demeurent dotées d'extrémités organo-métalliques très faiblement ionisées, mais réactives[25,26].

—la faible polydispersité des échantillons[27,28], imputable à la simplicité du mécanisme réactionnel et à la rapidité de l'équilibre paire d'ion $\rightleftarrows$ ions libres.

Nous examinerons dans ce qui suit l'application des processus de polymérisation anionique à la préparation de copolymères séquencés et de copolymères greffés.

1. Copolymères séquencés

Dès ses premiers travaux Szwarc[4] avait montré que la polymérisation de monomères vinyliques comme le styrène est rapide en milieu THF et que l'addition d'une nouvelle quantité de monomère conduit à une augmentation proportionnelle de la masse du polymère. Il suffit donc, théoriquement, de choisir un promoteur actif, qui réagisse rapidement et complètement avec un premier monomère, en amorçant la polymérisation de celui-ci, et d'ajouter à la solution du "polymère vivant" obtenu un 2e. monomère pour obtenir un copolymère séquencé. En fait, c'est bien ainsi qu'ont été préparés des copolymères séquencés styrène–isoprène[29] et styrène–oxyde d'éthylène[30] par Szwarc et ses collaborateurs, et dans d'autres laboratoires[31].

Mais cette méthode n'est cependant pas universelle, même parmi les les monomères polymérisables par voie anionique. Il faut en effet que la réaction d'amorçage de la polymérisation du 2e. monomère soit non seulement possible, mais encore rapide par rapport à la réaction de propagation subséquente. Ceci implique que le 2e. monomère possède un caractère plus nettement électrophile que le premier[32,33]. C'est ainsi que le carbanion styryle peut amorcer la polymérisation du méthacrylate de méthyle, mais l'inverse n'est pas vrai[34].

De très nombreux copolymères séquencés ont été préparés par voie anionique, soit en milieu tétrahydrofuranne avec des promoteurs organométalliques variés, soit en solution dans le toluène ou le benzène, le promoteur étant alors en général le butyl-lithium. Nous citerons quelques exemples:

—La préparation de copolymères séquencés styrène–butadiène et styrène–isoprène a fait l'objet de nombreux travaux[35–39] et connaît dès à présent des développements industriels. C'est l'un des rares systèmes où les 2 monomères possèdent des électroaffinités comparables: chaque monomère réagit sur le carbanion de l'autre.

—Les copolymères séquencés styrène–méthacrylate de méthyle ont fait l'objet de recherches pousées. Le carbanion styryle peut réagir non seulement sur la double liaison du méthacrylate de méthyle, mais également sur sa fonction ester, ce qui entraîne des complications. Pour éliminer cette réaction parasite, il suffit de faire réagir d'abord du diphényl-1,1 éthylène, ce qui donne naissance à un carbanion diphényl-alkyle moins réactif; ce dernier amorce cependant rapidement la polymérisation du méthacrylate de méthyle. On obtient aisni des copolymères séquencés exempts d'homopolymères, homogènes en composition, et de faible polydispersité[40,41].

—De nombreux autres copolymères séquencés ont pu être obtenus parmi lesquels nous citerons les systèmes polystyrène-polyvinyl-2 pyridine[42]; polystyrène-poly-p-bromostyrène; polystyrène-polyvinylnaphtalène; polyméthacrylate de méthyle-polyacrylonitrile[43].

—Dans le domaine des monomères cycliques, outre l'oxyde d'éthylène déjà mentionné, signalons les travaux de Sigwalt et Boileau[44] qui ont permis à ces auteurs de préparer des copolymères séquencés polystyrène-polysulfure de propylène et polysulfure de propylène-polysulfure d'éthylène[45].

D'autres recherches portent sur la copolymérisation séquencée anionique de styrène et d'octaméthylcyclotétrasiloxane, et ont conduit à des résultats prometteurs[46].

Avant de clore cette énumération il convient de rendre compte brièvement d'une méthode originale de synthèse d'un copolymère séquencé, méthode basée sur la réaction de désactivation réciproque d'un polystyrène "vivant" préparé par voie anionique et d'un polytétrahydrofuranne "vivant" préparé par voie cationique. Les réactions de polymérisation cationique sont sujettes à des réactions d'arrêt par transfert ou par désactivation, ce qui explique qu'elles ne se prêtent pas, sauf exception[47a], à la synthèse de copolymères séquencés ou greffés. Cependant, en utilisant le promoteur de Merwein dans le THF lui-même—car sa polymérisation n'affecte qu'une partie des molécules—M. Levy[47] a réussi une désactivation réciproque de polystyrène anionique et de poly THF cationique, et le produit de la réaction est effectivement un copolymère séquencé. Cette méthode originale de préparation pourrait être avantageusement étendue à d'autres systèmes dans lesquels une séquence polystyrène est associée à une séquence de motifs polymérisables par voie cationique seulement.

2. Procédés de greffage anionique

Tout comme dans le cas des méthodes procédant par voie radicalaire, le greffage anionique peut s'effectuer de deux manières différentes:
—par réaction d'une chaîne macromoléculaire ω-fonctionnelle sur les fonctions électrophiles portées par une autre chaîne macromoléculaire: c'est le greffage par désactivation carbanionique.
—par création, sur une chaîne polymérique de sites organométalliques et utilisation de ces derniers comme promoteurs pour la polymérisation "latérale" d'un 2e. monomère: c'est le greffage par amorçage carbanionique.

Nous examinerons successivement ces 2 procédés de greffage par voie anionique en phase homogène, qui exigent tous deux que le milieu réactionnel soit rigoureusement aprotique.

A. Greffage par désactivation carbanionique

Il est établi depuis longtemps que les fonctions organo-métalliques terminales d'un polymère "vivant" peuvent réagir non seulement sur tout composé donneur de proton, mais encore avec divers composés à caractère électrophile[25,26]: halogénures d'alkyle, chlorures d'acides, anhydrides, esters, anhydride carbonique, etc. . . . Ces réactions conduisent à la formation de sites fonctionnels terminaux variés:

$$\sim CH_2-\underset{\underset{\varphi}{|}}{CH^-} + \underset{\underset{}{|}}{\overset{\overset{Cl}{|}}{CH_2}}-CH_3 \rightarrow \sim CH_2-\underset{\underset{\varphi}{|}}{CH}-CH_2-CH_3 + Cl^-$$

$$\sim CH_2-CH^- + \overset{\overset{\displaystyle O}{\|}}{\underset{\underset{\displaystyle O-CH_3}{|}}{C}}-CH_3 \rightarrow \sim CH_2-\underset{\underset{\displaystyle \varphi}{|}}{CH}-CO-CH_3 + CH_3O^-$$

Si les fonctions halogénures, ou ester, ou anhydride sont elles-mêmes portées par une chaîne macromoléculaire, la réaction de désactivation doit conduire à la fixation d'une chaîne polymérique sur l'autre, c'est-à-dire au greffage.

Le greffage de polystyrène "vivant" sur le polyméthacrylate de méthyle a été étudié en détail par Gallot[48-50], et les produits obtenus ont pu être caractérisés de façon satisfaisante, car on peut déterminer séparément les masses moléculaire du squelette, du greffon individuel et du copolymère greffé.

La proportion de fonctions ester affectées par la réaction de greffage est faible ($<10\%$), ce qui n'empêche pas de préparer des produits de masse moléculaire élevée portant un grand nombre de greffons[50]. La séparation du polystyrène non greffé ne pose pas de problèmes particuliers.

Cette technique de greffage, si elle présente des avantages évidents—distribution statistique des greffons[51], identité des greffons, possibilité de caractérisation complète des échantillons—n'est malheureusement applicable qu'à un nombre limité de systèmes, parmi lesquels nous citerons: le greffage de polystyrène sur chlorure de polyvinyle[52]; le greffage de polystyrène sur la polyvinylpyridine[53], qui ne peut être réalisé qu'à basse température; le greffage de polystyrène sur des xylanes[54,55], effectué après méthylation totale par O'Malley et Marchessault. Signalons d'autre part les travaux de Korotkov[56] et de ses collaborateurs, sur le système styrène-méthacrylate de méthyle, dont les résultats corroborent ceux obtenus à Strasbourg.

B. Greffage par amorçage carbanionique

Pour créer des sites organo-métalliques sur une chaîne polymérique il faut que celle-ci soit dotée
—ou d'atomes d'hydrogène acides, aisément remplaçables par des atomes métalliques,
—ou de fonctions susceptibles de réagir avec un composé organométallique,
—ou de noyaux aromatiques pouvant donner naissance à des complexes radical–ioniques.

Mais il faut en outre que la chaîne polymérique ne contienne aucune fonction électrophile incompatible avec les sites organométalliques.

Nous donnerons des exemples de chacun des types de réactions de métallation et des réactions de greffage auxquels ces polymères métallés peuvent donner lieu, s'ils sont mis en présence d'un monomère adéquat.

(a) Métallation par remplacement d' hydrogène.
Le polydiphényl-propène[57] et le polyvinylfluorène[58] peuvent tous deux être métallés selon la méthode de Normant et Angelo[59]; l'agent de métallation est le naphtalène-sodium, ce qui permet à la réaction de se dérouler en milieu homogène.

$$-CH_2-CH- \xrightarrow{\text{(napht)}^-Na^+} -CH_2-CH-$$

$$| \qquad\qquad\qquad\qquad |$$

$$CH \qquad\qquad\qquad\qquad C$$

$$/\ \backslash \qquad\qquad\qquad /\ |\ \backslash$$

$$\varphi \qquad \varphi \qquad\qquad\qquad \varphi \ \ Na \ \varphi$$

Les sites métallés ainsi formés peuvent amorcer la polymérisation de styrène, d'isoprène, de vinylpyridine, de méthacrylate de méthyle, d'acrylonitrile, d'oxyde d'éthylène et conduire ainsi à la formation de produits greffés. Toutefois l'amorçage de la polymérisation du styrène est lent et le nombre de greffons formés est inférieur à celui des sites métallés sur la chaîne.

(b) La métallation d'une chaîne macromoléculaire par réaction de substitution ou d'addition avec un composé organométallique a également été étudiée dans divers laboratoires. Plusieurs exemples de réaction de ce type ont été donnée par Greber et ses collaborateurs[60,61], et des copolymères greffés ont été préparés à partir de polyvinylbenzophénone-sodium et à partir de polyacrylonitrile préalablement traité par du butyl-lithium.

La métallation de poly-p-bromostyrène a été réalisée par action de naphtalène-lithium—qui ne joue pas dans ce cas le rôle d'un réducteur[62]—et les motifs poly-p-lithiostyrène ainsi obtenus peuvent amorcer de nombreuses polymérisations vinyliques dans des conditions satisfaisantes[63]. La métallation et le greffage du poly-p-chlorostyrène ont été étudiés également[64].

Dans la même catégorie on peut ranger également les travaux de Zilkha et al.[65,66] qui se sont attachés à métaller des dérivés de la cellulose, soit par Na en milieu ammoniac liquide, soit par la naphtalène–sodium en milieu dioxane. Les sites alcoolates qu'ils ont obtenus leur ont permis d'amorcer la polymérisation de certains monomères cycliques tels que les anhydrides de Leuchs, et de l'acrylonitrile, dont l'électroaffinité est telle qu'un alcoolate suffit à en faire démarrer la polymérisation.

(c) La formation de complexes radical–ioniques par action d'un métal alcalin sur un polymère dont le substituant est un hydrocarbure aromatique à noyau condensé: naphtalène, acénaphtylène, biphényle, anthracène, et même diphénylanthracène, a été réalisée dans divers laboratoires[67–72], mais il a été établi que cette réaction s'accompagne assez fréquemment de ruptures parasites de la chaîne[68,69]. Toutefois il n'est pas possible d'utiliser les sites radical-ioniques formés pour la synthèse de copolymères greffés, car l'amorçage de la polymérisation des monomères vinyliques s'opère, dans ce cas, par transfert électronique et non par addition: la polymérisation s'effectue normalement mais il ne s'établit pas de liaison chimique entre le polymère formé et le squelette porteur des sites radical-ioniques[70]. Cependant la polymérisation de l'oxyde d'éthylène constitue une exception, car elle se produit par addition et des copolymères greffés polyvinylnaphtalène-POE ont pu être obtenus[69,71].

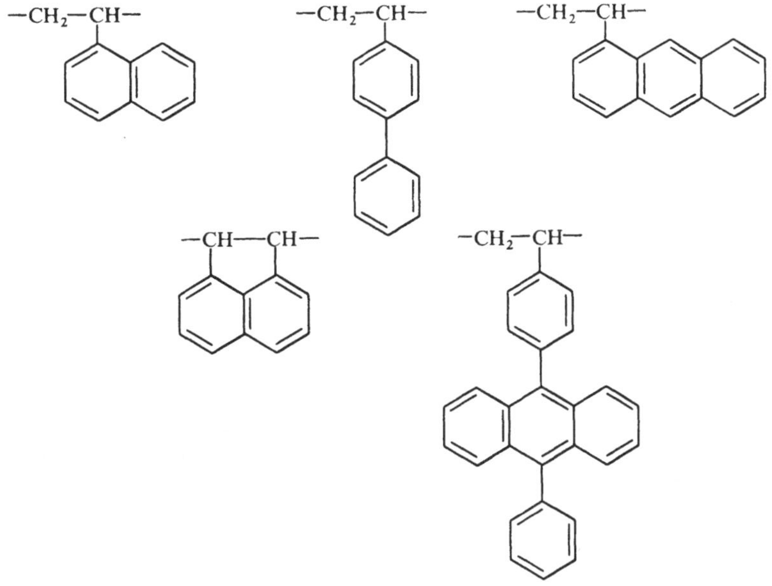

(d) Formation de catalyseurs Zeigler–Natta—Enfin, pour que l'énumération des méthodes de greffage soit complète, il convient de signaler les essais qui ont été réalisés en vue de doter des chaînes macromoléculaires de sites promoteurs de type Ziegler–Natta. On peut notamment chlorométhyler du polystyrène et faire réagir le produit avec de l'hydrure de diéthyl-aluminium; l'addition de $TiCl_4$ permet d'amorcer la polymérisation de l'éthylène ou du propylène avec semble-t-il un greffage partiel des chaînes formées sur le squelette de polystyrène. D'autres essais du même genre ont été effectués par Greber et ses collaborateurs[73] et ont également conduit à des résultats encourageants.

IV. HOMOPOLYMERES A RAMIFICATION CONTROLEE

Parmi les homopolymères préparés par voie radicalaire, certains sont plus ou moins ramifiés, les ramifications étant dues principalement à des réactions de transfert sur la chaîne macromoléculaire elle-même, au cours de la polymérisation. Cependant il est très difficile de préparer, par voie radicalaire, des homopolymères dont le degré de ramification soit connu avec précision. Dans ce domaine encore, les méthodes de polymérisation sans terminaison, par voie anionique, se sont révélées précieuses, car elles ont permis de préparer des polymères de structure bien définie, dont le nombre et la longueur des branches peuvent être déterminés avec une précision satisfaisante.

1. Polymère à structure en "peigne"

La préparation de polymères en peigne, dont le squelette et les greffons sont de même nature chimique peut être réalisée selon les méthodes utilisées pour la préparation de copolymères greffés, à condition de remplacer le

411

squelette homopolymère par un copolymère statistique contenant une faible proportion du monomère électrophile, ou de sites promoteurs, respectivement.

(a) Méthodes procédant par désactivation carbanionique

Des polystyrènes en peigne ont été préparés par réaction de polystyrène vivant sur les fonctions ester portées par un copolymère statistique styrène-méthacrylate de méthyle, contenant environ 10% de ce dernier monomère. Le greffage affecte de 60 à 80% des fonctions esters présentes et la teneur en fonction ester du produit brut est de l'ordre de 1 à 2%[74]. On peut donc considérer ce produit comme un homopolystyrène; mais il faut se rappeler que la jonction entre greffons et squelette est assurée par une fonction carbonyle, ce qui explique la photosensibilité de ces produits.

La chlorométhylation partielle de polystyrènes permet également de doter un squelette de groupes électrophiles, sur lesquels peuvent venir se greffer des polystyrènes "vivants"[75]. Dans les 2 cas les polymères "en peigne" obtenus sont souillés de polystyrène non fixé, de masse plus faible, et dont l'élimination quantitative ne pose pas de problème, et peut être contrôlée par GPC[76]. La détermination du nombre de greffons est possible car le polymère en peigne, le greffon et le squelette peuvent être caractérisés séparément Notons d'autre part que les greffons sont tous de taille comparable.

(b) Méthodes procédant par amorçage carbanionique

Si on traite un copolymère statistique styrène-p-bromostyrène contenant environ 10% de ce monomère par du naphtalène-lithium, à basse température, la métallation s'effectue sans pontages[62], et les sites métallés ainsi créés peuvent servir à l'amorçage de la polymérisation de styrène. Cependant le processus d'amorçage étant lent dans ce cas, on ne peut considérer les greffons comme identiques entre eux, et il n'est même pas assuré que tous les sites promoteurs aient rempli leur mission[74]. Ainsi, bien que chimiquement pur, le polystyrène en "peigne" préparé selon cette méthode ne peut être caractérisé intégralement.

De la polyvinyl-2 pyridine à structure en "peigne" a été préparée de façon analogue, par action de butyl–lithium sur une chaîne de polyvinylpyridine, suivie de l'addition de vinylpyridine monomère. La polymérisation se déroule rapidement et les produits obtenus présentent la structure haute-ramifiée attendue[77].

2. Polymères en étoile

Pour la préparation, par voie anionique, de polymères en étoile on peut a priori envisager plusieurs voies:
—L'utilisation d'un promoteur organo-métallique polyfonctionnel
—La réaction de plusieurs chaînes macromoléculaires ω-carbanioniques sur les fonctions électrophiles d'un composé "désactivateur" polyfonctionnel.
—La copolymérisation séquencée anionique d'un monomère vinylique avec un 2e. monomère possédant 2 fonctions insaturées.

Bien qu'envisagée par divers auteurs, la première de ces méthodes n'a jamais été utilisée: outre la difficulté de préparer un initiateur polyfonctionnel adéquat, les branches des polymères en étoile ainsi préparées ne seraient pas forcément identiques.

La 2e. méthode a été appliquée avec succès par divers auteurs[78-80] au cas du polystyrène, le promoteur utilisé étant en général le cumyl–potassium ou le butyl–lithium, de sorte que les chaînes élémentaires de polystyrène vivant peuvent être considérées en première approximation comme identiques entre elles. Parmi les composés électrophiles polyfonctionnels qui ont été utilisés citons le trichlorométhylbenzène, le tétrachlorure de silicium, ainsi que

le trimère du chlorure de phosphonitrile. Ces réactions sont loin d'être quantitatives et le produit brut de la réaction est constitué d'un mélange de produits linéaires et de polymères en étoile, le nombre maximum de branches étant égal à la fonctionnalité du désactivateur. Les opérations de fractionnement sont délicates, mais les polymères en étoile ont pu être caractérisés comme tels.

La préparation de polymères en étoile par copolymérisation séquencée anionique de styrène avec du divinylbenzène repose sur le principe suivant[81,82]; les chaînes de polystyrène ω-carbanioniques préparées dans la première étape sont utilisées comme macropromoteurs pour l'amorçage d'une petite quantité de divinylbenzène. La polymérisation de ce 2e. monomère conduit à la formation de nodules réticulés qui sont reliés chacun à n chaînes linéaires solvatées de polystyrène qui les protègent. L'efficacité de cette protection est démontrée par le fait que le milieu réactionnel demeure parfaitement homogène même si la proportion de divinylbenzène atteint 30%. Le nombre de branches fixées à un même nodule de DVB peut atteindre 20 ou 30. Ici encore, les branches peuvent être considérées comme identiques entre elles, mais les échantillons obtenus sont très fortement polydispersées, car le nombre de branches varie considérablement d'une molécule à l'autre. Ici encore le fractionnement se révèle délicat, et les fractions obtenues sont elles-mêmes très polydispersées. Néanmoins des études en solution, conduites sur ces échantillons se révèlent fructueuses[83].

Des polyméthacrylates de méthyle à structure en etoile[83] ont pu être obtenus par le même procédé, le promoteur étant ici le diphénylméthyl-sodium, et l'agent de réticulation le diméthacrylate de glycol.

CONCLUSION

De cette longue énumération des méthodes qui ont été mises au point—ou simplement essayées—en vue de la préparation de copolymères greffés et séquencés ou d'homopolymères ramifiés, il ressort que ce domaine de la Science macromoléculaire est en plein essor. Ces synthèses ont été abordées de multiples façons, et le souci de préparer des échantillons caractérisables, homogènes en masse et en composition, est de plus en plus évident.

413

Les méthodes de polymérisation anionique sans terminaison, en milieu solvant aprotique ont ouvert de nouvelles voies en cette matière, et ont permis la synthèse de copolymères effectivement homogènes en masse et en composition, et de structures ramifiées caractérisables. Malheureusement les problèmes liés à ces synthèses ne sont pas pour autant tous résolus, ne serait-ce que parce que les polymérisations anioniques sans terminaison ne s'appliquent qu' à certains monomères. D'autres monomères ne se polymérisent que par voie cationique, tels les éthers vinyliques, l'isobutène, le vinylcarbazole, de nombreux monomères cycliques oxygénés, tels que les dérivés de l'oxéthane, du tétrahydrofuranne, le trioxane, etc. . . . Il serait donc d'un grand intérêt si l'on pouvait préparer à volonté des polymères cationiques vivants, et utiliser les sites carbonium ou oxonium terminaux en vue de réactions ultérieures, soit sur des fonctions adéquates, soit sur des sites anioniques comme cela a déjà été fait. C'est dans cette direction que devront porter de nouveaux efforts, car les polymérisations cationiques sont souvent affectées par des réactions de transfert ou d'arrêt qui rendent impossible toute synthèse ultérieure. Mais il est permis de penser que l'on trouvera des systèmes permettant la préparation de polymères cationiques authentiquement "vivants".

Il ne nous appartient pas de discuter ici de l'application des méthodes anioniques que nous avons décrites à la préparation industrielle de copolymères séquencés ou greffés. Cependant l'utilisation de réactifs et de solvants d'un haut degré de pureté constitue une condition impérative, et toutes les opérations doivent être effectuées sous vide ou sous atmosphère inerte; cela constitue certes un handicap, mais non pas un obstacle majeur au développement industriel de ces méthodes.

Bibliographie

[1] T. Alfrey, J. J. Bohrer, et H. Mark. *Copolymerization*, New York, Interscience (1952).
[2] I. Skeist. *J. Amer. Chem. Soc.* **68**, 1781 (1946).
[3] J. A. Hicks et H. W. Melville. *Nature, Lond.* **171**, 300 (1953) et *J. Polymer Sci.* **12**, 461 (1954).
[4] M. Szwarc. *Nature, Lond.* **178**, 1168 (1956).
[5] M. Szwarc. *Makromol. Chem.* **35**, 132 (1960).
[6] G. Smets et G. Hart. *Adv. Polymer Sci.* **2**, 173 (1960).
[7] P. Molyneux. *Makromol. Chem.* **37**, 160 (1960); **43**, 31 (1961).
[8] G. M. Burnett, P. Meares, et C. Paton. *Trans. Faraday Soc.* **58**, 723 (1962).
[9] M. Van Beylen et G. Smets. *Makromol. Chem.* **69**, 140 (1963).
[10] J. Vuillemot, B. Babier, G. Riess, et A. Banderet. *J. Polymer Sci.* **A-3**, 1969 (1965).
[11] G. Smets, A. Poot, et G. L. Duncan. *J. Polymer Sci.* **54**, 65 (1961).
[12] G. Smets et W. Van Rillaer. *J. Polymer Sci.* **A-2**, 2417, 2423 (1964).
[13] G. Oster, G. Oster, et K. Prati. *Nature, Lond.* **173**, 300 (1954) et *J. Amer. Chem. Soc.* **79**, 595 (1957).
[14] G. Smets, W. De Winter, et G. Del Zenne. *J. Polymer Sci.* **55**, 767 (1961).
[15] G. Riess et A. Banderet. *Bull. Soc. Chim. Fr.* (1959) 51; (1959) 733; (1960) 1625.
[16] R. A. Wallace et K. L. Hadley. *J. Polymer Sci.* **A-1**, **4**, 71 (1966).
[17] F. S. Holahan, S. Stivala, et D. Levi. *J. Polymer Sci.* **A-3**, 3993 (1965).
[18] A. Chapiro. *J. Polymer Sci.* **48**, 109 (1960).
[19] A. Chapiro et Z. Mankovoski. *Europ. Polymer J.* **2**, 163 (1960).
[20] E. Turska et S. Polowinsky. *J. Polymer Sci.* **A-1**, 2085 (1963).
[21] A. Chapiro et A. Matsumoto. *J. Polymer Sci.* **57**, 743 (1962).
[22] K. H. Kahrs et J. W. Zimmermann. *Makromol. Chem.* **58**, 75 (1962).
[23] W. Kobryner et A. Banderet. *J. Polymer Sci.* **34**, 381 (1959).
[24] H. S. Kolesnikov et Tsen Khan Ming. *J. Polymer Sci.* **61** 497 (1962).
[25] M. Szwarc. *Adv. Polymer Sci.* **2**, 275 (1960).
[26] M. H. Loucheux et P. Rempp. *Bull. Soc. Chim.* 1497 (1958).
[27] L. Gold. *J. Chem. Phys.* **28**, 91 (1958).
B. D. Coleman et T. G. Fox. *J. Amer. Chem. Soc.* **85**, 1241 (1963).

SYNTHESES ET STRUCTURES NOUVELLES DE POLYMERES

28 F. M. Brower et H. W. McCormick. *J. Polymer Sci.* **A-1**, 1749 (1963).
29 M. Levy, M. Szwarc, et R. Milkovitch. *J. Amer. Chem. Soc.* **78**, 2656 (1956).
30 D. H. Richards et M. Szwarc. *Trans. Faraday Soc.* **55**, 1644 (1959).
31 G. Finaz, P. Rempp, et J. Parrod. *Bull. Soc. Chim.* 262 (1962).
32 E. Franta et P. Rempp. *C.R. Acad. Sci.* **254**, 674 (1962).
33 M. Shima, D. W. Battacharyya, J. Smid, et M. Szwarc. *J. Amer. Chem. Soc.* **85**, 1306 (1963).
34 C. G. Overberger et N. Yamamoto. *Polymer Letters* **3**, 569 (1965).
35 S. Schlick et M. Levy. *J. Phys. Chem.* **64**, 883 (1960).
36 S. E. Bresler, L. M. Pyrkov, S. Ya Frenkel, L. Laius, et S. I. Klenin. *Vysokomol. Soed.* **4**, 250 (1962).
37 A. F. Johnson et D. J. Worsfold. *Makromol. Chem.* **85**, 273 (1965).
38 D. N. Cramond, P. S. Lawry, et J. R. Urwin. *Europ. Polymer J.* **2**, 107 (1966).
39 M. Morton. ACS Rubber Div. Meeting (Montréal, 1967).
40 D. Freyss, M. Leng, et P. Rempp. *Bull. Soc. Chim.* 221 (1964).
41 D. Freyss, P. Rempp, et H. Benoit. *Polymer Letters* **2**, 217 (1964).
42 P. Sigwalt et M. Fontanille. *C.R. Acad. Sci.* **251**, 2947 (1960).
43 R. Graham, 0. Panchak, et M. Kampf. *J. Polymer Sci.* **44**, 411 (1960).
44 S. Boileau et P. Sigwalt. *C.R. Acad. Sci.* **261**, 132 (1965).
45 S. Boileau, G. Champetier, et P. Sigwalt. *Makromol. Chem.* **69**, 180 (1963).
46 M. Morton, A. Rembaum, et E. E. Bostick. *J. Appl. Polymer Sci.* **8**, 2707 (1964).
47 G. Berger, M. Levy, et D. Vofsi. *Polymer Letters* **4**, 183 (1966).
47a M. P. Dreyfuss et P. Dreyfuss. *Polymer* **6**, 93 (1965).
48 Y. Gallot, H. Benoit, et P. Rempp. *C.R. Acad. Sci.* **253**, 989 (1961).
49 Y. Gallot, P. Rempp, et J. Parrod. *Polymer Letters* **1**, 329 (1963).
50 Y. Gallot, Z. Grubisic, P. Rempp, et H. Benoit. Communication au Colloque IUPAC (Bruxelles, 1967).
51 Y. Gallot. Thèse, Strasbourg (1964).
52 P. Rempp, J. Parrod, G. Laurent, et Y. Gallot. *C.R. Acad. Sci.* **260**, 903 (1965).
53 A. Dondos et P. Rempp. *C.R. Acad. Sci.* **264**, 869 (1967).
54 J. J. O'Malley et R. H. Marchessault. *Polymer Letters* **3**, 685 (1965).
55 J. J. O'Malley et R. H. Marchessault. *J. Phys. Chem.* **70**, 3235 (1966).
56 S. P. Mitzengendler, G. A. Andreeva, K. I. Sokolova, et A. A. Korotkov. *Vysokomol. Soed.* **4**, 1366 (1962).
57 A. Dondos. *Bull. Soc. Chim.* 2762 (1963).
58 G. Goutière et J. Gole. *C.R. Acad. Sci.* **257**, 674 (1963).
Bull. Soc. Chim. 153 (1965).
59 H. Normant et G. Angelo. *Bull. Soc. Chim.* 354 (1960).
60 G. Greber et G. Egle. *Makromol. Chem.* **54**, 136 (1962); **59**, 174, (1963); **62**, 196 (1963).
61 G. Greber et G. Egle. *Makromol. chem.* **64**, 68, 207 (1963).
62 A. Dondos et P. Rempp. *C.R. Acad. Sci.* **258**, 4045 (1964).
63 M. B. Huglin. *Polymer* **5**, 135 (1964).
64 G. Greber, J. Tölle, et W. Burchard. *Makromol. Chem.* **71**, 47 (1964).
65 B. A. Feit, A. Bar Nun, M. Lahav, et A. Zilkha. *J. Appl. Polymer Sci.* **8**, 1869 (1964).
66 Y. Avny, S. Migdal, et A. Zilkha. *Europ. Polymer J.* **2**, 355, 367 (1966).
67 A. Rembaum et J. Moacanin. *Polymer Letters* **1**, 41 (1963); **2**, 117 (1964).
68 E. Cuddihy, J. Moacanin, et A. Rembaum. *J. Appl. Polymer Sci.* **9**, 1385 (1965).
69 A. Rembaum, J. Moacanin, et R. Haack. *J. Macromol. Chem.* **1**, 657, 673 (1966).
70 J. Gole, G. Goutiere, et P. Rempp. *C.R. Acad. Sci.* **254**, 3867 (1962).
71 J. E. Herz. Résultats non publiés.
72 J. Parrod et G. Meyer. *C.R. Acad. Sci.* **262**, 1244 (1966).
73 G. Greber et G. Egle. *Makromol. Chem.* **53**, 206, 208 (1962).
74 D. Decker et P. Rempp. *Polymer Symposia.*
75 T. G. Fox. Résultats non publiés.
76 H. Benoit, Z. Grubisic, P. Rempp, D. Decker, et J. G. Zilliox. *J. Chim. Phys.* **63**, 1507 (1966).
77 A. Dondos. *Bull. soc. Chim.* 910 (1967).
78 S. P. S. Yen. *Makromol. Chem.* **81**, 152 (1965).
79 J. A. Gervasi et A. B. Gosnell. *J. Polymer Sci.* A-1, **4**, 1391, 1400 (1966).
80 T. A. Orofino et F. Wenger. *J. Phys. Chem.* **67**, 566 (1963); M. Morton, T. E. Helminiak, S. D. Gadkary, et F. Bueche. *J. Polymer Sci.* **57**, 471 (1962).
81 D. Decker et P. Rempp. *C.R. Acad. Sci.* **261**, 1977 (1965).
82 J. G. Zilliox, D. Decker, et P. Rempp. *C.R. Acad. Sci.* **262**, 726 (1966).
83 J. G. Zilliox, P. Rempp, et J. Parrod. Communication au Colloque IUPAC Bruxelles, (1967).

NUCLEAR MAGNETIC RESONANCE AND OPTICAL STUDIES OF POLYPEPTIDE CHAIN CONFORMATION

F. A. Bovey

Polymer Chemistry Research Department, Bell Telephone Laboratories, Murray Hill, New Jersey, U.S.A.

1. INTRODUCTION

Everyone is aware of the intense effort that has been devoted to the study of the structure of proteins and polypeptides during the last two decades, culminating in the complete x-ray determination of the structures of *myoglobin, hemoglobin, lysozyme,* and *ribonuclease.* Few problems in natural science have been subjected to such a heavy assault with such a variety of weapons. The most important of these, of course, has been x-ray diffraction. One cannot imagine at present any other technique capable of providing the thousands upon thousands of individual parameters necessary to describe a complete protein structure. In view of the spectacular triumphs of this method, why do many of us who are concerned with polypeptide and protein structure continue to employ such relatively humble techniques as optical rotation, circular dichroism, and nuclear magnetic resonance, which are capable of giving at most only a handful of structural parameters? I think there are good answers to this question, and I would suggest a few as follows:

(*a*) If our main interest is in polypeptides of a single repeating residue, then only a relatively limited number of molecular parameters may be adequate for a good description.

(*b*) X-ray studies are limited for the most part to crystalline solids and, except in certain special circumstances, cannot inform us what biopolymers are doing in solution. (One such exception is the x-ray study of polypeptides in solution by Brady, Salovey and Reddy[1] in our laboratories; this requires labelling with heavy atoms such as bromine.)

(*c*) Optical and n.m.r. methods have an intrinsic interest in their own right. This may be the most important reason for using them.

2. THE α-HELIX

This lecture will centre chiefly on the α-helix and its conformational changes, as observed by both high resolution n.m.r. and circular dichroism (CD). In particular, we shall be concerned with the α-helix-to-random coil transition. This phenomenon has, of course, been intensively studied both experimentally and theoretically, but there nevertheless are some unresolved problems. In particular, the question of why the transition occurs at all appears to be quite unanswered for organic solvent systems. We may not be able to give a complete answer here either, but at least we will furnish some additional evidence.

3. THE HELIX–COIL TRANSITION

In *Figure 1* is shown the now very familiar structure of the α-helix. The structure represented is that of poly-L-alanine in the right-handed helical conformation. This representation claims two somewhat novel features: *first*, it is a pair of stereoscopic views; *second*, it was not drawn by a draftsman, but was generated directly on microfilm by the General Electric 645 computer, using a programme adapted for this purpose by R. L. Kornegay

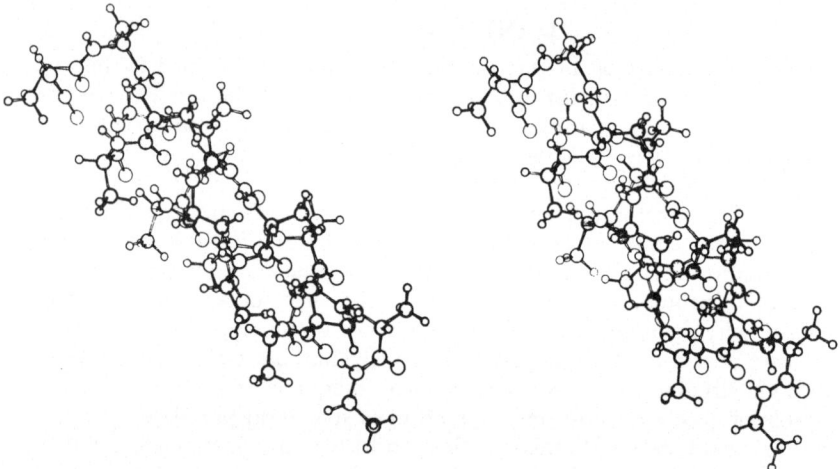

Figure 1. Stereo views of α-helix (must be viewed with a stereo lens system obtainable from Stereo-Magniscope, Inc., New York City, or from the author upon request)

(Bell Telephone Laboratories) from a more general programme devised by C. K. Johnson (Oak Ridge National Laboratories).

As it occurs in nature and in many synthetic polypeptides, the α-helix has approximately 18-fold symmetry, i.e., it repeats exactly every 18 amino-acid residues. There are 3·6 residues per turn and a residue translation of 1·49 Å, i.e., as we pass from a point in one residue to the corresponding point in the next, we move 1·49 Å along the helical axis.

In *Figure 2* are shown the absorption spectrum and the circular dichroism spectrum characteristic of a polypeptide in the α-helical conformation. These spectra are of poly-γ-methyl-L-glutamate (DP 400) in hexafluoroisopropanol solution. They are generally in agreement with those presented by Holzwarth and Doty[2], but the CD band intensities were found to be slightly smaller than they report. In *Figure 3*, the CD spectrum is decomposed into the constituent bands as now recognized, and the relative intensity of each is indicated.

The currently accepted interpretation of the circular dichroism bands, and of the corresponding bands of the absorption spectrum, is as follows:

PMLG. DP 400. in HFiP

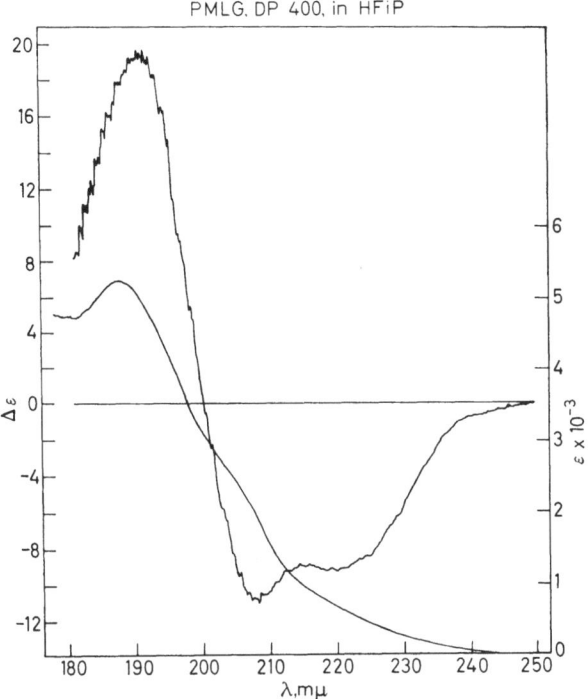

Figure 2. Absorption and CD spectra of poly-γ-benzyl-L-glutamate, 2 × 10⁻² M in hexa-fluoroisopropanol

(a) At 223 mμ is the negative $n-\pi^*$ peptide band, quite strong in di-chroism, but weak in absorption.

(b) At 190 mμ and 205 mμ are the opposite-sign, exciton-split $\pi-\pi^*$ bands of the peptide chromophor. The positive band is polarized perpen-dicularly to the helical axis; the negative band is polarized parallel to the helical axis. The positive band has twice the intensity of the negative band.

We are concerned also with the left-handed α-helical conformation which poly-L-aspartate chains are believed to prefer. There appear to be little if any published circular dichroism data for these polymers, the left-handed configuration being deduced from b_0 values obtained from optical rotatory dispersion measurements[4]. From our own limited data, we know that poly-β-methyl-L-aspartate and poly-β-benzyl-L-aspartate give *positive* $n-\pi^*$ CD bands in about the same position as the negative $n-\pi^*$ band of a right-handed α-helix.

When transition from the right-handed α-helix to the random coil occurs in aqueous solution, it has been reported[2, 3] that the negative $n-\pi^*$ band gives place to a very weak positive band. (As we shall see, this is not neces-sarily the case in organic solvents, where a weak negative $n-\pi^*$ band may persist in the random coil conformation.) The $\pi-\pi^*$ exciton-split bands are

419

P

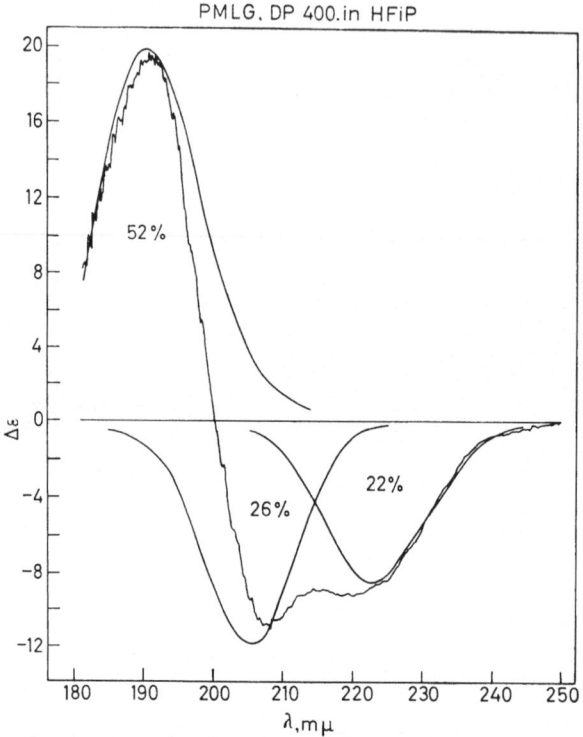

Figure 3. The CD spectrum of *Figure 2* decomposed into *n–π**, ‖ *π–π**, and ⊥ *π–π** bands

replaced by a single negative band at about 200 mμ. In the work to be described, the π–π* transition could not be observed because of high absorption, but the n–π* band furnished a satisfactory measure of the helix-coil transition.

4. NMR OBSERVATIONS OF THE α-HELIX–RANDOM-COIL TRANSITION

The first n.m.r. study of the helix–coil transition was reported in 1959 by Bovey and Tiers[5], who observed an extreme broadening of the n.m.r. peaks of poly-γ-benzyl-L-glutamate in trichlorethylene, but were able to see reasonably well-resolved spectra for both helix and coil on addition of appropriate quantities of trifluoroacetic acid. Several other authors have since examined this and closely related systems[6–9]. We report here a more detailed study of poly-γ-benzyl-L-glutamate (PBLG) than our previous one, and also a parallel study of poly-β-benzyl-L-asparate (PBLA). We have used chloroform as the helix-supporting solvent and trifluoroacetic acid as the helix-breaking solvent. For both studies we have used polymers of low molecular weight (DP *ca.* 50) and of high molcular weight (DP *ca.* 1000).

In *Figure 4* are shown 100 Mc/s spectra of PBLG of DP 55 in chloroform alone; with 15% TFA; and with 30% TFA. These spectra were run at 50°

and are representative of the many that have been run. All polymer solutions contained 10% (wt./vol.) of polymer. There is a marked broadening of all peaks in $CHCl_3$. The peaks of the backbone protons, NH and α-CH, are so broadened that they seem to have disappeared, as has been previously noted by Goodman and Masuda[6]. However, on increasing the spectrometer gain, they can be readily seen. The protons of the side-chain are progressively less broadened as we move out from the glutamyl methylenes (β-CH_2 at *ca.*

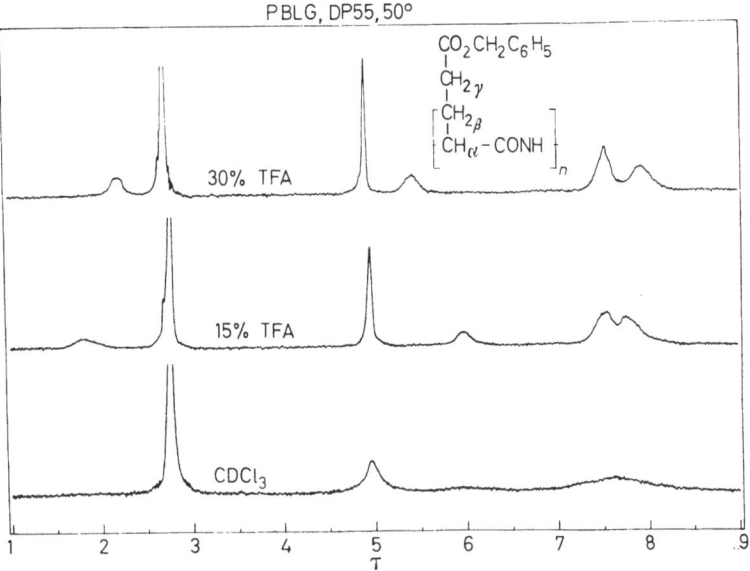

Figure 4. 100 Mc/s spectra of PBLG (10%; DP 55) in deuterochloroform alone; in 15% TFA in deuterochloroform; and in 30% TFA in deuterochloroform

$7 \cdot 7\tau$ and γ-CH_2 at *ca.* $7 \cdot 5\tau$) to the benzylic methylene group ($4 \cdot 96\tau$) and to the phenyl group ($2 \cdot 76\tau$). Upon addition of 15% of trifluoroacetic acid (as little as $2 \cdot 5\%$ has a nearly equal effect), all peaks narrow considerably, and the NH and α-CH resonances become clearly visible. However, their positions are the same and the chains must be still largely helical, for the circular dichroism measurements, which we shall discuss below, show an n–π* band at 224 mμ with an intensity, $\Delta\epsilon$, of $-6 \cdot 4$ l.-mole^{-1}.-cm^{-1}. At 30% TFA, all peaks narrow further, but the effect is not striking. Circular dichroism measurements show that the polymer is now a random coil, the n–π* band having nearly disappeared.

In *Figure 5* are shown corresponding spectra for the PBLG of DP 1000. In pure chloroform, the entire spectrum appears to have vanished completely when recorded over the usual 1000 c/s width. The smaller inset spectrum, recorded over a 5000 c/s sweep, shows that it is actually still there, but very greatly broadened. The only visible peak is the phenyl resonance, now about 250 c/s in width. Measurable intensity is spread out over about 4000 c/s, as integration of the spectrum clearly shows. In the presence of

2·5% TFA, the peaks are likewise extremely broad. At 15% TFA the polymer is still helical, but the peaks are very much narrower (probably by a factor of at least 100); they are, however, markedly broader than for the DP 55 polymer under the same conditions. The peaks of the random coil exhibit the same width as those of the low molecular weight polymer in the random-coil state; the spectra are indistinguishable.

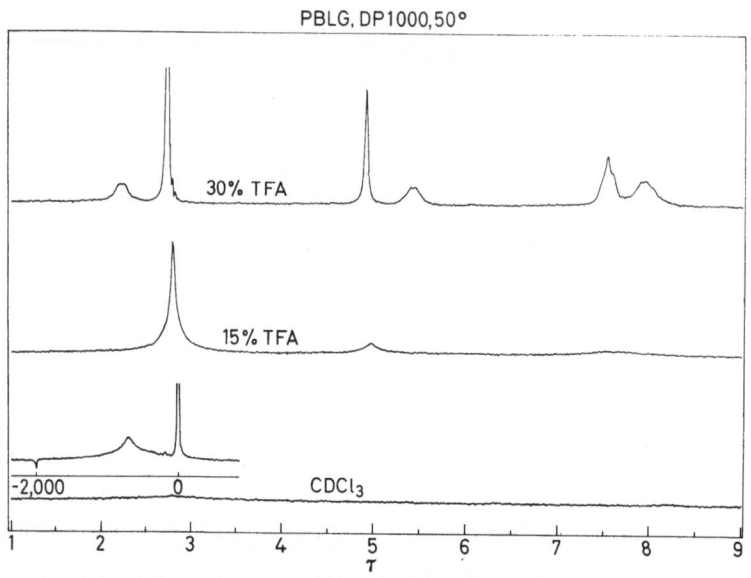

Figure 5. 100 Mc/s spectra of PBLG (10%; DP 1000) in deuterochloroform alone; in 15% TFA in deuterochloroform; and in 30% TFA in deuterochloroform

As Bovey and Tiers[5] pointed out, the extreme broadening in chloroform is very likely due to aggregation. The existence of PBLG in a liquid crystalline state under these conditions has been established[10]. The broadening is closely analogous to that exhibited by native proteins, but somewhat more extreme. It is strongly dependent upon molecular weight. Upon adding trifluoroacetic acid, these aggregates are broken up. A considerable dependence of the line-width of the free helices upon molecular weight is still noticeable, as might be expected for a rod-like macromolecule, but it is much less marked. In the random-coil state, there is no dependence upon molecular weight, since now local segmental motion determines the line-width. This last is the behaviour normally characteristic of vinyl polymers in solution.

Figures 6 and *7* show similar data for poly-β-benzyl-L-aspartate of low and high molecular weight. The spectral broadening in chloroform is comparable, but is not quite so great: both the phenyl and benzyl protons can now be discriminated. The onset of narrowing occurs at much lower acid concentration for both the low and high molecular weight polymer. In 30% TFA, the NH resonance is a doublet and the α-CH a binomial quartet, indicating approximately equal couplings (*ca.* 7 c/s) of the α-CH to the β-CH and NH protons.

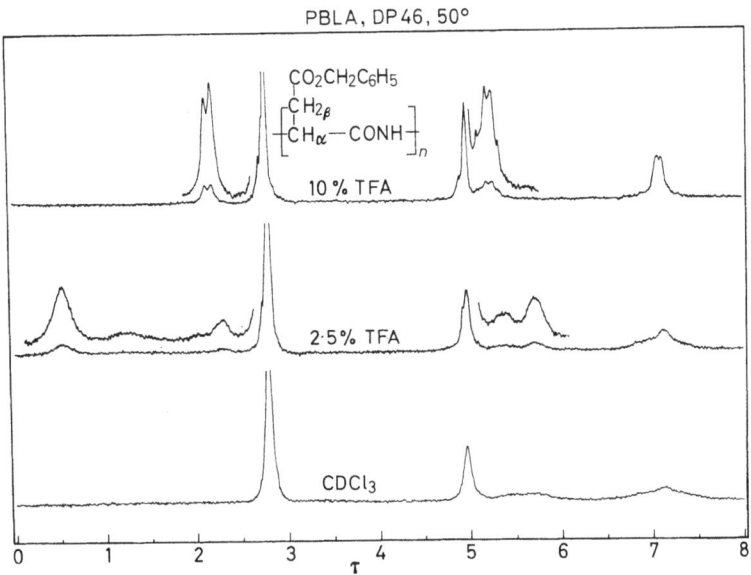

Figure 6. 100 Mc/s spectra of PBLA (10%; DP 46) in deuterochloroform alone; in 2·5% TFA in deuterochloroform; and in 10% TFA in chloroform

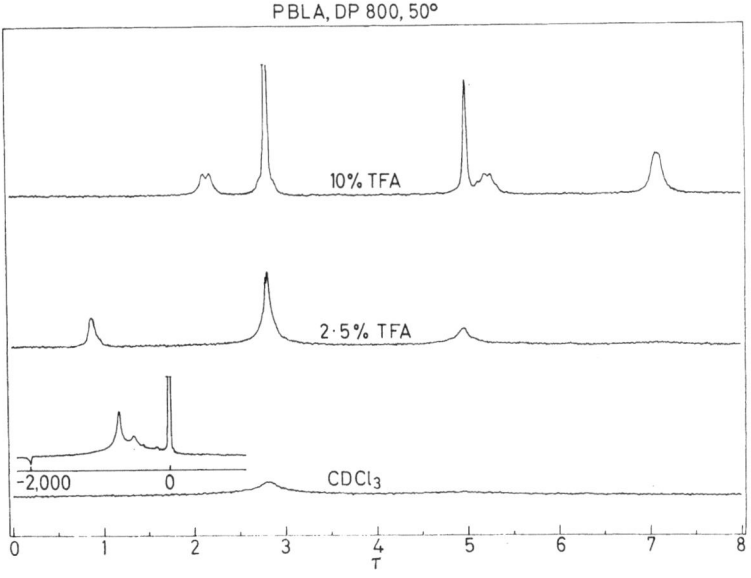

Figure 7. 100 Mc/s spectra of PBLA (10%; DP 800) in deuterochloroform alone; in 2·5% TFA in deuterochloroform; and in 10% TFA in chloroform.

PBLG, DP 55, 50°, CDCl₃

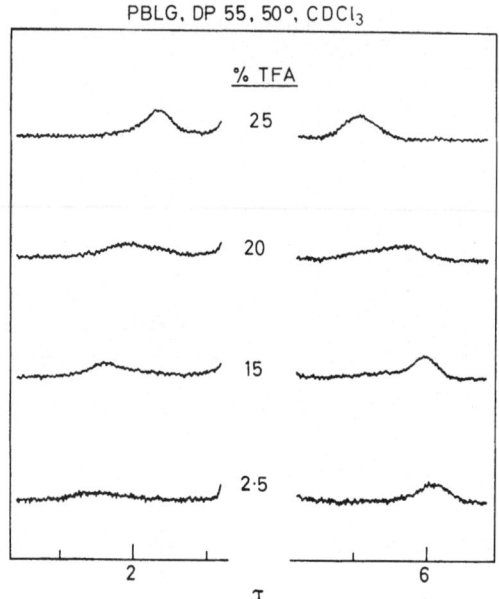

Figure 8. NH and CH peaks of PBLG (10%; DP 55) as a function of TFA conc. in CDCl₃.

PBLA, DP 46, 50°, CDCl₃

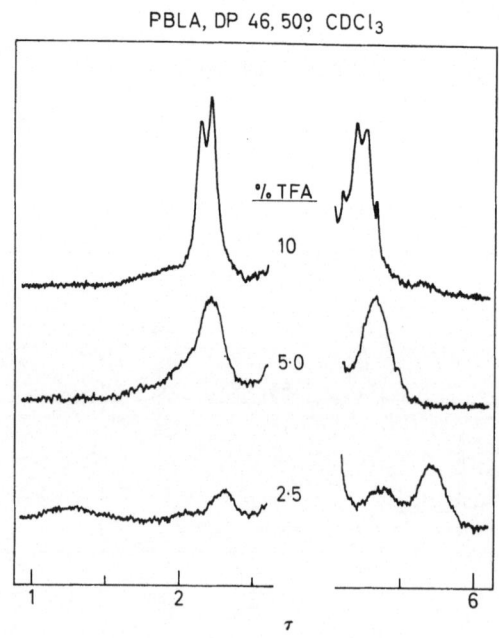

Figure 9. NH and CH peaks of PBLA (10%; DP 46) as a function of TFA conc. in CDCl₃

This clearly is a measure of the actual local conformation of the random coil, but unfortunately we do not at present know the dependence of the vicinal coupling upon the H—N—C—H dihedral angle.

Let us turn now to a consideration of the chemical shifts of the various polypeptide protons as a function of conformation. Certain obvious trends can be seen in *Figures 4–7*. The behaviour of the NH and α-CH protons is shown in greater detail in *Figures 8* and *9*, which represent the low-molecular-weight polymers at 50°. For PBLG, the α-CH peak remains unchanged at 6·00τ until about 20% TFA, when a down-field shift begins which is complete at 25% TFA. This corresponds to the helix–coil transition. At 20% TFA (50°), there is a clear indication of a smaller peak at 5·70τ. We believe this corresponds to those chains which are too short to sustain a helix under these conditions, and begin to undergo transition to the coil. The NH peak also undergoes a marked change in position at the transition, but it becomes *more* shielded. Similar observations have been reported by Stewart *et al.* for poly-L-alanine[8]. (*But see Note 1 added in proof on p. 432*).

These same general features are exhibited in the poly-β-benzyl-L-aspartate spectra shown in *Figure 9*. Here, however, the NH peak moves up-field more markedly when transition to the random coil occurs. We observe also that

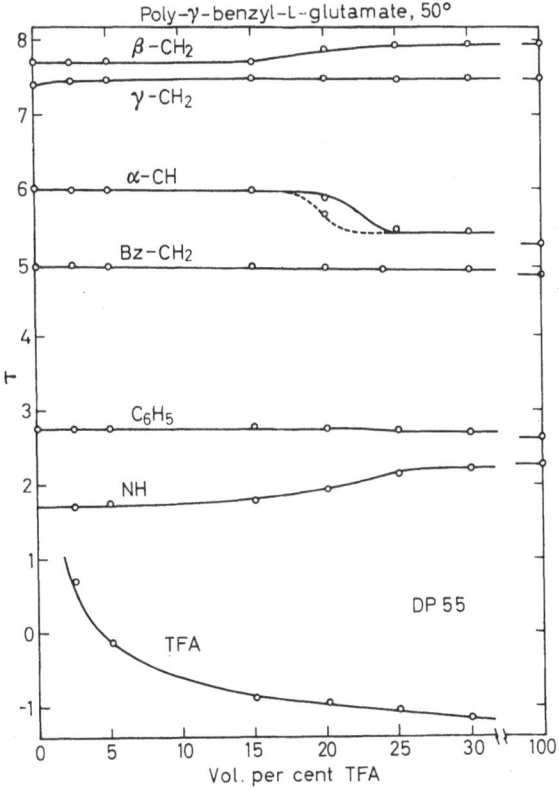

Figure 10. Chemical shifts of PBLG (10%; DP 55) plotted *versus* conc. of TFA in CHCl$_3$. Dotted curve corresponds to transition of low-molecular-weight fraction

even in chloroform alone, there is a substantial fraction of asparate chains, perhaps one-third, which presumably are too short to form a helix, and appear as a peak at *ca.* 5·3τ. Beyond the transition, i.e., at 5% or more of TFA, the helical peak disappears and this 5·3τ peak is the only one visible.

The appearance of these two α-CH peaks might at first suggest that we are about midway in the transition, and that the equilibration of magnetic nuclei between the coil and helical environments is so slow as to allow a separate peak for each to appear. This explanation raises questions however, for two reasons: (*a*) temperature jump measurements[11] indicate that the lifetimes of the helix and random-coil conformations in equilibration with each other are likely to be less than 10 microseconds—even a lifetime as long as a millisecond would still be short enough to merge the n.m r. peaks completely into a single peak, they must, therefore, correspond to non-exchanging spin populations; (*b*) the spectra of the high molecular weight PBLA (not shown here in detail) exhibit the same transition, but do not appear to show helical and coil peaks simultaneously. (*cf. Note 1 added in proof on p. 432*).

The chemical shift changes for all protons are plotted *versus* volume percent TFA in *Figure 10* for PBLG of DP 55. The high-molecular-weight

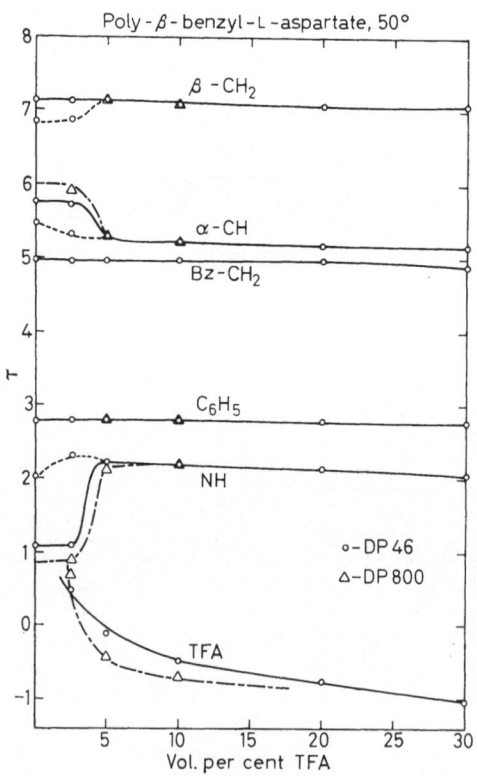

Figure 11. Chemical shifts of PBLA (10%) plotted *versus* conc. of TFA in CHCl₃; ○: DP 46, △: DP 800, – – – –: Low mol. wt. fraction in DP 46 polymer

polymer shows nearly the same behaviour, except that the separate transition for shorter helices (indicated by the dotted curve in the α-CH plot) does not appear. A similar plot for PBLA, DP 46, is shown in *Figure 11*. The PBLG transition between 20 and 25% TFA is clearly shown by the α-CH and NH peaks; those for side-chain protons show smaller and smaller changes the farther out they are. The β-protons show an appreciable shift in both spectra; the others exhibit small but easily observable changes.

We might at this point ask the question: how do we know that these changes in the n.m.r. spectra correspond to the helix–coil transition itself? Perhaps they are actually due to the protonation of the peptide oxygen atoms:

$$HA \quad + \quad \overset{O}{\underset{}{\overset{\|}{C}}} - N\underset{H}{\overset{}{\diagup}} \quad \rightleftharpoons \quad A^- \quad + \quad \overset{HO_{\cdots}}{\underset{}{C^{+}_{\cdots}}} = N\underset{H}{\overset{}{\diagup}}$$

Such protonation is known to occur for small-molecule amides, at least if the acid is strong enough[12-14]. This might precede but not necessarily coincide with the actual transition. To answer this question, let us look a little more closely at the circular dichroism measurements. Those shown in *Figure 12* were made at 33°, using the same solutions as used for the n.m.r. measurements. Because of solvent and polymer absorption in these concentrated solutions, it was not possible to make measurements beyond

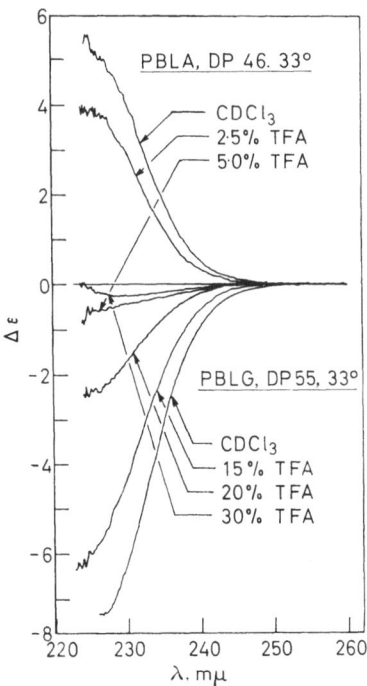

Figure 12. CD spectra of 10% solutions of PBLG (DP 55) and PBLA (DP 46) as a function of TFA conc. in CDCl₃

427

222 mμ, but this suffices to reach the *n–π** extrema for both the poly-glutamate and polyaspartate systems. In *Figure 13*, the α-CH peak positions and Δε are plotted *versus* the acid concentration. Both appear to depend on the acid concentration in the same way, and therefore one may logically (but not rigorously) conclude that both reflect the helix–coil transition.

It is well known that in organic solvent systems such as these, the helix is the form stable at elevated temperatures. If, therefore, we observe the

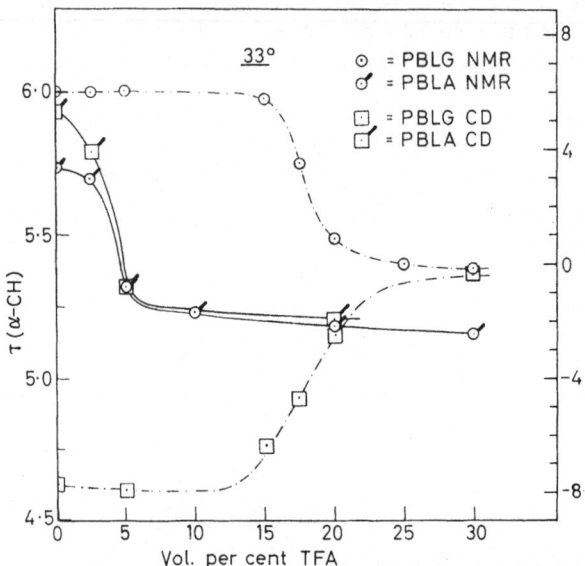

Figure 13. Δε and $\tau_{\alpha-CH}$ of 10% solutions of PBLG (DP 55) and PBLA (DP 46) plotted *versus* TFA conc. in CHCl₃; 33°

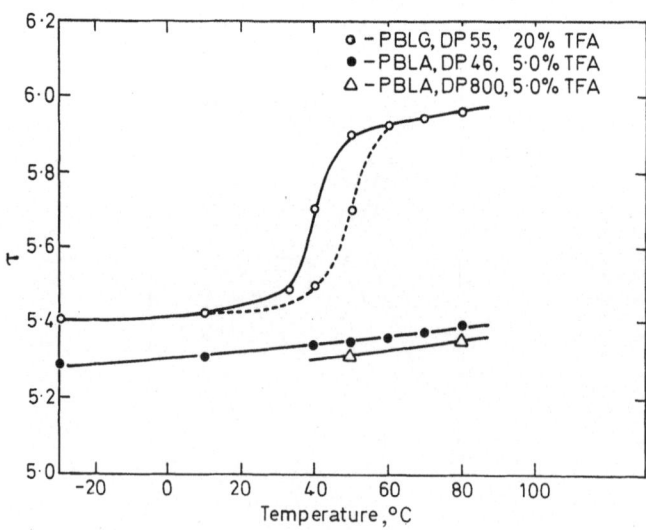

Figure 14. $\tau_{\alpha-CH}$ for PBLG (DP 55) and PBLA (DP 46 and 800) *versus* temperature

428

α-CH peak position as a function of temperature and if it reflects the helix–coil transition, we should see a marked up-field shift as we pass through the temperature region appropriate for the transition. This behaviour is indeed seen (*Figure 14*) for PBLG (DP 55) in 20% TFA, under which conditions the random coil is stable at *ca.* 25° and below. The transition is rather broad, as expected for a polymer of relatively low molecular weight, but seems to be complete above *ca.* 70–80°. In contrast, PBLA shows no such transition in the analogous solvent system (5% TFA). There appears to be no dependence of the aspartate helix–coil equilibrium upon temperature.

5. ORIGIN OF THE HELIX–COIL TRANSITION

These experiments raise once again a knotty and still-unanswered question: just what is it that causes the transition from helix to coil to occur in these systems as the concentration of acid is increased? The marked up-field shift of the TFA carboxyl protons caused by the polypeptide (*Figures 10* and *11*) clearly points to a strong interaction of some kind. This increased shielding has been observed for poly-L-alanine solutions by Stewart *et al.*[8] These investigators also studied the behaviour of small model molecules, *N*-methylacetamide and *N,N*-dimethylacetamide, in the chloroform-TFA system and found a very marked *deshielding* of the TFA protons[15]. They believed this pointed to the formation of an ion pair of protonated amide and trifluoroacetate ion. They further concluded that TFA does *not* protonate the polypeptide amide oxygens, since the change in carboxyl peak position is in the opposite direction. There is, however, infrared evidence for such polypeptide protonation. Klotz and Hanlon and their coworkers[16–18] have observed band changes in the NH overtone region which have been interpreted as indicating that protonation can occur for both model compounds and polymers, and they conclude that protonation causes the helix–coil transition. As we shall see shortly, however, there are grounds for doubting that trifluoroacetic acid is strong enough to cause protonation of either model amides or polypeptide chains.

We have found that perchloric acid is capable of causing the helix–coil transiton in poly-L-alanine (DP 25). In *Figure 15* are shown the $n-\pi^*$ CD spectra for poly-L-alanine in 10^{-3} M solution in hexafluoroisopropanol, a solvent similar to trifluoroethanol and capable of supporting the α-helical conformation of this polymer. In all solvents in which we have observed it, the $n-\pi^*$ band of poly-L-alanine is abnormally weak compared to other polypeptides, but we shall assume for present purposes that in hexafluoroisopropanol the polymer is nevertheless fully helical. It can be seen that the addition of only 0·003% of perchloric acid, corresponding to 0·30 molar equivalent per peptide residue, suffices to decrease the $n-\pi^*$ band markedly, and that the transition to the random coil is complete when 0·9 molar equivalent is present. In contrast, the poly-L-alanine transition requires over 75 vol. percent of trifluoroacetic acid, chloroform being the other solvent[8]. Since perchloric acid is known to be very strong, at least in solvents of low basicity, surely it would seem that here, at least, protonation of the peptide units is responsible for the disruption of the helix. But once again there are grounds for doubt.

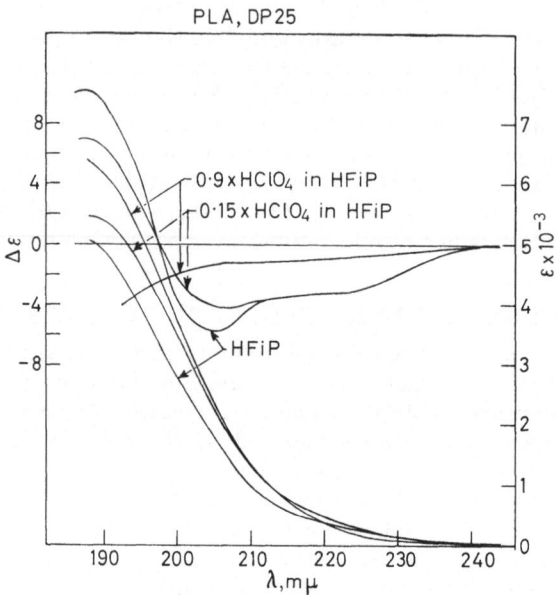

Figure 15. CD spectra of poly-L-alanine (10^{-3} M in hexafluoroisopropanol, HFiP) as a function of perchloric acid conc., expressed in moles per mole of peptide unit

The doubts have been generated by some preliminary studies of model compounds. These studies are still very incomplete, but the results appear significant. We need a model amide which has a strong n–π^* band, but which is rigid and incapable of conformational change. Small molecule amides giving observable n–π^* CD bands are very scarce; most simple amides show no observable dichroism in this region of the spectrum. Litman and Schellman[19] have observed a weak n–π^* CD band in L-3-aminopyrrolidone, and weak n–π^* bands have been found by us and by Balasubramanian and Wetlaufer[20] in a number of diketopiperazines. These latter compounds are subject, however, to possible conformational changes. A much more suitable compound is D-oxolupanine, which was observed some time ago by

Dr J. W. Longworth, then in our laboratory, to have an o.r.d. Cotton band in the n–π^* region. Its conformation appears from molecular models to be rigid. The CD spectrum of this diamide in the region of major interest is shown in *Figure 16* in a number of solvents. In chloroform, a band is seen at 230 mμ, comparable in intensity ($\Delta\epsilon = 5\cdot4$) to that of poly-β-benzyl-L-aspartate and remarkably strong for a nonpolymeric molecule. The CD spectrum in trifluoroethanol is similar, and even somewhat stronger

430

($\Delta\epsilon = 7\cdot2$), but shifted to 221 mμ. In addition to the positive band, there is a negative band (not shown) at 202 mμ, in the same position as the absorption maximum. It seems reasonable to assign these to n–π^* and π–π^* transitions, respectively, just as in the α-helix spectrum. If protonation were to occur to a substantial extent, the n–π^* band should decrease in intensity or disappear, for if the representation of the protonated form given above is correct, the nonbonded electrons of the unprotonated carbonyl oxygen are now bonded.

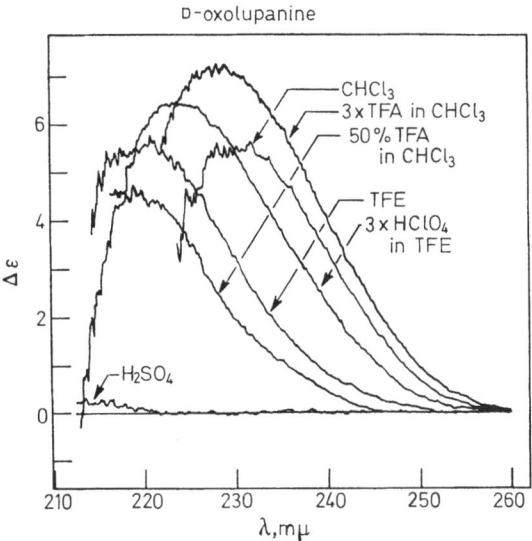

D-oxolupanine

Figure 16. The n–π^* region of the CD spectrum of D-oxolupanine in CHCl₃ and in CHCl₃ with added TFA; in TFE and in TFE with added HClO₄; and in conc. H₂SO₄

In concentrated sulphuric acid, this band does indeed disappear. (The π–π^* band and the absorption maximum shift in position, and the former changes sign.) But in the presence of a three-fold molar excess (based on two amide residues per molecule) of perchloric acid in trifluoroethanol (TFE) and in the presence of a 3-fold and 600-fold molar excess of trifluoroacetic acid, the band does *not* disappear and even increases slightly in intensity. It appears that the amide groups in oxolupanine should be at least as strongly basic as those of a polypeptide chain, and probably more so, since they are N,N-dialkyl amide groups, which are normally stronger than N-monoalkyl amides. The carbonyl groups do not appear to be any more sterically hindered in the model than in the polypeptide.

We therefore conclude that the helix–coil transitions reported so far for nonaqueous solvents do *not* involve proton loss and gain by the polymer chain, and that other interpretations must be sought for those experimental results which have been interpreted in this way. We thus must appeal in part to the familiar hydrogen bond competition as being responsible for the transition:

$$\text{HA} \cdots \text{HA} + \text{—C}=\text{O} \cdots \text{NH} \rightleftarrows \text{—C}=\text{O} \cdots \text{HA} + \text{HA} \cdots \text{NH}$$

431

We assume that this equilibrium runs to the right when HA is reasonably strong, and that increasing temperature pushes it to the left for the poly-γ-benzyl glutamate system, and probably for most right-handed α-helices. It further appears that for the poly-β-benzyl-L-aspartate system, the position of this equilibrium is unaffected by temperature.

Recent potential energy calculations by a number of authors have shown that α-helical conformations tend to be preferred even in the *absence* of intramolecular hydrogen bonding, as a result of van der Waals interactions and peptide dipole–dipole interactions. Therefore, the disruption of hydrogen bonding is itself not really a sufficient answer. It must be that there is in addition a substantial difference in solvation energy between the helix and the coil aside from that involving hydrogen bonding, and that more refined energy calculations should take this into account. Also, we must of course recall that, other things being equal, there is a substantial positive entropy term which encourages random coil formation whatever the heat terms involved may be.

NMR studies of polypeptide and other biopolymer systems are being pursued in this laboratory, and elsewhere, for it appears that this approach is an effective one, particularly when combined with optical measurements.

Acknowledgement
The author acknowledges with pleasure the experimental work performed by his collaborator, Mr F. P. Hood.

Notes added in proof
1. Recently Ferretti [*Chem. Commun.* 1030 (1967)] has observed NH and α-CH peak doubling in the helix-coil transition of poly-β-methyl-L-aspartate and poly-L-leucine.) Work now in progress at 220 Mc/s in our laboratory seems to show similar behaviour for DP 1000 PBLG.

2. After the completion of this work, CD results very similar to those shown in *Figure 13* were independently reported by F. Quadrifoglio and D. W. Urry [*J. Phys. Chem.* **71**, 2364 1967)].

References
[1] G. W. Brady, R. Salovey, and J. M. Reddy. *Biopolymers* **3**, 573 (1965).
[2] G. Holzwarth and P. Doty. *J. Amer. Chem. Soc.* **87**, 218 (1965).
[3] J. P. Carver, E. Schechter, and E. R. Blout. *J. Amer. Chem. Soc.* **88**, 2550, 2562 (1966).
[4] R. H. Karlson, K. S. Norland, G. D. Fasman, and E. R. Blout. *J. Amer. Chem. Soc.* **82**, 2268 (1960).
[5] F. A. Bovey, G. V. D. Tiers, and G. Filipovich. *J. Polymer Sci.* **38**, 73 (1959).
[6] M. Goodman and Y. Masuda. *Biopolymers* **2**, 107 (1964).
[7] D. I. Marlborough, K. G. Orrell, and H. N. Rydon. *Chem. Comm.* 518 (1965).
[8] W. E. Stewart, L. Mandelkern, and R. E. Glick. *Biochemistry* **6**, 143 (1967).
[9] K.-J. Liu, J. S. Lignowski and R. Ullman. *Biopolymers*, **5**, 375 (1967).
[10] C. Robinson, J. C. Ward, and R. B. Beevers. *Disc. Faraday Soc.* **25**, 29 (1958).
[11] R. Lumry, R. Legare, and W. G. Miller. *Biopolymers* **2**, 489 (1964).
[12] G. Fraenkel and C. Niemann. *Proc. Nat. Acad. Sci.* **44**, 688 (1958).
[13] G. Fraenkel and C. Franconi. *J. Amer. Chem. Soc.* **81**, 62 (1959).
[14] R. J. Gillespie and T. Burchall. *Can. J. Chem.* **41**, 148 (1963).
[15] W. E. Stewart, L. Mandelkern, and R. E. Glick. *Biochemistry* **6**, 150 (1967).
[16] S. Hanlon, S. F. Russo, and I. M. Klotz. *J. Amer. Chem. Soc.* **85**, 2024 (1963).
[17] I. M. Klotz, S. F. Russo, S. Hanlon, and M. A. Stake. *J. Amer. Chem. Soc.* **86**, 4774 (1964).
[18] S. Hanlon. *Biochemistry* **5**, 2049 (1966).
[19] B. J. Litman and J. A. Schellman. *J. Phys. Chem.* **69**, 978 (1965).
[20] D. Balasubramanian and D. B. Wetlaufer. *J. Amer. Chem. Soc.* **88**, 3449 (1966).

ON NEW CHEMICAL REACTIONS OF POLYMERS

Rolf C. Schulz

Institute of Organic Chemistry, University of Mainz, Germany

INTRODUCTION

Three major areas of work present themselves to a chemist who is concerned with polymer materials namely: synthesis and degradation of polymers; structure analysis of natural and synthetic polymers; and chemical transformations of polymers. Although these three fields are very closely related, I wish to concern myself only with the last of them.

Reactions with natural or synthetic polymers have played an important role in technology for a number of decades. I need only mention the nitration and acetylation of cellulose, the sulphochlorination of polyethylene, which leads to Hypalon, or, to give a more recent example, the use of reactive dyes. Chemical transformations of macromolecules are also of great importance in connection with basic research in polymer chemistry. I wish especially to call attention to the fact that many biochemical processes occur through reactions in dissolved macromolecules or on polymer surfaces.

Due to the great significance of polymers with reactive groups, several papers are devoted to this subject at each IUPAC symposium. I need only mention the main lectures of Smets on analogous polymer reactions[1], of Overberger on biologically active polymers[2] and of Manecke on enzyme resins[3] in Montreal, Prague and Tokyo respectively.

But in view of the great number of publications which appear annually on the subject, it is not possible to present a comprehensive survey within the compass of a single lecture and I shall be obliged to restrict myself to certain aspects. I have chosen topics which have not been treated at previous symposia and to which our group in Mainz has been able to make contributions. I wish to discuss the following items:

1. New reactions on polyacroleins
2. Reactions on polymers with anhydride groups, carbonate groups or lactone groups
3. Isomerizations on macromolecules
4. Asymmetric syntheses on polymers
5. Polyradicals

Certain other important aspects will be considered in detail by Professor Sakurada.

REACTIONS ON POLYACROLEINS

Under the influence of free-radical initiators on acrolein, only the vinyl group is polymerized, and polymers with aldehyde groups in the side chains are obtained[4]. [I, equation (1)]

$$\overset{1}{C}H_2 = \overset{2}{C}H - \overset{3}{C}H = \overset{4}{O}$$

R· / \ CN⁻

$$-CH_2-CH-$$
$$\quad\quad |$$
$$\quad\quad CH{=}O$$

$$-CH-O-$$
$$\quad |$$
$$\quad CH{=}CH_2$$

I

II

Poly-1,2-acrolein   Poly-3,4-acrolein   (1)

These aldehyde groups, although not in a free form, are nevertheless highly reactive. The numerous reactions which are possible with this type of poly-acrolein were extensively reported on at an earlier IUPAC symposium[5,6]. We have since been able to show that in the case of alkali cyanides at low temperature, polymerization takes place exclusively at the carbonyl group of acrolein. Polyacetals with vinyl groups in the side chains are obtained[7] (II). We call this polymer poly-3,4-acrolein. The physical and chemical proper-ties of this polyacrolein are of course quite different from those of polyacrolein obtained by free-radical polymerization.

Poly-3,4-acrolein softens at about 90°C and is soluble in various organic solvents. The double bonds of this polymer are highly reactive; hence, upon storage in air for an extended period of time, chemical changes take place, which lead to cross-linking and insolubility. But in solution or at low tem-peratures these polymers can be kept unchanged for some time.

In order to study the chemical properties and the structure of the polymers, we carried out various reactions with the double bonds.

$$-CH-O-$$
$$\quad |$$
$$\quad CH{=}CH_2$$

$\xrightarrow{H_2}$

$$-CH-O-$$
$$\quad |$$
$$\quad CH_2-CH_3$$

$\xrightarrow{\text{Depolym.}}$

$$CH{=}O$$
$$\quad |$$
$$\quad CH_2-CH_3$$

II   III   IV

$$-CH-O-$$
$$\quad |$$
$$\quad CH{=}CH_2$$

$\xrightarrow{Br_2}$

$$-CH-O-$$
$$\quad |$$
$$\quad CHBr-CH_2Br$$

II   V

$$-CH-O-$$
$$\quad |$$
$$\quad HC{=}CH_2$$

$\xrightarrow{Ph_3SnH}$

$$-CH-O-$$
$$\quad |$$
$$\quad CH_2-CH_2-SnPh_3$$

II   VI   (2)

Hydration of poly-3,4-acrolein yields polypropionaldehyde (III). But the latter compound is very unstable and depolymerizes, in part, even during the hydration. The monomeric propionaldehyde (IV) formed in this way can

434

be trapped as phenyl hydrazone and identified by thin-layer chromatography. Thus the structure of the polymer is proved.

Bromine can be added to the double bonds under very mild conditions. This reaction proceeds with yields of over 95% and can be used for the quantitative determination of the vinyl groups. Although the polymer (V) can be isolated, it is not very stable, since it readily cleaves off hydrogen bromide.

Addition of triphenyltin hydride is also possible; the maximum tin content, however, is 5%[8].

The vinyl groups in the side chains of poly-3,4-acrolein can undergo free-radical polymerization. If the polymer is dissolved in styrene or acrylonitrile and if AIBN is added, a grafting reaction takes place upon heating, which ultimately leads to cross-linking [equation (3)].

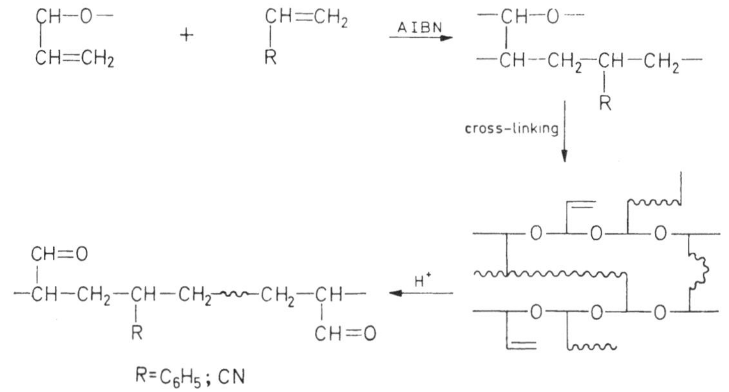

$$R = C_6H_5 ; CN \tag{3}$$

But the acetal links of the polyacrolein main chain can be cleaved by strong acids; hence it is possible to degrade the insoluble network hydrolytically and to isolate the network links of polystyrene or polyacrylonitrile. They are characterized by carbonyl groups[8]. From the molecular weight, the density of the network and the mesh size can be determined. However, these investigations have not yet been completed.

POLYMERS WITH ANHYDRIDE GROUPS, CARBONATE GROUPS OR LACTONE GROUPS

Anhydrides are highly reactive, but few polymers are known to contain this group. It was not till the discovery of cyclopolymerization[9] that poly (acrylic anhydride) and poly(methacrylic anhydride) became readily accessible[10]. These polymers show the usual reactions of the anhydride groups; for example, they form poly(acrylic acid) or poly(methacrylic acid) when hydrolysed. However, it is found that polyacids prepared in this way are not identical with the corresponding polymers obtained from acrylic acid or methacrylic acid by free-radical polymerization. The intra-intermolecular propagation steps during cyclopolymerization give rise to a particular stereo-regularity[11]; therefore the tacticity of the poly(acrylic acid) thus obtained differs from that of the conventional poly(acrylic acid).

At this point I wish to call attention to a peculiarity of reactions with

Q

cyclopolymers which has hitherto received little notice. Ring-opening reactions can produce 1:1 copolymers which frequently cannot be obtained by means of normal copolymerization. For example, by free-radical polymerization of allyl acrylate a cyclopolymer is obtained which, when hydrolyzed, yields a 1:1 copolymer of allyl alcohol and acrylic acid[12].

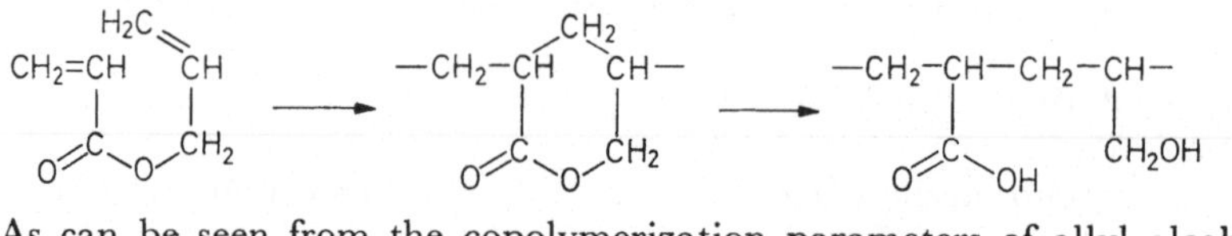

$$(4)$$

As can be seen from the copolymerization parameters of allyl alcohol or allyl acetate and acrylic acid, a copolymer of this type cannot be obtained by means of copolymerization.

Aminolysis of poly(acrylic anhydride) results in a copolymer consisting of monomer units of acrylic acid and acrylamide in a molar ratio which must necessarily be 1:1.

If it is true that ring opening always takes place in the same manner, an alternating copolymer results. This cannot, of course, be generally assumed; but even if the rings cleave at random and without neighbouring group effects, there cannot be more than two consecutive acid groups or amide groups. The maximum sequence length is thus 2. It is to be expected that such copolymers will differ from random copolymers.

In order to test this conclusion we allowed poly(acrylic anhydride) to react with (—)-phenylethylamine; an optically active polymer with the calculated nitrogen content results (5). By copolymerization of acrylic acid and (—)-N-α-phenylethylacrylamide a copolymer with the same nitrogen content can be obtained (6); but it certainly has a different distribution of sequence lengths. The rotatory powers of the two optically-active copolymers are in fact distinctly different[13].

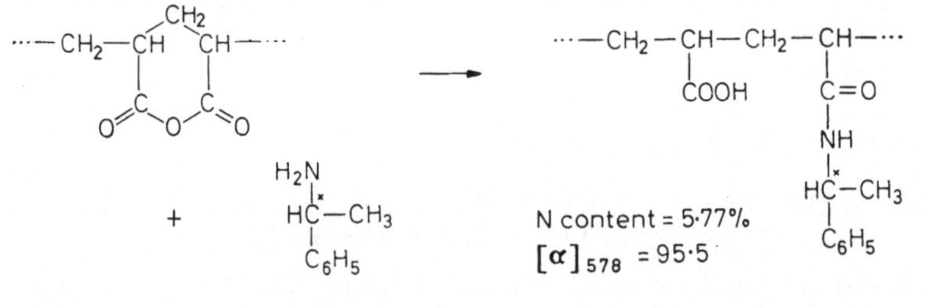

$$(5)$$

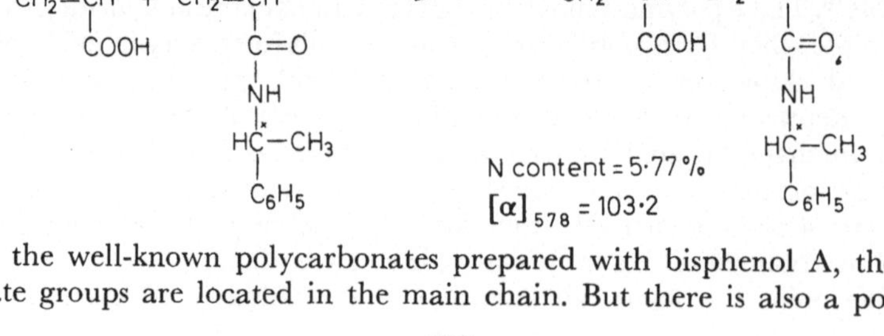

$$(6)$$

In the well-known polycarbonates prepared with bisphenol A, the carbonate groups are located in the main chain. But there is also a polymer with carbonate groups in the side chains. It is obtained by polymerization of vinylene carbonate[14] and has the following structure:

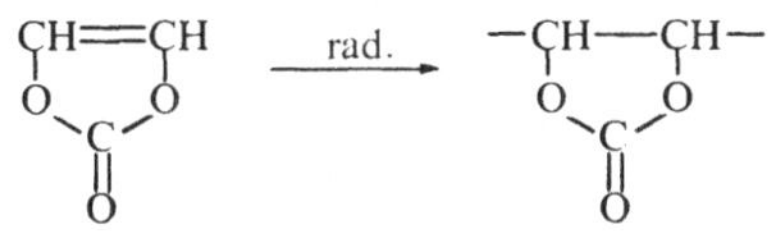

$$
\begin{array}{c}
\text{CH}=\text{CH} \\
| \quad | \\
\text{O} \quad \text{O} \\
\text{C} \\
\| \\
\text{O}
\end{array}
\quad \xrightarrow{\text{rad.}} \quad
\begin{array}{c}
-\text{CH}-\text{CH}- \\
| \quad | \\
\text{O} \quad \text{O} \\
\text{C} \\
\| \\
\text{O}
\end{array}
\tag{7}
$$

Melt-polymerization and polymerization in various solvents with free-radical initiators[14–16] or with γ-rays[17] have been thoroughly investigated by several authors. Cationic[18] or anionic catalysts were found to be unsuitable.

Vinylene carbonate can also be copolymerized with various vinyl compounds and with acrylic acid derivatives[18–20]. In most cases, however, only a small amount of vinylene carbonate is incorporated. Particularly favourable parameters were found for the system vinylene carbonate–vinylisobutyl-ether[21], namely $r_1 = 0.16$ and $r_2 = 0.18$. It follows that for a monomer feed ratio of 49:51 an azeotropic copolymerization occurs.

Both the homopolymers and the copolymers can be readily hydrolysed in an alkaline medium. From the homopolymer a polymer is obtained which consists solely of secondary hydroxyl groups and is therefore called poly-methylol.

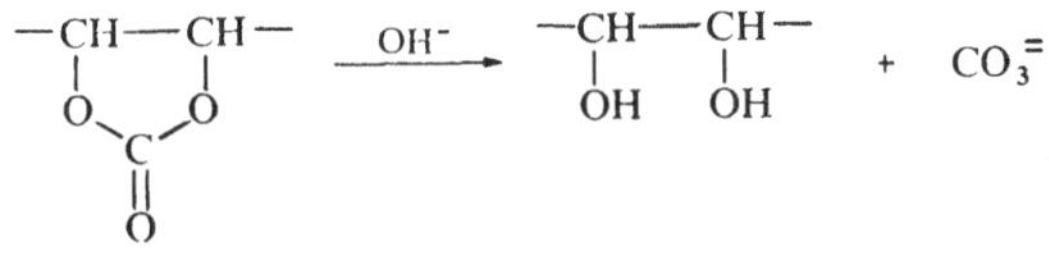

$$
\begin{array}{c}
-\text{CH}-\text{CH}- \\
| \quad | \\
\text{O} \quad \text{O} \\
\text{C} \\
\| \\
\text{O}
\end{array}
\quad \xrightarrow{\text{OH}^-} \quad
\begin{array}{c}
-\text{CH}-\text{CH}- \\
| \quad | \\
\text{OH} \quad \text{OH}
\end{array}
\quad + \quad \text{CO}_3^=
\tag{8}
$$

Polymethylol of high molecular weight is completely insoluble in the usual solvents, presumably because of strong hydrogen bonding. Reactions with the OH-groups of this polymer are therefore extremely difficult to carry out, and frequently only low yields result[15,22]. We found, however, that during the reaction with styrene oxide in an alkaline medium almost all OH-groups are alkylated (9)[17].

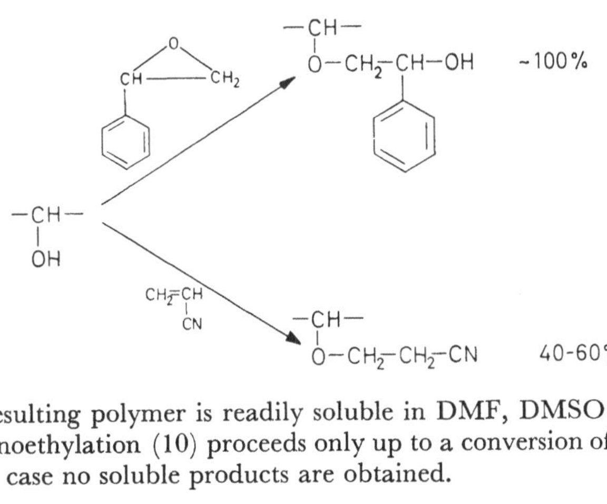

$$
\begin{array}{c}
-\text{CH}- \\
| \\
\text{O}-\text{CH}_2-\text{CH}-\text{OH} \quad \sim 100\%
\end{array}
\tag{9}
$$

$$
\begin{array}{c}
-\text{CH}- \\
| \\
\text{O}-\text{CH}_2-\text{CH}_2-\text{CN} \quad 40\text{-}60\%
\end{array}
\tag{10}
$$

The resulting polymer is readily soluble in DMF, DMSO and chloroform.

Cyanoethylation (10) proceeds only up to a conversion of about 60%; but in this case no soluble products are obtained.

It was then found to be far more advantageous to carry out such reactions not with polymethylol itself as starting material but directly with poly-(vinylene carbonate). For if it is saponified with 5N NaOH at room temperature, the resulting polymethylol remains in clear solution; it is not isolated, but the homogeneous solution is used for further reactions. Poly-(vinylene carbonate) also reacted with various other epoxides in the same manner[17].

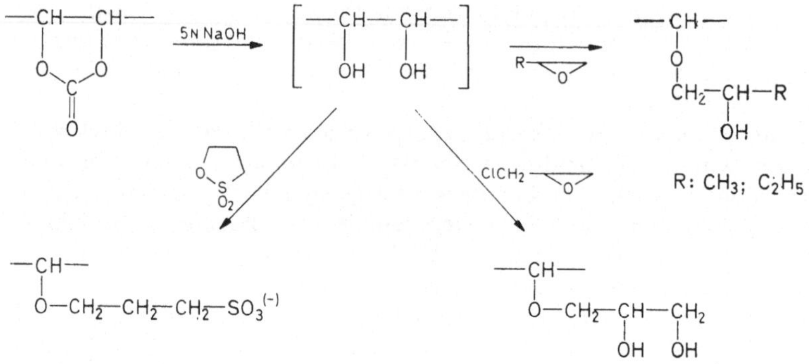

A conversion of up to 70% was obtained, the percentage depending on the experimental conditions. In most cases the reaction products were readily soluble in water.

When polymers containing OH-groups are treated with sulphones, polymeric sulphonic acids are obtained[23]. This reaction too can be applied to polymethylol under the conditions described above[17]. The conversion can be determined from the sulphur content or by titration and amounts to as much as 35%. The free acid and its alkali salts are soluble in water; they exhibit the viscosity behaviour which is characteristic for polyelectrolytes. The alkaline-earth salts and the lead salts are insoluble. [For further reactions of poly(vinylene carbonate) see ref. 69].

Recently polymers have been described which contain β- or γ-lactone groups along the main chain[24]. An example of the preparation of such polymers is shown in equation (12).

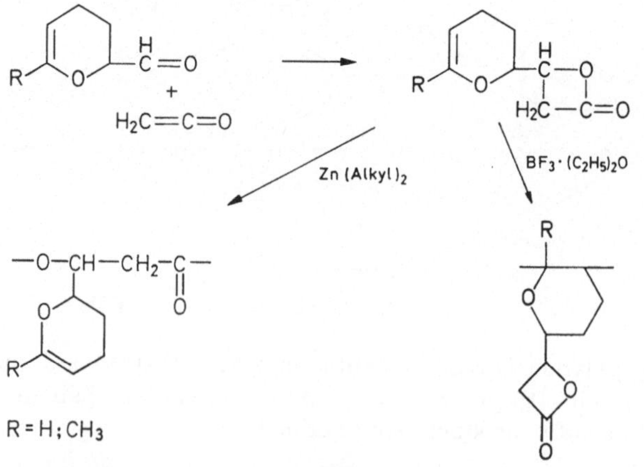

438

The starting material is the acrolein dimer or the methacrolein dimer, to which ketene is added. The resulting β-(2,3-dihydropyranyl-2)-β-lactone has two different groups which are capable of polymerization, namely the β-lactone group and the vinyl ether group; both can be polymerized selectively. Upon initiation with BF_3–etherate, polymers with β-lactone groups in the side chains are obtained. A second method of synthesis is based on the addition reaction of dihydropyran to the hydroxyl groups of a polymeric alcohol (Eq. 13).

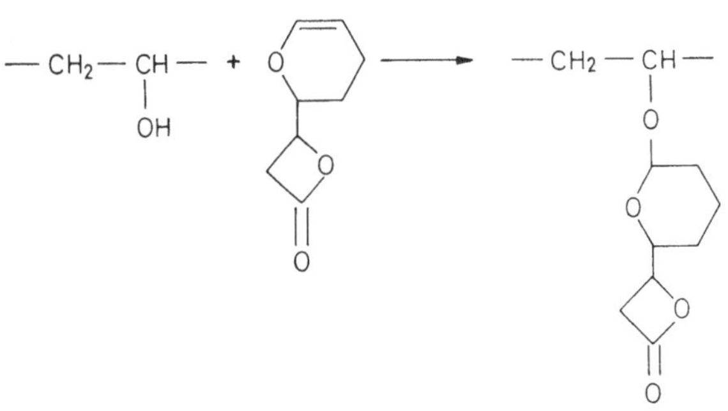

$$\text{(13)}$$

Polymers with γ-lactone groups can be obtained by base-catalysed polymerization of the lactone of γ-hydroxycrotonic acid.

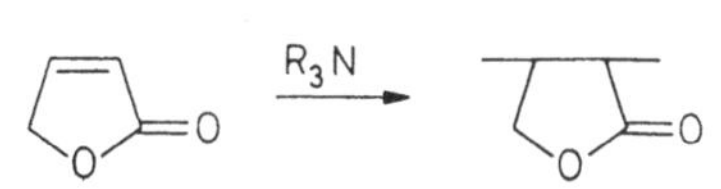

$$\text{(14)}$$

Both the β- and the γ-lactone groups are highly reactive and add water, alcohols, amines, acids etc. under ring opening and formation of the derivatives of the corresponding β- or γ-hydroxyacids.

ISOMERIZATIONS IN MACROMOLECULES

The work of Kennedy[25] has shown that in a number of cationic polymerizations isomerization can take place during the propagation reaction through hydride shifts or rearrangements. Therefore polymerizations of this kind are also called 'isomerization polymerizations'. But since isomerization here takes place only at the growing chain-ends, these investigations do not come within the scope of this paper. I wish to deal only with isomerizations which occur in previously-prepared polymers.

Cyclization reactions in natural rubber, polybutadiene and polyisoprene were the first to be observed and have been studied most thoroughly[26,27]. These reactions are also carried out on a technical scale because the resulting 'cyclo-rubber' is particularly suitable for certain uses[28]. It was not till recently that it was possible, with the help of i.r. and n.m.r. spectroscopy, to elucidate the exact structure of the cyclized polymers.

An example of cyclization of *trans*-1,4-polyisoprene initiated with sulphuric acid is shown in equation (15)[29].

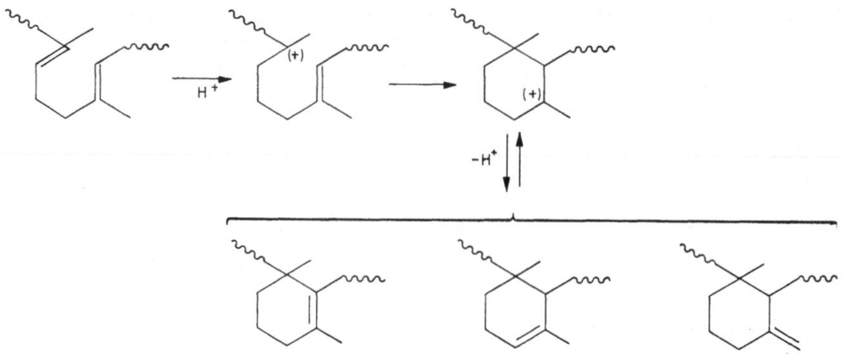

$$(15)$$

After completion of the reaction the cyclized polymer contains only 33% of the C═C double bonds originally present. Apart from the cyclization a shift of the double bonds must also be considered. But recent investigations with polyisoprene deuterated in the 3-position showed conclusively that this isomerization does not occur here[30,31]. On the other hand, Blatz and Johnson[32] have lately reported that, under the influence of acids, the isolated C═C double bonds of polycyclopentadiene shift in such a way that sequences of 3 to 6 conjugated C═C double bonds result.

Cyclization reactions of unsaturated polymers have also been investigated in connection with the preparation of double-strand polymers[33]. Starting materials used for this purpose were, for example, 1,2-polybutadiene and 3,4-polyisoprene. The nature of the cyclization steps and their number of course depend on the experimental conditions and particularly on the tacticity of the polymers used.

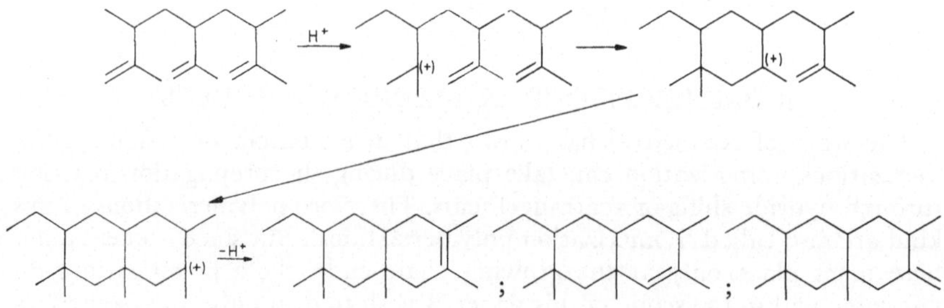

$$(16)$$

At first it was believed that the cyclization reactions spread over many consecutive monomer units and that ladder polymers are formed in the process. But the most recent results of Angelo et al.[34] showed that sequences of 2 to a maximum of 5 anellated rings are found; in addition, in accordance with statistical theory, there are about 18% of isolated vinylidene groups which are left over and remain unchanged.

A further example of cyclization in a polymer is the well-known pyrolysis of polyacrylonitrile, although in this case dehydrations and cleavage reactions also occur.

The *cis–trans* isomerization of unsaturated polymers has received much attention and has been the object of many investigations[26], which were essential for the elucidation of the structure of natural rubber, balata and gutta-percha. When, by means of Ziegler–Natta catalysts, sterically homogeneous polybutadienes and polyisoprenes became available by synthetic methods as well, *cis–trans* isomerization was studied in their cases too. Golub[35] and Cunneen[30] have made fundamental contributions in this field.

It was found that the effect in question is caused by a true thermodynamic equilibrium. In the case of polybutadiene the equilibrium at 25°C corresponds to a *cis–trans* ratio of about 20 : 80, regardless of whether the reaction was begun with a pure *cis*-isomer or a pure *trans*-isomer[36]. In the case of 1,4-polyisoprene at 140°C, an equilibrium with about 45% *cis* and 55% *trans* is reached. The isomerization is catalysed by ultraviolet light, γ-rays, sulphur dioxide (in the form of butadiene sulphone), disulphides etc. and doubtless proceeds according to a free-radical mechanism[37].

Although the chemical composition obviously remains unchanged during these isomerizations, very considerable changes occur in the physical properties. Hence these transformations are of particular interest for the elucidation of the relationship between structure and properties[36].

Cis–trans isomerizations are also possible in the case of unsaturated polyesters of maleic, fumaric or citraconic acids as starting material; such isomerizations have been thoroughly investigated[38].

In connection with the study of tactic polymers the question presented itself as to what extent the configuration of the tertiary carbon atoms in the main chain of vinyl polymers is stable. Braun, Hintz and Kern[39] showed that the tacticity of isotactic polystyrene is not perceptibly altered by mechanical degradation through the effect of aluminium chloride or even by metallization at the α-carbon atom. Thus it was found that the configurational stability of this polymer is considerably higher than that of comparable low-molecular-weight model compounds.

As has already been mentioned, the isomerization of polymers can best be demonstrated by physical methods. I now wish to report on some experiments in which we determined the rate of isomerization by measuring the optical rotations. It is the purely thermal racemization of an optically-active polymer that here concerns us.

For our experiments we chose a polymer whose monomer units are molecularly asymmetric because of hindered rotation and are therefore optically active[40]. The synthesis of a monomer which seemed suitable is shown in equation (17). By Ullmann condensation of o-iodobenzoic acid and 2-iodo-3-nitrotoluene, 2-methyl-6-nitrobiphenyl-2'-carboxylic acid is obtained in 50% yield. When allowed to react with vinyl acetate it yields the corresponding vinyl ester. But, although a variety of conditions were tried, it was not possible to obtain homopolymers or copolymers with styrene from this compound because of the strong inhibiting effect of the aromatic nitro-group.

We therefore introduced the biphenyl groups by reaction with a polymer.

For this purpose the quinidine salts of the acid were separated into the anti-podes, and from the dextrorotatory form the acid chloride was prepared. From it the desired polymer can be obtained through Schotten–Baumann reaction with polyvinyl alcohol. Esterification with methanol, ethanol or

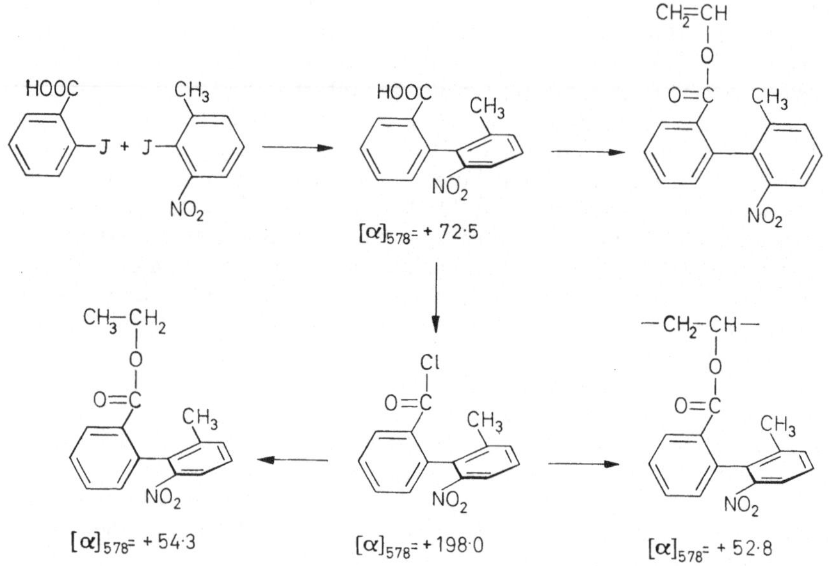

$$[\alpha]_{578} = + 72 \cdot 5$$

$$[\alpha]_{578} = + 54 \cdot 3 \qquad [\alpha]_{578} = + 198 \cdot 0 \qquad [\alpha]_{578} = + 52 \cdot 8$$

(17)

1,4-butanediol yields the low-molecular-weight model compounds which we need for reference measurements. The conversion, as determined by the nitrogen content of the polymer, was found to be 97% by weight; in accordance with this value, the rotations of the polymer also amount to 95% of the molar rotations of the model compounds. This polymer, it should be noted, is the first example of an optically-active polymer without asymmetric carbon atoms.

It is known that atropisomeric biphenyl derivatives racemize when heated. At elevated temperatures the restriction of free rotation is reduced, and when the phenyl rings can rotate freely the optical activity disappears. The racemization can be readily observed by following the gradual decrease of the rotations at elevated temperatures. This is a first-order reaction; the length of the half-life depends on the reaction medium and on the kind of substituents. The results are shown in *Figure 1*. The ethyl ester as well as the methyl ester racemize exactly according to first-order kinetics. At 120°C in dioxane the half-life is 24 hours in solutions of 1% as well as of 0·1%.

The polyvinyl alcohol ester, on the other hand, shows a totally different behaviour. In 1% solutions a curved line results, i.e. there are distinct deviations from first-order kinetics. Presumably this is due to disturbances caused by cross-linking reactions. In 0·1% solutions these deviations do not occur: racemization here proceeds according to first-order kinetics but at a considerably lower rate than in low-molecular-weight model compounds. It might be objected that the ethyl ester represents only one monomer unit and is therefore not an adequate model. But the racemization of the diester of butanediol proceeds with the same half-life; hence no neighbouring-group effect occurs here.

We also checked for a possible effect of the viscosity of the polymer solutions. By adding poly(vinyl acetate) we raised the viscosity of the reaction medium by a factor of 100. This does not influence the racemization of the model compound perceptibly[41].

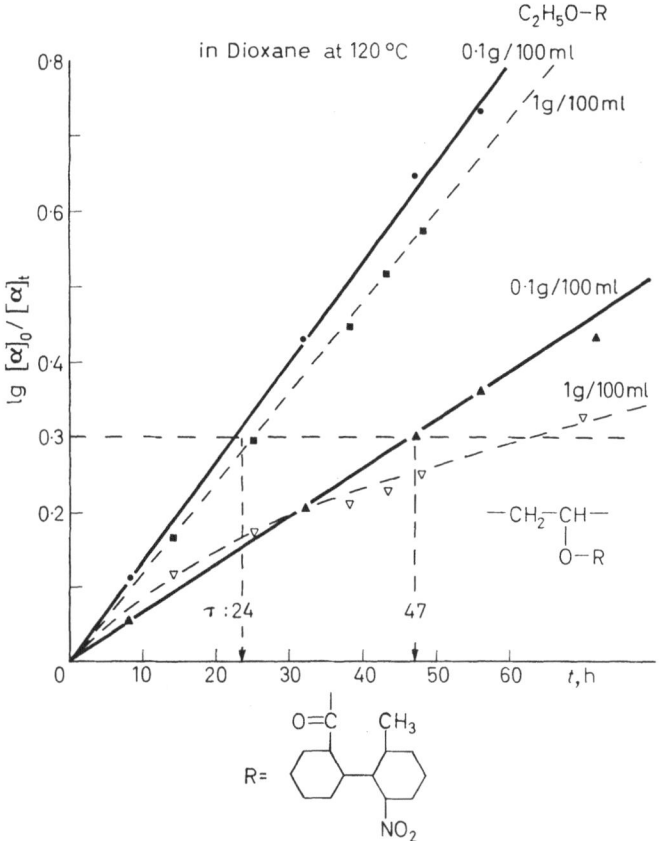

Figure 1. Kinetics of racemization of atropisomeric esters of ethanol and polyvinyl alcohol

We therefore conclude that the low rate of racemization of the poly(vinyl ester) is a result of its macromolecular structure. The free rotation of the phenyl rings is in this case impeded not only by the substituents but by the tight coil shape of the macromolecules as well.

We have since extended our investigations to polymers in which the atropisomeric groups are located in the main chain[42]. Without going into details I merely wish to state that the measurement of optical rotations is a useful method of studying certain isomerization reactions in polymers.

ASYMMETRIC SYNTHESIS ON POLYMERS

In an asymmetric synthesis a new asymmetric centre is formed in the molecule, the reaction proceeding in such a way as to cause preferential formation of one of the antipodes. This means that from a symmetric optically-inactive compound used as starting material an asymmetric optically-active product is obtained. Ideally only one of the antipodes should be formed at all, as a pure product. But in most cases the other antipode is also formed to a greater or lesser extent. The ratio of the optical rotation of the mixture to that of the pure antipodes is the so-called optical yield. In the case of low-molecular-weight compounds numerous asymmetric synthesis have already been described[43]. It is not necessary here for me to enter into a discussion of the steric, kinetic and thermodynamic problems; there are detailed reports and reviews on these subjects[44,45]. Recently Ugi and Ruch have also set up a mathematical model for calculating asymmetric syntheses[46].

The following example is intended to illustrate the principle and the chemical process involved in an asymmetric synthesis[47].

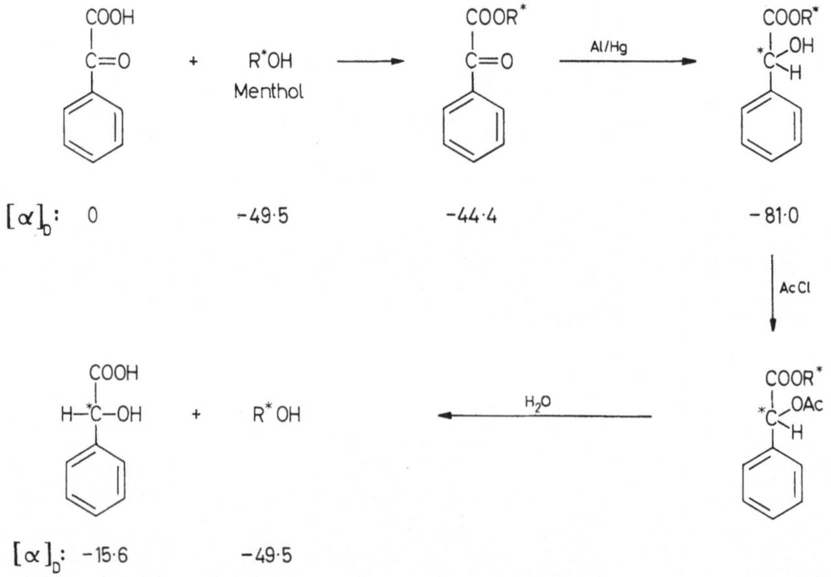

$$(18)$$

Phenylglyoxylic acid, an optically-inactive compound, is esterified with (−)-menthol and then reduced. After the reduction menthol is cleaved off hydrolytically and mandelic acid is obtained, which contains an excess of the laevo-rotatory form. The optical yield is in this case 10%. Menthol here serves only as an auxiliary substance whose purpose is to influence the reduction sterically. Other optically-active compounds can also serve as auxiliary materials; they are frequently employed as catalysts or as solvents[43]. But in each case they are removed when the asymmetric synthesis has taken place.

In polymer chemistry two different kinds of asymmetric syntheses can be distinguished. In the first kind the propagation steps during polymerization of unsaturated compounds are affected in such a way that a monomer without asymmetric centres is converted into an optically-active polymer. Synthesis by this method were first attempted by Marvel and his coworkers in 1943[48]. Later a number of other authors also dealt with this problem, but in no case did the asymmetric synthesis of a polymer succeed in the respect detailed above[49].

It was not until 1961 that Natta[50] was able to report on the first successful synthesis of macromolecules.

$$CH_2=CH-CH=CH \xrightarrow[]{-50\,^{\circ}C} \quad -CH_2-CH=CH-\overset{*}{C}H- $$

COOCH$_3$	COOCH$_3$

$-OC_2H_5$

$+ BuLi$

$[\alpha]_D:$ 0 $+ 7\cdot2$

Natta *et al.* 1960   (19)

An example of this reaction is illustrated in equation (19). β-Vinylacrylic acid esters contain no asymmetric carbon atoms. But during 1,4-polymerization the α-carbon atom becomes asymmetric. If stereospecific catalysts are used together with asymmetric cocatalysts to effect the propagation steps in such a way that *one* of the configurations of this asymmetric carbon atom is preferred in all monomer units, an optically active polymer results. It was possible to show that the optical activity is not the result of built-in catalyst fragments. Through ozonolysis of these polymers, optically-active di- or tri-carbonic acids are obtained[50]; this proves conclusively that we have here an asymmetric synthesis of macromolecules.

Another example is the polymerization of benzofuran with cationic catalysts which are complexed with optically-active compounds[51]. If (−)-β-phenylalanine is used as the asymmetric catalyst, a laevo-rotatory polymer results; if (+)-β-phenylalanine is used, a dextro-rotatory polymer is obtained.

Thus in the cases described so far the asymmetric synthesis depends on the sterically-controlled propagation reaction during the formation of the macromolecules. But in the case of polymers there is yet another kind of asymmetric synthesis. It involves the formation of asymmetric carbon atoms through reactions on functional groups of a previously-prepared optically-inactive polymer. As nothing has yet been published on this problem, to our knowledge, I would like to report on the work we have done so far in this field.

We studied the Reformatsky reaction between acetophenone and menthyl bromoacetate[52] (*see* Eq. 20). In this reaction an asymmetric carbon atom is formed. Menthol again serves merely as an auxiliary material and is cleaved off after the reaction. An optically-active β-hydroxy-β-phenylbutyric acid is obtained whose specific rotation is $[\alpha]_D^{25} = 2\cdot384$. This corresponds to an optical yield of 30%, which is in accordance with published data.

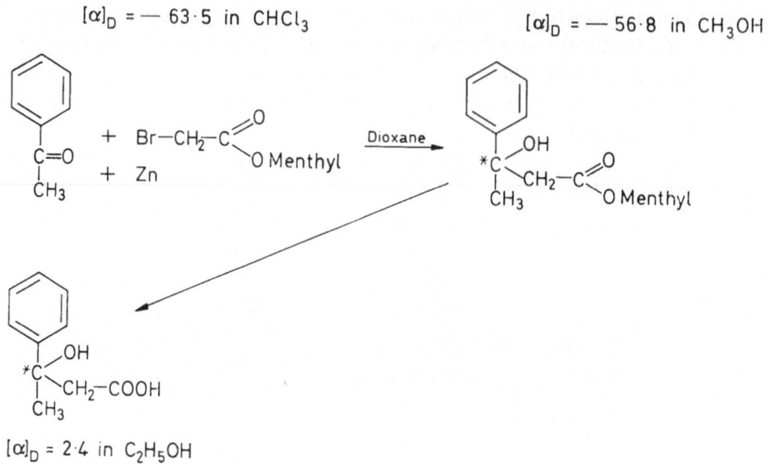

$[\alpha]_D = -63 \cdot 5$ in $CHCl_3$ $[\alpha]_D = -56 \cdot 8$ in CH_3OH

$[\alpha]_D = 2 \cdot 4$ in C_2H_5OH

(20)

After analogous preliminary experiments we applied the same reaction conditions in the case of polyvinylacetophenone[53].

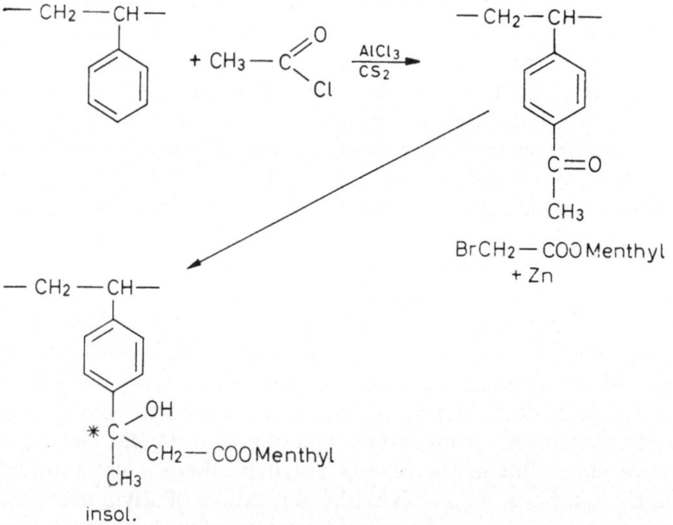

insol.

(21)

In dioxane at 90°C it was in fact possible to carry out a Reformatsky reaction. In the infrared spectrum the bands for the OH-groups and ester groups appear. But the highly reactive zinc–organic intermediates lead to cross-linking of the polymer. Although the reaction conditions were varied, our attempt to obtain a soluble polymer was not successful. Therefore it was not possible to measure the optical rotations or to determine the optical yield. Nor was the application of the Reformatsky synthesis to poly(methyl vinyl ketone) successful.

446

We therefore studied the reaction conditions for an asymmetric reduction of the keto groups. The reduction of polyvinylacetophenone with lithium aluminium hydride has long been known and results in good yields of the corresponding secondary alcohol[54]. On the other hand, it is also known that low-molecular-weight ketones can be asymmetrically reduced if the lithium aluminium hydride has been previously exchanged with an equivalent amount of an optically-active alcohol. Certain glucose derivatives[55] or quinine[56] are suitable for this purpose (22). The structure of the complexes has not yet been elucidated. For instance, optically active α-phenylethanol is obtained from acetophenone. The optical yield amounts to 15 per cent or 48 per cent, depending on the experimental conditions.

$$\text{LiAlH}_4 \; + \; Y \; \longrightarrow \; \left[\text{LiAlH}_{4-n} \cdot Y \right] \; + \; \frac{n}{2}\,\text{H}_2 ; \tag{22}$$

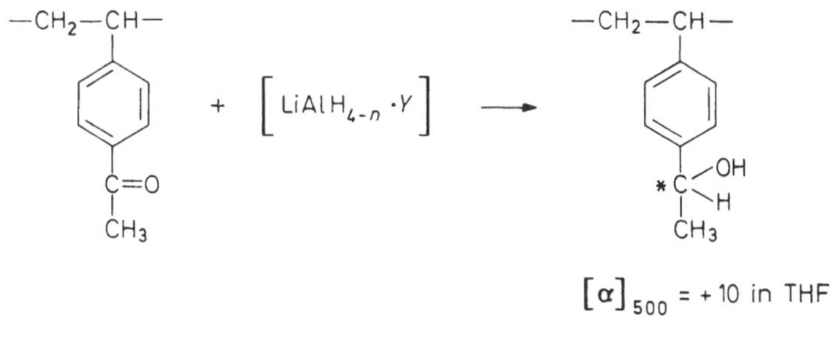

$$[\alpha]_{500} = +10 \text{ in THF}$$

Y: Chinin, n=1; 1,2 Dicyclohexyliden-D-glucofuranose, n=3; .

$$\tag{23}$$

We applied this reaction to polyvinylacetophenone (23)[53]. But because of the solubility properties of the polymer the reaction conditions must be somewhat altered. Nearly all keto-groups were reduced, and readily-soluble polymers resulted. It was possible to separate quantitatively the asymmetric auxiliary agent, i.e. the glucose derivative or the quinine, by reprecipitation or by dialysis. The polymers prepared and carefully purified in this way were in fact optically active. When the glucose derivative is used, a laevo-rotatory polymer results. But the reduction with lithium aluminium hydride–quinine proved to be more advantageous. The polymer resulting from this reduction is dextro-rotatory; for instance, at 500 mμ in THF [α] amounts to +10° (C = 1·404). If it is assumed that the optically-pure polymer has the same specific rotation as (+)-α-phenylethanol, the resulting optical yield should be 15 per cent. This yield is lower than those of the model reactions; possibly it can be yet increased. In any case, this shows for the first time that asymmetric syntheses can also be carried out with macromolecules. We plan to investigate whether other asymmetric syntheses from the chemistry of low-molecular-weight compounds can likewise be adapted to polymers.

MACROMOLECULES WITH NUMEROUS ODD ELECTRONS ALONG THE CHAIN [POLY-RADICALS]

Polymers with a positive or a negative charge at each base unit—as for example polyacrylic acid or polyvinyl pyridinium chloride—are called polyanions or polycations respectively. Consequently, polymers with unpaired electrons in the side chain have to be labelled as "poly-radicals".

These poly-radicals should be distinguished from growing chain-ends in radically-initiated polymerizations; they carry a free electron, but they are termed "macro-radicals" because they contain just *one* radical position per macromolecule. In the following I should like to treat only uncharged poly-radicals, i.e. macromolecules with many unpaired electrons along the main chain.

A well-known reaction to generate O-radicals is based on the homolytical scission of peroxides. Free radicals are also formed by a redox-reaction of hydroperoxides and reducing agents. Application of this reaction to polymers with hydroperoxide side-groups should result in poly-radicals.

In 1955 such polymers were prepared by Hahn and Lechtenböhmer[57] and by Metz and Mesrobian[58].

$$
\begin{array}{ccc}
-CH_2-CH- & \xrightarrow{O_2} & -CH_2-CH- \\
| & & | \\
C_6H_4 & & C_6H_4 \\
| & & | \\
H_3C-\overset{\displaystyle H}{\underset{\displaystyle}{C}}-CH_3 & & H_3C-\overset{\displaystyle OOH}{\underset{\displaystyle}{C}}-CH_3 \\
\end{array}
$$

$$
\begin{array}{ccc}
-CH_2-CH- & \xleftarrow{\qquad} & \\
| & & \\
C_6H_4 & & \\
| & & \\
H_3C-\overset{\displaystyle}{\underset{\displaystyle O\cdot}{C}}-CH_3 & \xrightarrow{nM} & H_3C-\overset{\displaystyle}{\underset{\displaystyle O-M_x-}{C}}-CH_3 \\
\end{array}
\qquad (24)
$$

Autoxidation of poly-*p*-isopropyl styrene yields a polymer with 1–4 per cent hydroperoxide groups, from which, by the action of heat or reducing agents, O-radicals are produced. They are very unstable and cannot be isolated in this form. However, intermediate radical formation can be proved with a highly-sensitive method: these polymeric peroxides initiate radical-chain polymerization of vinyl- or acrylic monomers. Thus, graft copolymers are formed confirming the intermediate existence of poly-radicals.

Hahn and Fischer[59] and more recently, Smets *et al.*[60] also prepared polymers with perester side groups. The reactions of copolymers containing 10 per cent acrylic acid chloride with tertiary butyl hydroperoxide (26) or benzoic peracid (25) have been studied. These polymers, too, yield poly-radicals on thermocracking; Smets and his coworkers studied the kinetics of the peroxide decomposition and the graft copolymerization.

$$-(CH_2-CH)_{\overline{m}}(CH_2-CH)_n- \qquad (25)$$
$$\begin{array}{cc} | & | \\ R & C=O \\ & | \\ & O-O-C-C_6H_5 \\ & \qquad \| \\ & \qquad O \end{array}$$

$$-(CH_2-CH)_{\overline{m}}(CH_2-CH)_n-$$
$$\begin{array}{cc} | & | \\ R & C\!\!\stackrel{O}{\diagdown}\!\!Cl \end{array}$$

$C_6H_5\,COOH$

$(CH_3)_3\,COOH$

R: C_6H_5; COOMe

$$-(CH_2-CH)_{\overline{m}}(CH_2-CH)_n-$$
$$\begin{array}{cc} | & | \\ R & C=O \\ & | \\ & O-O-C(CH_3)_3 \end{array}$$

$$(26)$$

$$R-CH_2-C\!\!\stackrel{\overline{O}}{\diagdown}\!\!\overline{O}\!^{(-)} \quad \xrightarrow{-e} \quad \left[R-CH_2-C\!\!\stackrel{\overline{O}}{\diagdown}\!\!\overline{O}\cdot \right]$$

$$\downarrow$$

$$R-CH_2\cdot + CO_2 \longrightarrow R-CH_2-CH_2-R \qquad (27)$$

$$\begin{array}{c} CH_3 \\ | \\ -CH_2-C- \\ | \\ COO^{(-)} \end{array} \quad \xrightarrow{-e} \quad \left[\begin{array}{c} CH_3 \\ | \\ -CH_2-C- \\ \cdot \\ +CO_2 \end{array} \right] \quad \longrightarrow \begin{array}{l} \text{Disproportionation} \\ \text{Cyclization} \\ \text{Degradation} \end{array}$$

$$(28)$$

Another, well-known method to create radicals is the so-called Kolbe-synthesis (27). Smets[61] has transferred this reaction to polymethacrylic acid (28). The reaction process can be followed by carbon dioxide formation. Obviously, intermediate poly-radicals do exist; however, due to numerous side-reactions, insoluble and non-uniform polymers are formed.

Recently, Braun and Faust[62] described a polymer carrying the Gomberg triphenylmethyl radicals as side-groups (29). From p-vinylphenyl-Mg-bromide and benzophenone the corresponding carbinol is obtained. It is easily polymerized by free radicals. Copolymerization with styrene is also possible. The hydroxy-groups of the polymer are chlorinated by acetyl chloride. Reaction with zinc powder or metallic potassium yields poly-radicals. Intermolecular dimerization leading to cross-linking can be circumvented by working in dilute solution (for example 5 g/l.). The e.s.r.-spectra show indeed that free radicals are present. Quantitative determination revealed on the average 100 unpaired electrons per macromolecule. With a molecular weight of ca. 30 000 this means 1 radical per 3–4 base units. Naturally, the poly-radical is extremely sensitive towards oxygen as is its parent monomeric radical.

Most of the radicals with the unpaired electron located at C- or O-atoms, are unstable. Certain N-radicals, however, are known for high stability, as for example DPPH. Braun et al.[63] were also able to introduce this group into a macromolecule and they obtained a completely stable poly-radical. The synthesis is shown in equation (30).

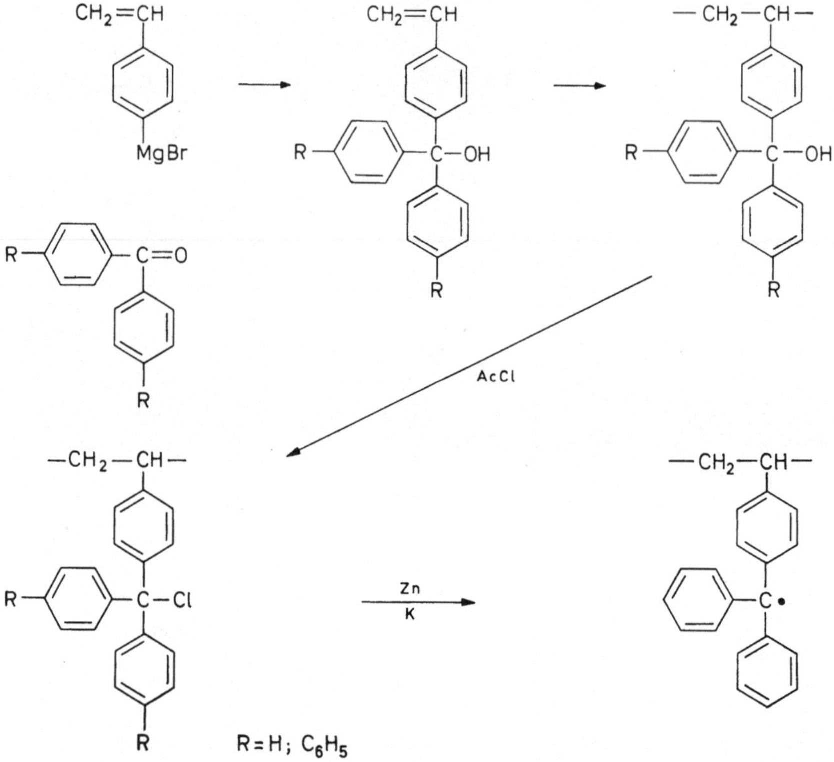
CH2=CH
MgBr
R
C=O
C—OH
—CH2—CH—
AcCl
C—Cl
Zn
K
C•
R=H; C6H5

(29)

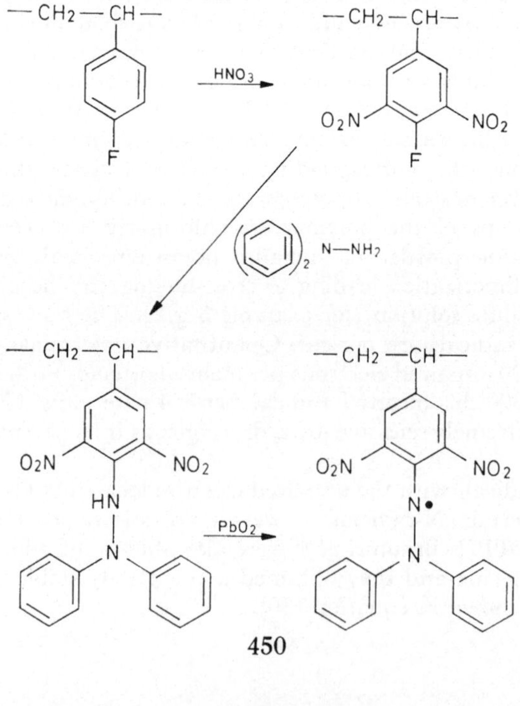
—CH2—CH—
F
HNO3
O2N
NO2
N—NH2
HN
N
PbO2
N•

(30)

By nitration of poly-p-fluoro-styrene the fluorine atom becomes highly activated and nucleophilic substitution of asymmetric diphenyl hydrazine is possible. On shaking with lead dioxide, a dark-purple solution is obtained, from which the poly-radical can be precipitated as a black powder. It is soluble in several organic solvents and quite stable in air and humid conditions. The absorption and the e.s.r spectra are consistent with the low-molecular-model compound. The radical character, determined after different methods amounts to 20–30 per cent. For further poly-radicals see ref. 64.

In 1963 Kuhn[65] discovered a new stable radical which he named verdazyle. The synthesis and structure are given in equation (31).

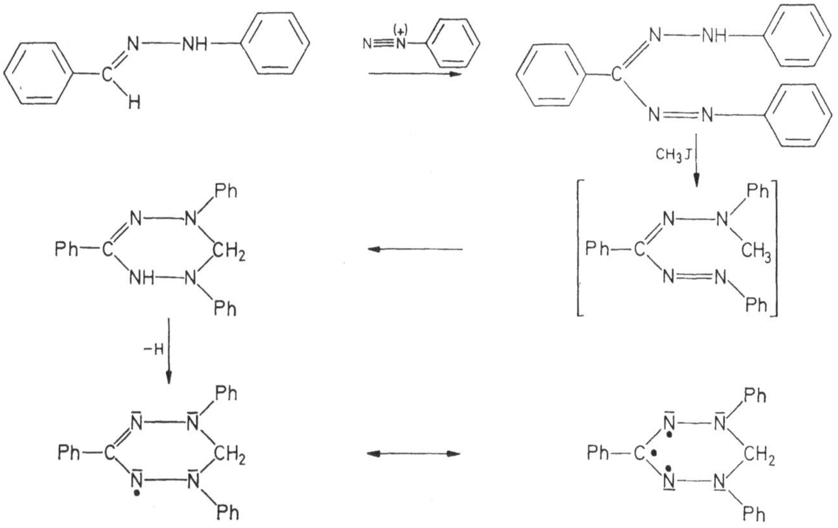

$$(31)$$

Aldehyde phenyl hydrazone and diazonium salt solution form the well-known formazanes. On methylation a tetraaza-cyclohexane is formed. Oxidizing agents remove an H-atom and a solution of deep green colour is obtained from which black crystals can be separated. According to the e.s.r. spectrum, a free radical exists in solution and in the crystal. The unpaired electron might be distributed evenly on the 4 N-atoms.

Some years ago, starting from polyacrolein, we prepared a polymer with formazane side-groups[66]. We thought it would be easy to methylate this and to prepare the polymer verdazyle by oxidation. Unfortunately, all attempts

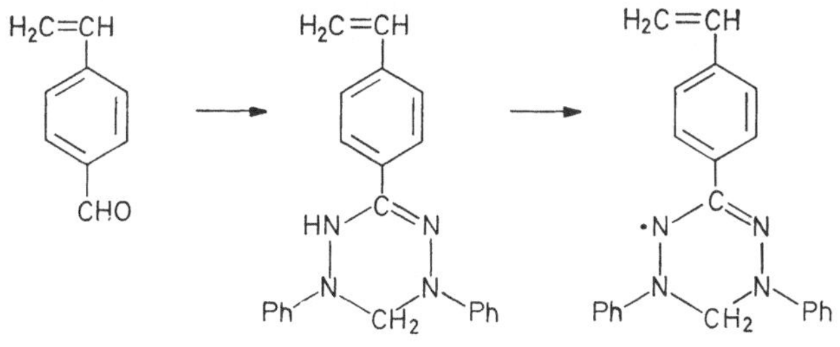

$$(32)$$

451

failed. Too many side reactions occurred and radicals—if any were formed—
were too unstable.

Therefore we prepared a verdazyle carrying a vinyl group at one of the
benzene rings[67] (32). This shows a typical absorption spectrum with a maximum
at 720 mμ and 9 lines in the e.s.r. spectrum (*Figure 2*). However, being a
strong radical-quenching agent, it cannot be polymerized radically.
Polymerization with anionic catalysts is possible, but the investigations on

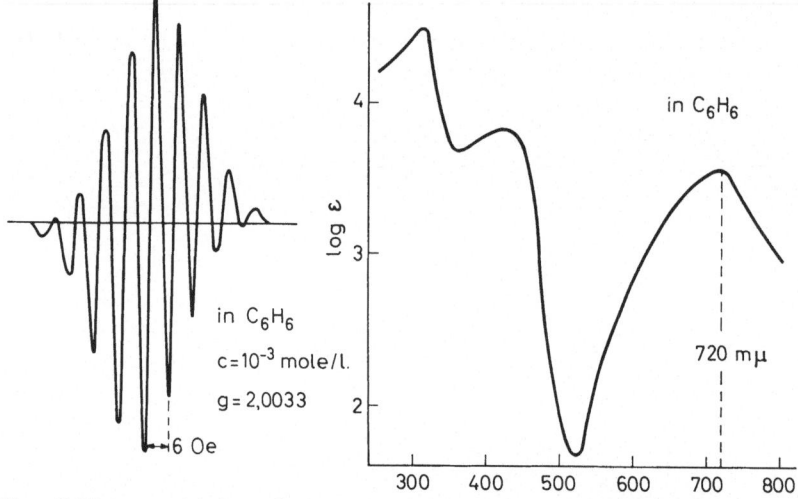

Figure 2. The e.s.r. and absorption spectra of monomeric vinyl verdazyle (see equation 32)

this topic are not yet finished. We then attempted the polymerization of the
precursors, but here insoluble polymers resulted which did not react further.
So we had to start with poly-*p*-vinyl benzaldehyde and conduct all steps
of synthesis, one after the other, with the polymer. In this way it was
possible to obtain a polymer with verdazyle radicals along the main chain.

$$\cdots-CH_2-CH-\cdots \qquad \cdots-CH_2-CH-\cdots$$

(33)

Proof for the presence of free radicals comes from the e.s.r. spectra (*Figure 3*).
The absorption spectrum reveals the agreement in chemical structure of the
polymer radical with the low-molecular-weight verdazyle. From the
spectra it is also possible to judge the radical-concentration. In our best
samples it was *ca.* 45 per cent, that means each second to third base unit is a
verdazyle radical.

The polymer is a green powder, soluble in DMF and stable in air for many months. From the solution green films can be cast in which radicals are also found. A specific property of the verdazyles is the loss of the unpaired electron on the action of oxidizing agents, as for example chlorine or

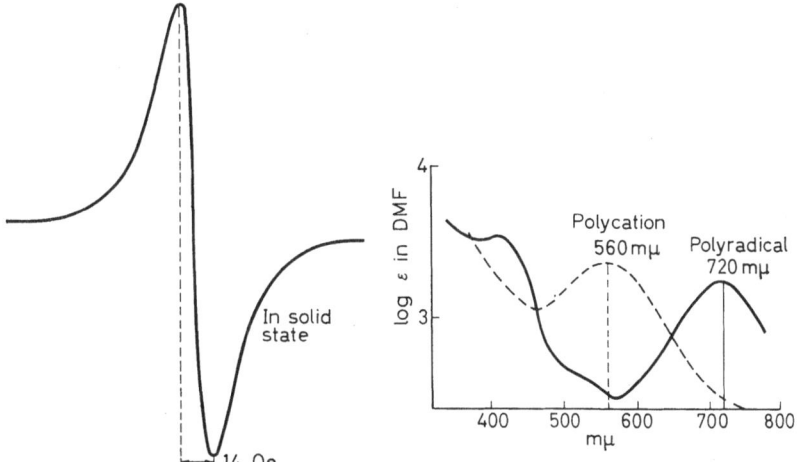

Figure 3. The e.s.r. and absorption spectra (solid lines) of polyverdazyle radical (see equation 33). Broken line: absorption spectrum of poly-cation (see formula 34)

bromine and creation of a purple cation. The maximum at 720 mµ disappears and a new one occurs at 550 mµ. The same reaction is also given by the polymeric verdazyle, in solution as well as in cast film. Again this reaction confirms the structure of the polymer and proves the nature of the poly verdazyle.

$$----CH_2---CH---\cdots \tag{34}$$

This has brought me to the end of my lecture. It is my hope that the examples I have selected are representative of the present stage of development. It was also my purpose to give you an impression of the diversity of chemical transformations of polymers. It is difficult to make predictions about future development in this area of polymer chemistry; but some major fields of interest may be expected to emerge:

1. The preparation of polymers with particularly reactive groups, for example new macromolecular metal–organic compounds.

2. The application of particularly selective reactions to polymers, for example new Redox reactions, complex formations, models of enzymatic reactions and further development of the so-called "matrix reactions"[68].

3. The application of selective reactions to biopolymers.

4. Researches about the relations between tacticity or secondary structure and reactivity of the polymers.

In conclusion, I wish to offer warm thanks to my coworkers for their industrious assistance and cooperation in the course of these investigations. Furthermore I am indebted to the "Deutsche Forschungsgemeinschaft" and to "Fonds der Chemischen Industrie" for supporting our research.

References

[1] G. Smets. *Makromol. Chem.* **34**, 190 (1959); *Pure and Applied Chemistry* **12**, 211 (1966); I.U.P.A.C. *Symposium on Macromolecular Chemistry*, Prague, 1965.

[2] C. G. Overberger, J. J. Ferraro, P. V. Bonsignore, F. W. Orttung, and N. Vorchheimer. *Pure and Applied Chemistry* **4**, 521 (1962); *I.U.P.A.C. Symposium on Macromolecular Chemistry*, Montreal, 1961.

[3] G. Manecke. *Pure and Applied Chemistry* **4**, 507 (1962).

[4] R. C. Schulz. *Kunststoffe* **47**, 303 (1957).

[5] W. Kern, R. C. Schulz, and D. Braun. *J. Polymer Sci.* **48**, 91 (1960).

[6] R. C. Schulz. *Encyclopedia of Polymer Sciences and Technology* Vol. I, p. 160 Interscience, New York, 1964.

[7] R. C. Schulz and W. Passmann. *Makromol. Chem.* **60**, 139 (1963); R. C. Schulz. *Chimia* **19**, 143 (1965); R. C. Schulz, G. Wegner, and W. Kern. *J. Polymer Sci. c* **16**, 989 (1967). *Makromol. Chem.* **100**, 208 (1967).

[8] R. C. Schulz and D. Margotte. unpublished results.

[9] G. B. Butler in *Encyclopedia of Polymer Science and Technology*, Vol. 4, p. 568, Interscience, New York, 1966.

[10] G. B. Butler, A. Crawshaw, and W. L. Miller in *Macromolecular Syntheses* (Ed. C. G. Overberger), Vol. 1, p. 38 Wiley, New York, 1963.

[11] W. L. Miller, W. S. Brey, and G. B. Butler. *J. Polymer Sci.* **54**, 329 (1961); G. B. Butler. *Pure and Applied Chemistry* **4**, 299 (1962); J. C. H. Hwa. *J. Polymer Sci.* **60**, S12 (1962); M. Reinmöller and T. G. Fox. *Polymer Preprints* **7**, 1005 (1966); J. Mercier and G. Smets. *J. Polymer Sci.* **A1**, 1491 (1963).

[12] R. C. Schulz, M. Marx, and H. Hartmann. *Makromol. Chem.* **44/46**, 281 (1961).

[13] J. Schwaab and R. C. Schulz, unpublished results, Mainz, 1963.

[14] M. S. Newman and R. W. Addor. *J. Amer. Chem. Soc.* **75**, 1263 (1953).

[15] N. D. Field and J. R. Schaefgen. *J. Polymer Sci.* **58**, 533 (1962).

[16] G. Smets and K. Hayashi. *J. Polymer Sci.* **29**, 257 (1958).

[17] N. Vollkommer and R. C. Schulz, unpublished results, Mainz, 1966.

[18] H. C. Haas and N. W. Schuler. *J. Polymer Sci.* **31**, 237 (1958).

[19] H. L. Marder and C. Schuerch. *J. Polymer Sci.* **44**, 129 (1960); J. M. Judge and C. C. Price. *J. Polymer Sci.* **41**, 435 (1959).

[20] K. Hayashi and G. Smets. *J. Polymer Sci.* **27**, 275 (1958).

[21] R. C. Schulz and R. Wolf. *Kolloid Z.* **220**, 148 (1967).

[22] C. C. Unruh and D. A. Smith. *J. Org. Chem.* **23**, 625 (1958).

[23] E. J. Goethals and G. Natus. *Makromol. Chem.* **93**, 259 (1966).

[24] R. Palm, H. Ohse and H. Cherdron. *Angew. Chem.* **78**, 1093 (1966); *Angew. Chem. Internat. Edit.* **5**, 994 (1966).

[25] J. P. Kennedy and A. W. Langer. *Advances in Polymer Sci.* **3**, 523 (1964).

[26] For a summary of the work see: M. A. Golub in *Chemical Reactions of Polymers* (Ed. E. M. Fettes), pp. 110, 116, Interscience, New York, 1964.

[27] J. Scanlan. *ibid* p. 125.

[28] *Ullmann's Encyclopädie der techn. Chemie* Vol. **9**, pp. 319–321 (1957).

[29] M. Stolka, J. Vodehnal and I. Kössler. *J. Polymer Sci.* **A2**, 3987 (1964).

[30] J. I. Cunneen, G. M. C. Higgins, and R. A. Wilkes. *J. Polymer Sci.* **A3**, 3503 (1965).

[31] M. A. Golub. *J. Polymer Sci.* **B4**, 227 (1966).

[32] P. E. Blatz and R. H. Johnson. *Polymer Preprints* **7**, 616 (1966).

[33] W. De Winter. *Reviews in Macromolecular Chemistry* **1**, 329 (1966).

[34] R. J. Angelo, M. L. Wallach, and R. M. Ikeda. *Polymer Preprints* **8**, 221 (1967).

[35] M. A. Golub. *J. Polymer Sci.* **25**, 373 (1957); *J. Amer. Chem. Soc.* **81**, 54 (1959); *J. Polymer Sci.* **B4**, 227 (1966).

[36] M. Berger and D. J. Buckley. *J. Polymer Sci.* **A1**, 2945 (1963).

[37] G. R. Seely. *J. Amer. Chem. Soc.* **84**, 4404 (1962).

[38] G. Bier. *Makromol. Chem.* **4**, 41 (1949); H. Batzer and B. Mohr. *ibid.* **8**, 217 (1952); H. Batzer. *Angew. Chem.* **66**, 513 (1954); see also W. Brügel and K. Demmler preprint 10/25 of this symposium.

[39] D. Braun, H. Hintz and W. Kern, *Makromol. Chem.* **62**. 108 (1963).

[40] R. C. Schulz and R. H. Jung *Makromol. Chem.* **96**, 295 (1966).

[41] R. C. Schulz and R. H. Jung, *Tetrahedron Letters* 4333 (1967).

[42] R. C. Schulz and R. H. Jung *Angew. Chem.* **79**, 423 (1967); *Internat. Edit.* **6**, 461 (1967).

[43] Je. I. Klabunowski *Asymmetrische Synthese* Deutscher Verlag der Wissenschaften, Berlin 1963. W. Theilacker in *Houben-Weyl-Müller*, 4. Aufl. Vol. **IV/2**, p. 535, Thieme Verlag, Stuttgart, 1955.

[44] V. Prelog *Helv. Chim. Acta* **36**, 308, (1953).

[45] E. L. Eliel (transl. by A. Luttring and R. Cruse) *Stereochemie der Kohlenstoffverbindungen* Verlag Chemie, Weinheim, 1966.

[46] E. Ruch and I. Ugi *Theoret. Chim. Acta* **4**, 287 (1966).

[47] A. McKenzie and H. B. P. Humphries. *J. Chem. Soc.* **95**, 1105 (1909).

[48] C. S. Marvel, R. L. Frank and E. Prill. *J. Amer. Chem. Soc.* **65**, 1647 (1943).

[49] For further information see R. C. Schulz and E. Kaiser. *Advances Polymer Sci.* **4**, 289 (1965).

[50] G. Natta. *Pure and Applied Chemistry* **4**, 363 (1962).

[51] G. Natta, M. Farina, M. Peraldo and G. Bressan. *Makromol. Chem.* **43**, 68 (1961); M. Farina and G. Bressan. *ibid.* **61**, 79 (1963).

[52] J. A. Reid and E. E. Turner. *J. Chem. Soc.* 3365 (1949); 3694 (1950).

[53] H. Mayerhöfer and R. C. Schulz, *Angew. Chem.* **80** (1968), in the press.

[54] J. A. Blanchette, U.S.P. 2,915,511 (1959); *Chem. Abs.* **54**, 6199c (1960).

[55] S. R. Landor, B. J. Miller, and A. R. Tatchell. *Proc. Chem. Soc.* 227 (1964); *J. Chem. Soc. C*, 1822 (1966).

[56] O. Cervinka. *Coll. Czech. Chem. Comm.* **30**, 1684 (1965).

[57] W. Hahn and H. Lechtenböhmer. *Makromol. Chem.* **16**, 50 (1955).

[58] D. J. Metz and R. B. Mesrobian. *J. Polymer Sci.* **16**, 345 (1955).

[59] W. Hahn and A. Fischer. *Makromol. Chem.* **16**, 36 (1955).

[60] G. Smets. *J. Polymer Sci.* **54**, 65 (1961); **A2**, 2417, 2423 (1964); *Makromol. Chem.* **91**, 160 (1966).

[61] G. Smets, X. van der Borght and G. van Haeren. *J. Polymer Sci.* **A2**, 5187 (1964).

[62] D. Braun and R. J. Faust. *Angew. Chem.* **78**, 905 (1966).

[63] D. Braun, I. Löflund and H. Fischer. *J. Polymer Sci.* **58**, 667 (1962).

[64] D. Braun. *Polymer Preprints* **8**, 683 (1967).

[65] R. Kuhn and H. Trischmann. *Angew. Chem.* **75**, 294 (1963); *Mh. Chem.* **95**, 457 (1964).

[66] R. C. Schulz, R. Holländer, and W. Kern. *Makromol. Chem.* **40**, 16 (1960).

[67] M. Kinoshita and R. C. Schulz, *Makromol. Chem.* **111**, 137 (1968).

[68] H. Kämmerer, J. Shukla, N. Oender, and G. Scheuermann, Preprint 1/128 of this symposium.

[69] V. D. Nemirovskii, M. A. Pavlovskaja, V. V. Stepanov, and S. S. Skorokhodov. *Vysokomol. Soedin* **7**, 1580 (1965); M. G. Krakovyak, S. J. Klenin and S. S. Skorokhodov. *Vysokomol. Soedin* **7**, 1576 (1965). V. D. Nemirovski and S. S. Skorokhodov. *J. Polymer Sci.* **C16**, 1471 (1967).

CATIONIC POLYMERIZATION OF
α, β-DISUBSTITUTED OLEFINS

A. Mizote, T. Higashimura, and S. Okamura

Kyoto University, Kyoto, Japan

INTRODUCTION

It is well known that α,β-disubstituted olefins cannot usually be polymerized to high polymers, especially by free-radical-type polymerization. The difficulty of polymerization has been attributed to the steric repulsion induced by the β-substituted group in the transition state of the propagating reaction[1].

It can be presumed, however, that the introduction of an alkyl group at the β-position of the vinyl double bond increases the electron density of the carbon–carbon double bond and thus the olefins become more reactive, especially by cationic polymerization.

Experimental results, however, showed these monomers to have rather low reactivity. For instance, C. G. Overberger and his colleagues (1958)[2] concluded that the steric effect of the β-alkyl group in styrene derivatives was so large that these monomers could not be polymerized into high polymers even by the cationic mechanism, in spite of the electron-donating property of the β-alkyl substituent. Usually, the electronic effect of the β-substituted group combines with the steric effect in the monomer reactivity and separation has not been previously tried. We should like to discuss first the effect of the β-methyl group of β-methylstyrenes and propenyl ethers (β-methyl vinyl ethers) in the copolymerization technique.

It is also interesting to study the reactivities of *cis*- and *trans*-isomers in the polymerization of α,β-disubstituted olefins. In the free-radical polymerization of fumaric and maleic esters, as already reported by F. R. Mayo and his colleagues (1948)[3], the *trans*-isomer was more reactive than the *cis*-isomer. This fact was explained by the large resonance stabilization of the propagating radical formed from the *trans*-isomer at the transition state of the propagating reaction.

In the cationic polymerization of α,β-disubstituted olefins a few experimental results have been reported. C. G. Overberger found that in the cationic polymerization of β-methylstyrenes the *trans*-isomer was a little more reactive than the *cis*-isomer. P. H. Plesch and his colleagues (1958)[4] observed that the less stable *cis*-stilbene was more reactive than the *trans*-isomer in cationic polymerization. Recently, J. Furukawa and his coworkers[5] studied the reactivity of α,β-unsaturated ether by hydrolysis and suggested that the *trans*-isomer was more stable than the *cis*-isomer. Many

457

different results have now been reported for the reactivity difference between cis- and trans-isomers.

We should like to discuss, secondly, the reactivity difference between isomers in the cationic polymerization of α,β-disubstituted olefins. It is also important to know the type of double bond opening for the investigation of the propagation reaction in ionic polymerization.

G. Natta and his colleagues[6] have concluded from x-ray diffraction on the di-tactic structure of polymers that the double bond of trans-propenyl ether usually opened in cis-type in the cationic polymerization. From x-ray examination, however, it is very difficult to know the steric structure of a polymer, or to discuss quantitatively the type of opening of a double bond, especially when the polymer is amorphous.

Finally, we should like to discuss the type of opening of the monomer double bond in the cationic polymerization of α,β-disubstituted olefins by n.m.r. investigation.

All these three factors are very important, we believe, in discovering the general characteristics of ionic polymerization of vinyl monomers.

CHANGE OF REACTIVITIES BY INTRODUCTION OF β-METHYL GROUP

Here the effects of the β-methyl group of β-methylstyrenes and propenyl ethers (β-methyl vinyl ethers) are studied by comparison of the monomer reactivity ratios in copolymerization with the corresponding β-unsubstituted monomers.

Styrenes[7]

We chose monomer pairs for which the propagating cations would be quite similar to each other in structure. Here the copolymerizations are made between styrene derivatives and β-methylstyrene derivatives in which nuclear substituents are the same, as shown in *Figure 1*. As seen in this *Figure* the propagating cations of styrenes and β-methylstyrenes can be

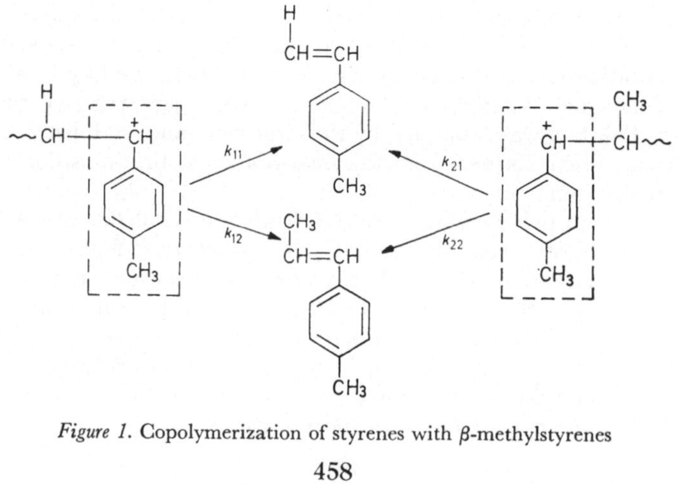

Figure 1. Copolymerization of styrenes with β-methylstyrenes

considered to be quite similar, because the inductive effect of the β-methyl group for the growing cation may be small for the charge stabilization.

The values of monomer reactivity ratios obtained here are summarized in *Table 1*. If the β-methyl group had no steric repulsion in the propagating

Table 1. Cationic copolymerization of styrenes (M_1) and β-methylstyrene (M_2)

M_1	M_2	R	r_1	r_2
CH$_2$=CH 丨 ⬡ 丨 R	CH$_3$ 丨 CH=CH 丨 ⬡ 丨 R	—H —CH$_3$ —OCH$_3$	$1{\cdot}8 \pm 0{\cdot}2$ $1{\cdot}3 \pm 0{\cdot}3$ $1{\cdot}2 \pm 0{\cdot}2$	$0{\cdot}07 \pm 0{\cdot}02$ $0{\cdot}04 \pm 0{\cdot}04$ $0{\cdot}04 \pm 0{\cdot}02$

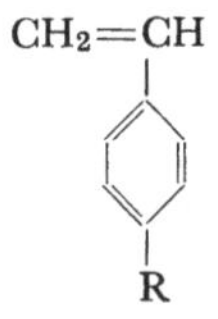

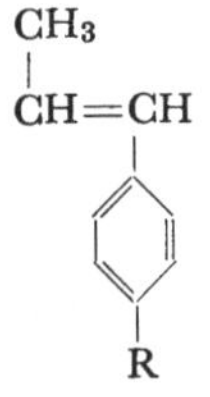

step, then $r_1 \, (= k_{11}/k_{12})$ should be equal to $1/r_2 \, (= k_{21}/k_{22})$. From the observed values, however, all r_2-values are found to be much smaller than expected when $r_2 = 1/r_1$. The results in this table showing all $r_1 > 1$ and $1/r_2 > 1$, can be interpreted as indicating that the lowered reactivity of β-methylstyrenes is due to the steric effect of the β-methyl group.

Now we should like to make a quantitative examination of the lowering of reactivities of β-methylstyrenes. The influence of monomer structure on the free energy of the propagating reaction can be expressed as the sum of the polar, resonance and steric factors, given by Branch and Calvin[8] as:

$$-\Delta F^{\ddagger} = P + R + S = RT \ln (k/k_0)$$

$$\log (k/k_0) = F_P + F_R + F_S$$

where F_P is the polar factor, F_R the resonance factor, and F_S the steric factor. Applying this relation to copolymerization, we obtain

$$\log 1/r_1 = \log k_{12}/k_{11} = F_P + F_R + F_S$$

$$\log r_2 = \log k_{22}/k_{21} = F_P' + F_R' + F_S'$$

Here propagating cations, M_1^+ and M_2^+ might be the same, in resonance and polar factors, then,

$$F_R + F_P = F_R' + F_P'$$

$$\therefore \quad \log r_2 - \log 1/r_1 = F_S' - F_S$$

$$F_S \ll F_S'$$

$$\therefore \quad \log r_2 - \log 1/r_1 = F_S'$$

Then, the steric factor can be expressed by ($\log r_2 - \log 1/r_1$) being between $-1\cdot0 \sim -1\cdot3$, calculated in this table. This means that the rate of homo-polymerization is depressed by a factor of $1/10 \sim 1/20$ compared with the corresponding styrenes, due to the steric repulsion of the β-methyl group.

Vinyl ethers[9]

For the estimation of reactivity, propenyl ethers were copolymerized with vinyl ethers, in which monomer reactivity ratios were determined by measuring the amount of residual monomers by gas chromatography. As shown in *Table 2*, the monomer reactivity of propenyl n-butyl ether, here (M_1), was found to be surprisingly higher than that of vinyl ether, here (M_2).

Table 2. Monomer reactivity ratio in the copolymerization of propenyl ether $(CH_3 \cdot CH = CHOR)$, (M_2), with the corresponding vinyl ether $(CH_2 = CHOR)$, (M_1), by $BF_3 \cdot O(C_2H_5)_2$ ([M]$_0$: 10 vol. %, [C]: 2 mmole/l., -74 to $-78°C$)

Monomer R	Geometric structure	Toluene[a] r_1	Toluene[a] r_2	Methylene chloride[b] r_1	Methylene chloride[b] r_2
Ethyl	cis	$0\cdot35 \pm 0\cdot1$	$4\cdot0 \pm 0\cdot5$	—	—
	trans	$0\cdot94 \pm 0\cdot1$	$0\cdot94 \pm 0\cdot1$	—	—
n-Butyl[3]	cis	—	—	$0\cdot50 \pm 0\cdot2$	$4\cdot0 \pm 1\cdot0$
	trans	—	—	$0\cdot80 \pm 0\cdot3$	$2\cdot3 \pm 0\cdot3$
Isopropyl	cis	$1\cdot1 \pm 0\cdot2$	$0\cdot80 \pm 0\cdot4$	—	—
	trans	$4\cdot9 \pm 0\cdot4$	$0\cdot19 \pm 0\cdot05$	—	—
tert-Butyl	cis	$2\cdot2 \pm 0\cdot4$	$0\cdot28 \pm 0\cdot08$	—	—
Isobutyl[4]	cis	—	—	$0\cdot29 \pm 0\cdot05$	$2\cdot20 \pm 0\cdot14$
	trans	—	—	$1\cdot04 \pm 0\cdot04$	$0\cdot90 \pm 0\cdot03$

a Toluene contains 5 vol. % methylene chloride as an internal standard for the gas-chromatographic measurement.
b Methylene chloride contains 5 vol. % of toluene for the same reason.

In both the *cis*- and *trans*-isomers, propenyl ethyl and n-butyl ethers have larger reactivities than corresponding vinyl ethers. In the copolymerization of β-methylstyrene with unsubstituted styrene, the steric hindrance makes the polymerization of β-methylstyrene difficult. This is in contradiction with the results for propenyl ether. However, propenyl ethers having a branched alkoxy group at the first carbon atom after ether oxygen (isopropyl and t-butyl propenyl ethers) were less reactive than the corresponding vinyl ethers.

β-Alkoxystyrenes[10]

It is interesting to study the behaviour of β-alkoxystyrenes in cationic polymerization. If the β-alkoxystyrenes act as the styrene type monomers, the polymerizability should decrease with the increase of size of the β-alkoxy group due to the increase of steric hindrance. On the other hand, if the β-alkoxystyrenes act as the vinyl ether type monomers, the reactivity is expected to rise as the size of alkoxy group increases.

Here the monomers used are β-methoxy-, β-ethoxy-, β-n-propoxy- and β-n-butoxy-styrenes. From the copolymer composition curves of β-alkoxy-styrenes with vinyl ether, the monomer reactivity ratio is calculated. The order of reactivity found in these monomers is shown in *Table 3*. We can consider that β-alkoxystyrenes act as the vinyl-ether-type monomers.

Table 3. Monomer reactivity ratios of the copolymerization of vinyl-n-butyl ether (M_1) with β-alkoxystyrenes (M_2)

	β-Methoxy	β-Ethoxy	β-n-Propoxy	β-n-Butoxy
r_1	0·55	0·45	0·40	0·25
r_2	0·15	0·20	0·25	0·45

As reported previously, in the case of cationic polymerization of vinyl ethers, the molecular weight of the resultant polymer increases with the decreasing polarity of solvent but decreases in the case of styrene derivatives. In the homopolymerization of β-methoxystyrene catalysed by $SnCl_4$, the high molecular weight polymer was obtained in a non-polar solvent such as toluene. A similar tendency is found in the polymerization of vinyl ethers.

These results suggest that the growing end of β-alkoxystyrene is a similar structure to that of vinyl ether. Now β-alkoxystyrene can be regarded as β-phenyl vinyl alkyl ether in the propagation reaction.

REACTIVITIES OF *cis*- AND *trans*-ISOMERS

In cationic polymerization the *trans*-isomer of β-methylstyrenes[2] was reported to be a little more reactive than the *cis*-isomer. However in the polymerization of stilbene, the *cis*-isomer was recognized to be more reactive than the *trans*-isomer[4].

As already shown in *Table 2*, *cis*-propenyl ether is more reactive than the *trans*-isomer in the copolymerization with vinyl ether. We should next like to clarify the difference in reactivities of *cis*- and *trans*-isomers in a,β-disubstituted olefins.

Styrene derivatives[11]

The residual monomer composition in the polymerization of the mixture of two isomers of β-methylstyrene and of β-methyl-p-methoxystyrene (i.e. anethol) is measured by gas chromatography. In this experiment, no isomerization could occur during polymerization. If the nature of the growing end is assumed to be the same in both *cis*- and *trans*-isomers, the gradients of first-order plots represent the polymerizability of both monomers. Polymerization was carried out in toluene or ethylene dichloride with stannic chloride or boron fluoride etherate at 0°C. It was found that the *trans*-β-methylstyrene was $1\cdot3 \sim 1\cdot5$ times more reactive than the *cis*-β-methylstyrene against styrene–carbonium ion. The copolymerization between the *trans*- and the *cis*-isomer of β-methylstyrene showed little difference in their monomer reactivities.

In contrast to β-methylstyrene, the *cis*-anethole was $1 \cdot 5 \sim 2 \cdot 0$ times more reactive than the *trans*-anethole, estimated by the copolymerization between the *trans*- and the *cis*-isomer of anethole.

Propenyl ethers[9]

Propenyl ethers having various molar ratios of *cis*- and *trans*-isomers were polymerized in toluene with BF_3OEt_2 at $-78°C$, and the gradients of monomer consumption of first-order plots are compared as the polymerizabilities. As shown in *Figure 2*, *cis*-isomers have larger reactivities than those of *trans*-isomers.

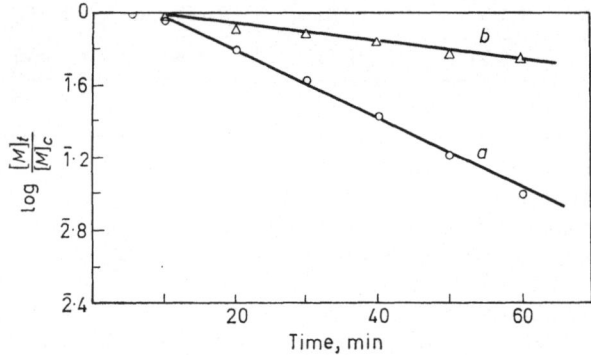

Figure 2. First-order plot for the monomer concentration in the copolymerization of *cis*- and *trans*-isopropyl propenyl ether.

a ($\bigcirc$): *cis*-isopropyl propenyl ether
b ($\triangle$): *trans*-isopropyl propenyl ether

$$\frac{d[M_c]}{d[M_t]} = \frac{k_{pc}[M_c]}{k_{pt}[M_t]}$$

Table 4. Relative reactivity of *cis*- and *trans*-propenyl ether by $BF_3 \cdot O(C_2H_5)_2$. ($[M]_0$: 10 vol. %, $[C]$: 2–4 mmole/l., -74 to $-78°C$)

Monomer	Solvent	$(k_p)_{cis}/(k_p)_{trans}$
Methyl	Toluene[a] Methylene chloride[b]	3·7 —
Ethyl	Toluene Methylene chloride	2·4 1·4
Isopropyl	Toluene Methylene chloride	3·7 2·2
n-Butyl	Toluene Methylene chloride	1·9 1·5
Isobutyl	Ethyl benzene Methylene chloride	6·4 2·2

[a] Toluene contains 5 vol. % of methylene chloride as an internal standard for the gas-chromatographic measurement.
[b] Methylene chloride contains 5 vol. % of toluene for the same reason.

The copolymerization of the *cis*- and *trans*-isomers of propenyl *n*-butyl ether was examined and the monomer reactivity ratios of the *cis*-isomer, $r_c = 1\cdot35 \pm 0\cdot1$, and the *trans*-isomer, $r_t = 0\cdot74 \pm 0\cdot1$, were obtained. The *cis*-isomer has a larger reactivity ratio than the *trans*-isomer and these two values satisfy the relationship $r_1{\cdot}r_2 = 1\cdot0$. The nature of the growing end is therefore considered to be the same in both *cis*- and *trans*-isomers.

The reactivity ratios of the *cis*- and *trans*-isomers of propenyl ethers are summarized in *Table 4*. The *cis*-isomer is more reactive than the *trans*-isomer irrespective of the kind of monomer and the polarity of the solvent.

STEREOREGULATED POLYMERIZATION

G. Natta and his coworkers[6] have already investigated the stereo-regulated polymerization of propenyl alkyl ether and found that the *trans*-isomer was polymerized into crystalline threo-di-isotactic polymer, but the *cis*-isomer was polymerized into an amorphous polymer whose stereo-structure was not revealed by x-ray examination.

We have studied the polymerization of propenyl ethers to crystalline polymers by homogeneous and heterogeneous catalysts.

Homogeneous and heterogeneous catalysts[12]

BF_3OEt_2 and $AlEtCl_2$ were used as homogeneous catalysts at $-78°C$, and the $Al_2(SO_4)_3/H_2SO_4$ complex catalyst was used as a heterogeneous catalyst at $0°C$ in *n*-hexane, toluene or methylene chloride as solvents. The experimental results are summarized in *Tables 5* and *6*.

Table 5. Stereospecific polymerization of $CH_3{\cdot}CH{=}CHOR$ by $BF_3{\cdot}OEt_2$ at $-78°C$. ([*M*]₀; 5 vol. %, [*C*]; 2–4 mmole/l., toluene)

Monomer		Cryst.	m.p. (°C)
Methyl	*cis*	×	
	trans	○	150
Ethyl	*cis*	×	
	trans	○	207
Isopropyl	*cis*	×	
	trans	○	211
n-Butyl	*cis*	×	
	trans	○	*ca.* 100
tert-Butyl	*cis*	○	>250

○: Crystalline, ×: Amorphous

By the homogeneous catalyst system the *trans*-isomer was polymerized into a crystalline polymer. By the heterogeneous catalyst, on the other hand, the *cis*-isomer was converted into crystalline polymer.

X-ray diffraction patterns of both crystalline polymers obtained by homogeneous and heterogeneous catalysts, are recognized to be completely different as shown in the example of poly(propenyl isopropyl ether) in

Table 7. As **G.** Natta[6] has shown the *trans*-isomer produced threo-di-isotactic polymer, then the poly-*cis*-crystalline polymer obtained here might have erythro-di-isotactic structure.

On the other hand, the x-ray diffraction pattern and n.m.r. spectrum of poly(methyl propenyl ether) obtained from the *cis*-isomer by $Al_2(SO_4)_3/H_2SO_4$ complex are the same as that obtained from the *trans*-isomer by $BF_3 \cdot O(C_2H_5)_2$. This means that threo-di-isotactic polymer is produced from *cis*-methyl propenyl ether.

As shown in *Table 5*, the only exception is the case of propenyl *t*-butyl ether in which the homogeneous system can produce crystalline polymer even in the *cis*-isomer. This is considered to be due to the steric hindrance induced both by β-methyl and bulky *t*-butoxy groups.

Table 6. Stereospecific polymerization of $CH_3 \cdot CH = CHOR$ by $Al_2(SO_4)_3/H_2SO_4$ complex at 0°C.
([M]$_0$; 20 vol. %, [C]; 0·4 g/100 ml, toluene)

Monomer		Cryst.	m.p. (°C)
Methyl	cis	O	230
	trans	—	
Ethyl	cis	O	191
	trans	—	
Isopropyl	cis	O	204
	trans	—	
n-Butyl	cis	O	ca. 100
	trans	oily	

O: Crystalline,   —: No polymerization

Table 7. X-ray diffraction pattern of crystalline poly($CH_3CH = CHOiPr$)

trans $\sim$ BF₃·OEt₂		cis $\sim$ Al₂(SO₄)₃/H₂SO₄	
d (A)	Strength	d (A)	Strength
9·93	W		
8·78	VS	8·02	VS
6·33	S	6·00	M
5·06	M	5·38	W
4·29	S	4·68	W
3·69	W	4·17	S
2·83	W	2·89	W
2·20	W	2·23	W

Type of opening of double bond[13]

It is very important to know the type of double bond opening (i.e., *cis*- and *trans*-openings) for the investigation of the propagation reaction in the ionic polymerization. In the cationic polymerization of *trans*-propenyl ethers, G. Natta and his colleagues have concluded that the double bond of monomers opens in *cis*-type. However, in the x-ray examination adopted, quantitative evaluation of the steric structure is very difficult. When an amorphous polymer is obtained, the x-ray diffraction method is of little use.

In anionic polymerization, the double bond opening has been studied by the n.m.r. spectrum in the case of α,β-deuterium-2-acrylates by T. Yoshino (1964)[14] and C. Schuerch and his coworkers (1964)[15]. Here the ditacticity of poly(methyl propenyl ether) obtained by the cationic polymerization has been studied by n.m.r. spectra.

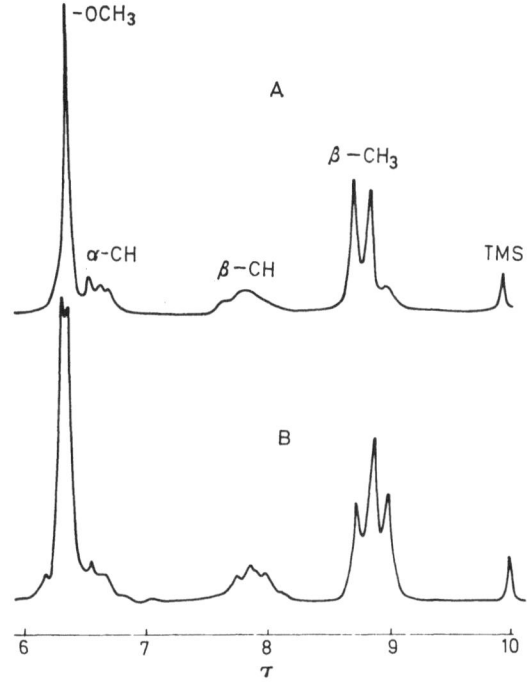

Figure 3. N.m.r. spectra of poly(methyl propenyl ethers) obtained by BF$_3$·O(C$_2$H$_5$)$_2$ in toluene at $-78°$C ([*M*]$_0$: 10 vol.%, [*C*] = 3 mmole/l.)
      A cis/trans ratio in monomer mixture: 1/9
      B cis/trans ratio in monomer mixture: 8/2

Two kinds of monomer mixture with different mole ratios of *cis*- and *trans*-isomers were used, that is, the *cis/trans* ratio equals to 1/9 and 8/2. N.m.r. spectra of polymer were measured in *o*-dichlorobenzene solution (10 w/v %) in a sealed tube at 160°C using the Varian HR-60 instrument. The n.m.r. spectra of β-methyl protons of the polymer were decoupled from β-methine proton by the side bond method, to know the amount of the di-tactic fraction in polymer. *Figure 3* shows the n.m.r. spectra of polymers obtained by BF$_3$OEt$_2$ at $-78°$C. The spectrum of β-methyl protons of polymer obtained from *cis/trans* = 1/9 mixture is clearly different from that of 8/2. To study the di-tactic fractions of a polymer quantitatively, the n.m.r. spectra of β-methyl protons are decoupled from the β-methine proton. As shown in

465

Figure 4 the spectra of β-methyl protons consist of two signals at τ8·78 and τ8·89. The intensity of τ8·78 is much stronger than that of τ8·89, in the polymer obtained from *trans*-rich mixture. The intensity of the single at

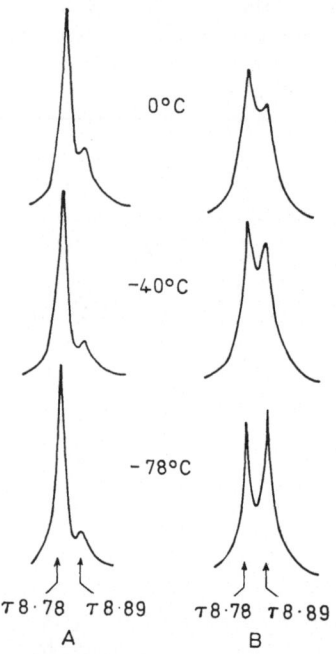

τ8·78   τ8·89   τ8·78   τ8·89

A   B

Figure 4. N.m.r spectra of β-methyl protons decoupled from β-methine proton in poly(methyl propenyl ethers) obtained by BF₃·O(C₂H₅)₂ in toluene at various polymerization temperatures. ([M]₀: 10 vol.%, [C] = 3 mmole/l)
A *cis/trans* ratio in monomer mixture: 1/9
B *cis/trans* ratio in monomer mixture: 8/2

τ8·89 for the polymer obtained from *cis*-rich monomer mixture increases with lowering polymerization temperature.

G. Natta has shown that *trans*-alkenyl ethers produce the threo-di-isotactic polymer, therefore, poly (*trans*-methyl propenyl ether) obtained by BF₃OEt₂ at a low temperature should be the threo-di-isotactic structure. From these results, the signals of β-methyl protons at τ8·78 and τ8·89 are assigned as spectra based on threo- and erythro-di-isotactic diads, respectively.

On the basis of the assignment of β-methyl protons, the content of the threo- and erythro-di-isotactic diads can be determined quantitatively. These results are shown in *Table 8* together with the polymerization conditions and properties of the polymers.

The polymers obtained from a *trans*-rich mixture are highly crystalline and contain more than 80 per cent of the threo-di-isotactic diad. On the other hand, the polymer obtained from a *cis*-rich mixture is amorphous and contains a mixture of the threo- and erythro-di-isotactic diads.

On the basis of polymer structure and composition, the type of double-bond opening may be discussed quantitatively. If the probability of *trans-*

466

Table 8.Polymerization conditions and properties of poly(methyl propenyl ether)
([M]₀ = 10 vol.%, [BF₃OEt₂] = 3 mmole/l., t = 2 h, solvent: toluene)

cis/trans in Monomer mole ratio	Polymeri-zation temp. °C	Conversion %	[η] 100ml/g	Crystal-linity	Threo-di isotactic fraction %	Erythro-di isotactic fraction %
1/9	−40	55	0·10	cryst.	82·6	17·4
1/9	−78	32	0·30	cryst.	85·7	14·3
8/2	−40	68	0·09	amorph.	51·9	48·0
8/2	−78	66	0·27	amorph.	44·4	55·6

opening in *cis*-monomer is defined as A (then, the probability of *cis*-opening is $1 - A$) and the probability of *trans*-opening in *trans*-monomer is defined as B (then, the probability of *cis*-opening is $1 - B$), we can obtain:

$$\frac{d[M_c]}{d[M]} A + \frac{d[M_t]}{d[M]} (1 - B) = \text{(Threo-di-isotactic fraction)}$$

where $d[M_c]$, $d[M_t]$ and $d[M]$ are the mole concentrations of *cis*-, *trans*- and total monomer in the polymer, respectively. At a constant temperature, the type of opening is considered to be constant in different compositions and the chain end formed does not rotate freely. Therefore A and B can be calculated from the experimental results. The percentages of *cis*- and *trans*-openings thus obtained are summarized in *Table 9*.

From these results, it is concluded that the double bond in a *trans*-monomer is opened exclusively in the *cis*-type while in a *cis*-monomer, *cis*- and *trans*-opening takes place at almost the same rates in a homogeneous catalyst system.

Table 9. Fraction of double bond opening at various polymerization temperatures

Polymerization temperature °C	cis-*Isomer*		trans-*Isomer*	
	cis-*Opening* %	trans-*Opening* %	cis-*Opening* %	trans-*Opening* %
0	47	53	80	20
−40	50	50	89	11
−78	60	40	*ca.* 100	*ca.* 0

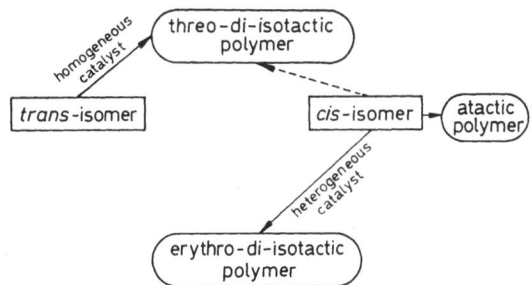

Stereospecific polymerization of *cis*- and *trans*-isomers in propenyl ethers

467

As a preliminary conclusion, we would like to emphasize the following six points.

First, by quantitative examination of the lowering of reactivities of β-methylstyrenes, the effect of steric repulsion of the β-methyl group is 1/10–1/20, compared to the styrenes.

Second, in the case of propenyl n-alkyl ethers, the reactivities of β-methyl-substituted monomers have been found to be larger than that of unsubstituted monomer by electronic effect. However, in branched-alkyl ether, the relationship was similar in styrene derivatives for which the steric effect as well as electronic effect must be considered both in α- and β-positions.

Third, from the reactivity behaviour, β-alkoxystyrenes are considered to be β-phenyl alkyl vinyl ethers.

Fourth, a *cis*-isomer has a larger reactivity than a *trans*-isomer, and, again styrene derivatives and branched alkyl vinyl ethers both have the tendency for reactivity to be larger in *trans*- than in *cis*-isomers, or the differences become smaller.

Fifth, not only *trans*-isomers but also *cis*-isomers could be polymerized into crystalline polymers, using homogeneous and heterogeneous catalysts, respectively. Thus, crystalline analysis by the x-ray diffraction method is made possible.

Sixth, by n.m.r. spectra of β-methyl protons, the type of opening was quantitatively measured. In the case of amorphous polymers, the proportion of *cis*- to *trans*-opening is approximately 50 : 50.

References

[1] e.g. T. Alfrey, Jr, J. J. Bohrer, and H. Mark. *Copolymerization*, Interscience, New York, 1952, p. 49.
[2] C. G. Overberger, D. Tanner, and E. M. Pearce. *J. Am. Chem. Soc.* **80**, 4566 (1958).
[3] F. M. Lewis, C. Walling, W. Cummings, E. R. Briggs, and F. R. Mayo. *J. Am. Chem. Soc.* **70**, 1519 (1948).
[4] D. S. Brackmann and P. H. Plesch. *J. Chem. Soc.* 3563 (1958).
[5] T. Fueno, T. Okuyama, O. Kajimoto, and J. Furukawa, Preprint of International Symposium on Macromolecular Chemistry, Tokyo-Kyoto, I-58 (1966).
[6] G. Natta. *J. Polymer Sci.* **48**, 219 (1960) and related literature.
[7] A. Mizote, T. Tanaka, T. Higashimura, and S. Okamura. *J. Polymer Sci.* **A, 3**, 2567 (1965).
[8] G. E. K. Branch and M. Calvin. *Theory of Organic Chemistry*, Prentice-Hall, New York, 1941, p. 192.
[9] A. Mizote, S. Kusudo, T. Higashimura, and S. Okamura. *J. Polymer Sci.* **A-1**, in the press.
[10] A. Mizote, T. Matsui, T. Higashimura and S. Okamura. *J. Macromol. Chem.* to be published.
[11] A. Mizote, T. Higashimura, and S. Okamura. *J. Polymer Sci.* **A-1**, to be published.
[12] T. Higashimura, S. Kusudo, Y. Ohsumi, and S. Okamura. *J. Polymer Sci.* **A-1**, to be published.
[13] Y. Ohsumi, T. Higashimura, R. Chūjō, T. Kuroda, and S. Okamura. *J. Polymer Sci.* **A-1**, in the press.
[14] T. Yoshino and J. Komiyama. *J. Am. Chem. Soc.* **86**, 4482 (1964) and related literature.
[15] C. Schuerch, W. Fowells, A. Yamada, F. A. Bovey, F. P. Hood, and E. W. Anderson. *J. Am. Chem. Soc.* **86**, 4481 (1964).

OPTICAL ACTIVITY AND OPTICAL ROTATORY DISPERSION IN SYNTHETIC POLYMERS

P. Pino, P. Salvadori, E. Chiellini and P. L. Luisi

Istituto di Chimica Organica Industriale dell'Università di Pisa, Centro Nazionale di Chimica delle Macromolecole del C.N.R., Pisa, Italy

Among the large number of methods proposed for the investigation of macromolecular conformation in solution, optical activity is in principle one of the most powerful since, in general, optical activity is strongly dependent on conformational equilibria[1]. Its use, however, is seriously hindered by the difficulty in calculating the optical activity even of small molecules theoretically. This difficulty can be partially solved in some cases by using a semi-empirical calculation of the optical rotation[2]; furthermore some information concerning the relationships between optical activity and conformation in polymers can be obtained by comparing the optical activity of polymers with that of low-molecular-weight models in which intermolecular interactions between the chromophoric systems responsible for the optical rotation can often be excluded.

From the experimental point of view the difficulties often encountered in preparing optically-active polymers are largely compensated by the fact that much information can be obtained using small quantities of polymer samples which are non-homogeneous with respect to molecular weight. In fact it has been shown[3] that the molar rotatory power in synthetic polymers, if referred to one monomeric unit, is in general independent of molecular weight and molecular-weight distribution, at least for macromolecules containing more than about 20 monomeric units in which the influence of the terminal groups on the rotation can be neglected. For this reason the molar rotation at each wavelength in the papers concerning optically-active polymers is referred to single monomeric units instead of to entire macromolecules, ignoring all the problems connected with the molecular-weight distribution.

In the present paper we shall consider for some polymers and low-molecular-weight models investigated in our laboratory, the origin of rotatory power, the influence of structure of monomeric units and stereoregularity on optical rotation, and finally, for the case of the poly-α-olefins in hydrocarbon solution, we shall consider some relationships between molar optical rotation and conformation.

We shall not attempt to make a complete review of the data on this subject as reviews covering both experimental data[4] and theoretical aspects[5] have been published quite recently.

1. GENERAL ASPECTS OF OPTICAL ACTIVITY IN POLYMERS

As in low-molecular-weight compounds, the optical rotation in polymers is connected with optically-active electronic transitions in definite chromophoric systems, which may be the same as in low-molecular-weight models, or can be modified by mutual interactions among the chromophoric systems existing in different monomeric units. In the first case the optical activity of the polymer, referred to a single monomeric unit, as far as electronic factors are concerned, should be about the same as in low-molecular-weight models, and the differences eventually found in optical rotation must be substantially attributed to different positions of the conformational equilibria in low-molecular-weight models and in the monomeric units of the polymers. No direct information can be obtained in this case from the optical activity on the main-chain conformation; however, in some cases[6] the respective position of the monomeric unit atoms inserted in the principal chain and respectively in the lateral chains can be established and hence the general features of the prevalent conformation of the main chain can be inferred.

In the second case the modification of the chromophoric systems present in the monomeric units can be directly related[7] to the conformation of the main chain of the polymers; in this case the optical activity of the polymer, referred to a single monomeric unit, and of the low-molecular-weight models must be different even if the conformational equilibrium assumes in both cases similar positions.

A mutual interaction among the chromophoric systems present in different monomeric units of synthetic polymers has so far been detected principally in polyamino acids and has been related to the presence of a helical conformation of the main chain in solution[8]; however, this topic has been discussed elsewhere both from theoretical and experimental points of view and will not be considered here. Poly-(S)-4-methyl-1-hexyne[9] (Ia) represents an extreme case of this type in which the chromophoric system in low-molecular-weight model Ib (isolated double bond) and in polymer (partially-conjugated double bonds) is substantially different. However, in this case the polymer is not crystalline at room temperature and its structure has not been investigated by x-ray diffraction; optical rotatory dispersion (o.r.d.) cannot be investigated below 450 mμ because the ratio of optical rotation to absorption coefficient is too small; no circular dichroism (c.d.) measurements have been carried out until now and the only conclusion that can be drawn is that the main chain is not planar, the type of folding being unknown.

$$
\left[\begin{array}{c} CH{=}C{-} \\ | \\ CH_2 \\ | \\ CH_3{-}\overset{*}{C}H \\ | \\ C_2H_5 \end{array}\right]_n
\qquad\qquad
\begin{array}{c} H{-}CH{=}C{-}H \\ | \\ CH_2 \\ | \\ CH_3{-}\overset{*}{C}H \\ | \\ C_2H_5 \end{array}
$$

$\lambda_{max} = 230$ mμ log $\epsilon_{max} = 3\cdot46$

$\lambda_{max} = 323$ mμ log $\epsilon_{max} = 3\cdot48$

$[\Phi]_D^{25} - 22\cdot9$

$\lambda_{max} < 200$ mμ

$[\Phi]_{D\ max}^{25} - 2\cdot97$

Ia   Ib

Two other cases are known in which the electronic transitions at the longest wavelength seem to be different in synthetic polymers and in the corresponding low-molecular-weight models; in the isotactic polyacrylamides a transiton at about 270 mμ seems to exist in the polymer[10] and not in the models[10]; in the isotactic polymethylmethacrylate, a maximum at 207 mμ, the intensity of which is strongly temperature-dependent, and two shoulders at 216·5 and 211·5 mμ exist[11] in the polymer while in the low-molecular-weight saturated esters a band at about 204 mμ without shoulders is normally present[12].

However no optically-active compounds of this type have been investigated from the above point of view and therefore the possible relationships between molar rotation and conformation in the above series are unknown (*see Note* 1 *added in proof on page* 489).

In the series of vinyl polymers investigated by our group[6, 13, 14] (II, III, IV) the wavelength of the optically-active transitions in the near u.v. seems not to be very different in the polymers (independent of their stereoregularity) and in the low-molecular-weight models, and no new bands at longer wavelengths have been found in the polymers.

The polymers which can be better investigated from this point of view are the polyvinyl ketones (II). As shown in *Table 1* the $n \rightarrow \pi^*$ electronic transition connected with the existence of the $>$C$=$O group is at the same wavelength for polymers having different stereoregularity, only the absorption

$$\begin{array}{cc}
\text{---CH}_2\text{---CH---} & \text{---CH}_2\text{---CH---} \\
| & | \\
\text{C}=\text{O} & \text{O} \\
| & | \\
(\text{CH}_2)_n & (\text{CH}_2)_n \\
{}^*| & {}^*| \\
\text{CH}_3\text{---CH} & \text{CH}_3\text{---CH} \\
| & | \\
\text{C}_2\text{H}_5 & \text{C}_2\text{H}_5 \\
n = 0, 1, 2 & n = 0, 1 \\
\text{II} & \text{III}
\end{array}$$

$$\begin{array}{cc}
& \text{H---CH}_2\text{---CH---H} \\
& | \\
\text{---CH}_2\text{---CH---} & \text{C}=\text{O} \\
| & | \\
(\text{CH}_2)_n & (\text{CH}_2)_n \\
{}^*| & {}^*| \\
\text{CH}_3\text{---CH} & \text{CH}_3\text{---C---H} \\
| & | \\
\text{C}_2\text{H}_5 & \text{C}_2\text{H}_5 \\
n = 0, 1, 2 & n = 0, 1, 2 \\
\text{IV} & \text{V}
\end{array}$$

coefficient being lower for the more stereoregular fraction. In comparison with the low-molecular-weight models the transition occurs in the polymers,

Table 1. Ultraviolet spectra and features of Cotton effect in polyvinyl ketones[a] and low-molecular-weight model compounds

Polymer	U.V. λ_{max} (mµ)	ε_{max}[b]	Cotton effect λ_0[c] (mµ)	Amplitude[a] ($>$C=O $n\to\pi^*$)	Sign	Model compound	U.V. λ_{max} (mµ)	ε_{max}	λ_0[c] (mµ)	Cotton effect Amplitude[a] ($>$C=O $n\to\pi^*$)	Sign
II (n=0) (i) (e)	292	66·8	292	77·0	−	V (n=0) [l]	283	29·3	283[g]	5·3	+
(f)	291	56·0	292	221·0	−				285[h]	n.d.	−
II (n=1) (m) (e)	290	67·5	290	33·6	−	V (n=1) [p]	282	26·5	282[g]	9·5	−
(f)	290	60·0	290	66·0	−						
II (n=2) (n) (e)	289	64·6	288	5·5	−	V (n=2) [p]	283	26·7	283[g]	1·0	−
(f)	288	53·5	288	10·8	−						

(a) In CHCl₃ solution.
(b) Referred to one monomeric unit.
(c) Value taken from experimental o.r.d. curve as (λ trough + λ peak)/2.
(d) Calculated from o.r.d. curve obtained by subtracting the background rotation from the experimental o.r.d. curve.
(e) Atactic, obtained by spontaneous polymerization.
(f) Stereoregular, obtained by anionic polymerization, initiator LiAlH₄.
(g) In methanol solution.
(h) In vapour phase.
(i) Optical purity of the polymerized monomer 68%.
(l) Optical purity 81%.
(m) Optical purity of the polymerized monomer 96%.
(n) Optical purity of the polymerized monomer 95%.
(p) Optical purity 95%.

according to the u.v. spectra, at a wavelength 6–8 mμ higher; the absorption coefficient is about twice; practically independent of temperature between 25 and 60°C. The above wavelength differences might be independent of interactions among $\diagup C = O$ groups in the polymer and hence from the main-chain conformation; in fact ketonic compounds having the $\diagup C = O$ transition at about 290 mμ are known in the literature[15]. However, as far as the values of ε found in the polymer are concerned, they are close to the

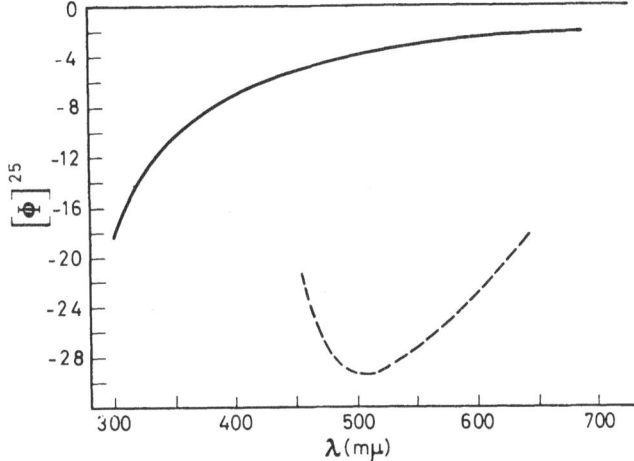

Figure 1. ————o.r.d. curve of (S)-4-methyl-1-hexene (optical purity 92%). — — — o.r.d. curve of poly-(S)-4-methyl-1-hexyne (optical purity 89%). Acetone ins., diethyl ether sol. fraction.

values found for acetonylacetone and could be connected with some type of unknown interaction between the keto groups, which might be dependent on the conformation of the principal chain.

In any case the large differences in optical activity observed in the more stereoregular fractions of polyvinyl ketones II ($n=0,1$) and low-molecular-weight models are mainly connected with the amplitude of the Cotton effects and hence with the position of the conformational equilibrium in the monomeric units of the polymers and in the models respectively.

In the case of polyvinyl ethers, Cotton effects cannot be detected with the spectropolarimeters at present available, and the deduction of the Cotton effect wavelength from the Drude equation is not advisable, as at least two chromophoric systems exist giving contributions of opposite sign to the rotation[16] (*Figure 2*). However u.v. spectra have shown that the band at the longest wavelength, which in [(S)-1-methylpropyl]-ethyl ether is optically active[16], is in the same region for the polymers (III, $n=0$, $\lambda_{max} = 190$; III, $n=1$, $\lambda_{max} = 191$)[13] (1966), and for the corresponding low-molecular-weight models[16]. Therefore, in this case, also interactions between the different chromophores present in the macromolecule should not play a very important role in determining the large rotation differences observed in some cases between polymers and models.

For poly-α-olefins neither Cotton effect nor u.v. maxima could be detected

473

with the experimental techniques used until now in our Institute; however in this case o.r.d. curves are simple up to 200 mμ and the wavelength corresponding to the Cotton effect in polymers and models can be evaluated by using a one-term Drude equation.

As shown in *Table 2*, all the λ_0 found for polymers and models are between 165 mμ and 179 mμ showing that in this case also modifications of chromo-

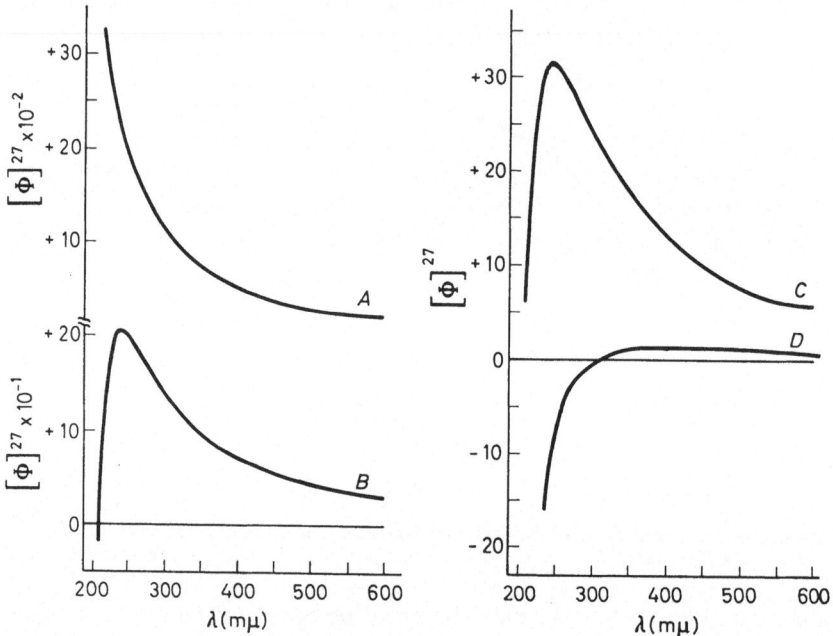

Figure 2. (*A*) o.r.d. curve of poly-[(S)-1-methyl-propyl]-vinyl-ether (monomer optical purity ~90%). Acetone ins., ether sol. fraction. (*B*) o.r.d. curve of [(S)-1-methyl-propyl]-ethyl ether (optical purity 80%). (*C*) o.r.d. curve of poly-[(S)-2-methyl-butyl]-vinyl-ether (monomer optical purity 99%). Acetone ins., diethyl ether sol. fraction. (*D*) o.r.d. curve of [(S)-2-methyl-butyl]-ethyl ether (optical purity 99%).

phoric systems should not be mainly responsible for the large rotation differences observed in some cases between polymers and models.

No systematic investigation has been carried out until now on the origin of rotatory power in polyacrylic derivatives. For atactic poly-[(S)-2-methyl-butyl]-methacrylate we have found, contrary to some published data[17], a plain o.r.d. curve up to 250 mμ; the Cotton effect, corresponding to the known $n \rightarrow \pi^*$ transition in the low-molecular-weight esters, has been detected in the polymers[20], λ_0 being located at about 215 mμ.

Also in the case of condensation polymers, with the exception of poly-amino acids which are not considered in the present paper, no systematic investigations on o.r.d. have been published until now, and we shall consider only two examples which are at the present under investigation in our laboratory. Among the polyethers the polypropylene oxide VI has an anomalous o.r.d. curve with a maximum not corresponding to a Cotton

Table 2(a). Molar optical rotatory power and optically-active electronic transition at the longest wavelength estimated from the one-term Drude equation[b] for some poly-α-olefins having (S) asymmetric carbon atom in the lateral chains, and for some model compounds of the same absolute configuration

Polymer	n	Polymerized monomer optical purity %	$[\Phi]^{25}_D$ (c)	λ_0 (f) (mμ)	Model compound	n	$[\Phi]^{25}_D$ (d)	λ_0, (mμ)
~CH—CH₂~ | (CH₂)ₙ | CH₃—C*—H | C₂H₅	0	91	+161	167	CH₃—CH—CH₃	0	−11·4(e)	176
	1	93	+288	165	(CH₂)ₙ	1	+21·3	170
	2	95	+ 68·1	169	CH₃—C*—H	2	+11·7	n.d.
	3	95	+ 20·4	n.d.	C₂H₅	3	+14·4	173

(a) For references see P. Pino *Adv. Polymer Sci.* **4**, 393 (1965).
(b) $[\Phi]_\lambda^T = \kappa/(\lambda^2 - \lambda_0^2)$.
(c) Referred to one monomeric unit of the most stereoregular fraction.
(d) Maximum value.
(e) At 20°C.
(f) ±10 mμ.

475

T

effect—at about 230 mμ in diethyl ether[18] analogous to that observed for the polyvinyl ethers. The u.v. spectra have not been investigated in this case.

Optical rotatory dispersion of poly-(—)-lactide (VII) has been investigated by R. C. Schulz[19] and more recently by our group[20] which has confirmed the maximum found by Schulz at 275 mμ.

$$-\left[CH_2-\overset{*}{\underset{|}{CH}}-O\right]_{\overline{n}} \qquad \qquad -\left[O-\overset{*}{\underset{|}{CH}}-CO\right]_{\overline{n}}$$

$$\overset{CH_3}{} \qquad \qquad \overset{CH_3}{}$$

VI   VII

Furthermore a maximum at 220 mμ has been found which corresponds to the first extremum of a Cotton effect having λ_0 at about 215 mμ and attributable therefore to the known $n \rightarrow \pi^*$ optically-active transition of the —COOR* group. In this case too, no investigation of the u.v. spectra of polymers and models has been carried out (*see Note 2 added in proof on page 490*).

2. RELATIONSHIPS BETWEEN OPTICAL ROTATION AND STRUCTURE IN SOME SYNTHETIC POLYMERS

As optical rotation cannot at present be calculated theoretically, relationships between optical rotation and structure in polymers can be drawn only on an empirical or semi-empirical basis in series of homologous compounds.

In the field of optically-active synthetic polymers this discussion must be limited to the series of vinyl polymers[6,13,14] in which at least two or three members of the series have been investigated as far as o.r.d., [Φ]–stereoregularity and [Φ] polymers–[Φ] models relationships are concerned (*Table 3*).

As emphasized in previous publications[6], in the case of poly-α-olefins the following facts appear clearly: (1) the sign of rotation of the polymer is related to the absolute configuration of the asymmetric carbon atom present in the lateral chains; (2) the rotation referred to one monomeric unit at 589 mμ and o.r.d. curves being simple, at all the wavelengths in the range investigated (above 200 mμ), is much higher in isotactic polymers than in the models when the asymmetric carbon atom of the lateral chains is in the α or β position with respect to the principal chain. The rotation is higher but of the same order of magnitude, in the two cases, when the asymmetric carbon atom is in the γ position and is practically the same when the asymmetric carbon atom is in the δ position with respect to the principal chain (*Table 3*).

As we shall discuss later these facts can be interpreted on the basis of conformational analysis which shows that, at least in the case of the poly-α-olefins, when the asymmetric carbon atom of the lateral chains is in α or β position with respect to the principal chain, few conformations of the monomeric units having high optical activity of the same sign prevail in the conformational equilibria. The conformational equilibrium position is entirely different in the case of the models shown in *Table 3* in which no large prevalence of conformations having high rotatory power of the same sign can exist. The situation is the same in polymer monomeric units and

476

Table 3. Molar optical rotatory power of the most stereoregular fractions of some optically active poly-α-olefins, polyvinyl ethers, polyvinyl ketones, and of some low-molecular-weight model compounds

	Polymers					Models		
Type	n	Position of the C^* in the lateral chain of the polymer	Polymerized monomer optical purity (%)	$[\Phi]^{25}$ (a)		Compound	n	$[\Phi]^{25}$ (b)
CH_3 $CH_2{\sim}$ (d) H–$\overset{*}{C}$–$(CH_2)_n$–CH C_2H_5	0	α	91	$+161$ (f)		CH_3 CH_3 (d) H–$\overset{*}{C}$–$(CH_2)_n$–CH C_2H_5 CH_3	0	-11.4 (c)
	1	β	93	$+288$ (f)			1	$+21.3$
	2	γ	95	$+68.1$ (f)			2	$+11.7$
	3	δ	95	$+20.4$ (f)			3	$+14.4$
CH_3 $CH_2{\sim}$ (d) H–$\overset{*}{C}$–$(CH_2)_n$–O–CH C_2H_5	0	β	90	$+312$ (f)		CH_3 CH_3 (d) H–$\overset{*}{C}$–$(CH_2)_n$–O–CH_2 C_2H_5	0	$+34.5$
	1	γ	>99	$+6.5$ (f)			1	$+1.1$
CH_3 O $CH_2{\sim}$ (e) H–$\overset{*}{C}$–$(CH_2)_n$–$\overset{\|}{C}$–CH C_2H_5	0	β	68	-118 (g)		CH_3 O CH_3 (e) H–$\overset{*}{C}$–$(CH_2)_n$–$\overset{\|}{C}$=CH_2 C_2H_5	0	$+34.8$
	1	γ	96	-43 (g)			1	$+11.5$
	2	δ	95	$+11.7$ (g)			2	$+15.2$

(a) Referred to one monomeric unit.
(b) Maximum observed value.
(c) At 20°C.
(d) For references see P. Pino, *Adv. Polymer Sci.* **4**, 393 (1965).
(e) See O. Pieroni, F. Ciardelli, C. Botteghi, L. Lardicci, P. Salvadori, P. Pino, paper presented at Symposium on Macromolecular Chemistry, Bruxelles, 1967. *J. Polymer Sci., C,* in the press.
(f) In aromatic hydrocarbon solution.
(g) In CHCl₃.

477

models when the asymmetric carbon atom in the lateral chains is in the δ position with respect to the principal chain while the case in which the asymmetric carbon atom is in the γ position is an intermediate one.

The relationships between optical rotation and structure in the other two series are more complicated: in fact the rotation originated by the hydrocarbon backbone which can be connected in a rather simple way with the conformation, is strongly altered by the presence of other chromophoric groups such as ethereal oxygen or keto-groups. These chromophores absorb at longer wavelengths than the paraffins, and the related electronic transitions occurring in the asymmetric environment make remarkable contributions to the observed rotatory power. Unfortunately the relationships between the optical rotation connected with the oxygen-containing chromophores and conformation are not very well known in aliphatic compounds and this lack of knowledge makes the interpretation of the experimental data even more difficult.

This situation is clearly shown in the case of polyvinyl ketones in which the positive background at 589 mμ arising from the transitions connected with the hydrocarbon backbone and from the $n \rightarrow \sigma^{*}$[21] transition of the keto groups is completely obscured by the negative Cotton effect corresponding to the $n \rightarrow \pi^{*}$ transition of the keto groups in II $(n=0)$ and II $(n=1)$, but is still apparent in II $(n=2)$. In this case the intensity of the above Cotton effect is much smaller because of the larger distance of the asymmetric carbon atom of the lateral chains from the carbonyl group and from the principal chain.

A similar situation exists in polyvinyl ethers in which the Cotton effect related to the presence of ethereal oxygen is not detectable by the available spectropolarimeters, but is certainly negative both in III $(n=0)$ and III $(n=1)$. In this case however, if we admit that the background rotation is substantially dependent on the hydrocarbon skeleton, we can conclude that a relationship between monomeric unit structure and conformational equilibria, similar to that observed in poly-α-olefins, also exists in polyvinyl ethers.

Despite the above difficulties we believe that from the above data it can be concluded that, in optically-active linear vinyl polymers, a secondary butyl group in the α or β position with respect to the principal chain considerably enhances the absolute value of the optical activity in comparison to the low-molecular-weight models. The enhancement is mainly related, at least in the cases examined up to now, to the largely different conformational equilibrium positions in the polymers and in the models.

3. RELATIONSHIPS BETWEEN ROTATORY POWER AND STEREOREGULARITY IN VINYL POLYMERS

The more detailed data on the relationship between stereoregularity and optical rotation concerns the series of poly-α-olefins IV $(n=0,1,2,3)$. In this case $[\Phi]$ increases by increasing stereoregularity as evaluated by melting point, infrared analysis, and solubility data (*Table 4*).

As o.r.d. curves are plain and λ_0 of the Drude equation is independent of stereoregularity, the increase of $[\Phi]_D$ must be connected with K values of

Table 4. Relationship between optical rotation and stereoregularity in some optically-active poly-α-olefins

Polymer	Polymerized monomer optical purity (%)	Fraction[a]	m.p.[b] (°C)	$[\Phi]_D^{25}$ [c, d]	One-term Drude equation[e] constants	
					λ_0[e] (mμ)	$\kappa . 10^{-6}$
IV (n=0)	89	Acetone ins., diethyl ether sol.	93–96[f]	+127	167	+40·5
		Diethyl ether ins., isooctane sol.	187–193[f]	+146	167	+46·5
IV (n=1)	93	Acetone ins., [h] diethyl ether sol.	[g]	+174	179	+59·8
		Acetone ins., ethyl acetate sol.	138–143	+243	165	+77·7
		Diethyl ether ins., diisopropyl ether sol.	210–215	+288	165	+92·0
IV (n=2)	95	Acetone sol.	[g]	+27·0	169	+8·6
		Acetone ins., diethyl ether sol.	54–55	+68·1	169	+21·7

(a) Obtained by extraction with solvent at boiling point.
(b) Determined by x-ray method if not otherwise indicated.
(c) Referred to one monomeric unit.
(d) In aromatic hydrocarbon solution.
(e) $[\Phi]_\lambda^{25} = \kappa/(\lambda^2 - \lambda_0^2)$.
(f) Determined by a Kofler m.p. apparatus.
(g) Amorphous.
(h) Obtained by hydrogenation of poly-(S)-4-methyl-1-hexyne derived from a (S)-4-methyl-1-hexyne sample of optical purity 89%.
(i) ± 10 mμ.

479

the Drude equation. As conformational analysis and semi-empirical calculation of optical activity indicate that configurational inversions in the main
chain should not substantially affect $[\Phi]_D$, at least for stereoblock polymers,
the dependence experimentally found of $[\Phi]_D$ on stereoregularity could be
attributed to the fact that, in samples with different stereoregularity,
differences in conformational equilibria exist, in agreement with the statistical
model assumed for the conformation of these relatively simple macromolecules as we shall discuss later[3] (Luisi 1968).

The above considerations, drawn for the chromophoric system responsible
for the rotation in the poly-α-olefins on the basis of the Drude equation, hold
also for $n \rightarrow \pi^*$ transition of the $>C{=}O$ chromophoric system in the α
position with respect to the main chain in polyvinyl ketones.

As shown in *Table 5*, λ_0 of the Cotton effect is independent of stereoregularity but the amplitude of the Cotton effect is strongly affected by
stereoregularity estimated on the basis of crystallinity in the case of II ($n{=}0$)

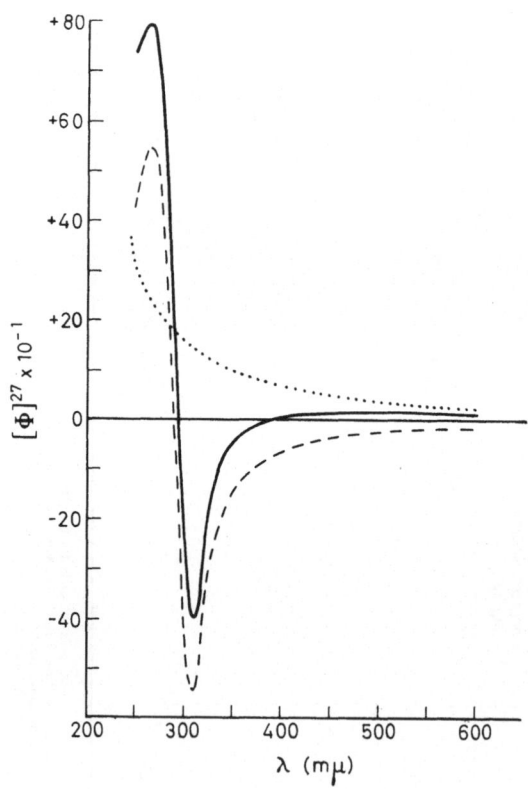

Figure 3. o.r.d. of poly-[(S)-3-methyl-pentyl]-vinyl-ketone
———— Experimental curve; . . . . Background rotation
— — — Cotton effect calculated for $>C{=}O$ $n \rightarrow \pi^*$ transition.

Table 5. Relationship between optical rotation(a) and stereoregularity in poly-vinyl-ketones

Polymer	Polymerized monomer optical purity (%)	Initiator	$[\varPhi]_D^{25}$ (b)	Features of the Cotton effect (a)			Crystallinity(e)
				Sign	λ_0 (mμ)	Amplitude(c)	
II (n=0)	68	(f) LiAlH₄	−42.5 −118.0	− −	292 292	77.0 221.0	none moderate
II (n=1)	96	(f) LiAlH₄	−10.0 −43.0	− −	290 290	33.6 66.0	none none
II (n=2)	95	(f) LiAlH₄	+15.6 +11.7	− −	288 288	5.5 10.8	none none

(a) In CHCl₃ solution.
(b) Referred to one monomeric unit.
(c) Calculated subtracting the background rotation from the experimental O.R.D. curve.
(d) Related to the carbonyl $n \rightarrow \pi^*$ transition.
(e) At room temperature.
(f) By spontaneous polymerization.

and on the basis of the stereospecificity of the polymerization process used for II ($n=1$) and II ($n=2$). The exception noted for II ($n=2$) in which the absolute value of $[\Phi]_D$ is higher for the less-stereoregular fractions is only apparent: in fact in this case the rotatory power measured is prevailingly given by the positive background rotation from which the contribution of the negative Cotton effect due to the $n \rightarrow \pi^*$ transition of the $>$C$=$O group is subtracted. The absolute $[\Phi]_D$ values found show that, by decreasing stereoregularity, the negative contribution given by the $n \rightarrow \pi^*$ transition of the $>$C$=$O group decreases more than the positive contribution by the other chromophoric systems absorbing at a much lower wavelength (*Figure 3*).

In fact separating the contribution by the Cotton effect connected with the presence of $n \rightarrow \pi^*$ transition of the $>$C$=$O from the background[21], the contribution to the rotation by the $n \rightarrow \pi^*$ transition of $>$C$=$O group is negative and its absolute value decreases by decreasing stereoregularity (*Table 6*).

In the case of polyvinyl ethers, $[\Phi]_D$ decreases by decreasing stereoregularity[22] (*Table 7*) as in the poly-α-olefins; the agreement is, however,

Table 6. Contributions to $[\Phi]_D^{25\,(a)}$ by $n \rightarrow \pi^*$ electronic transition of the $>$C$=$O chromophoric system and by background rotation in poly-[(S)-3-methyl-pentyl]-vinyl-ketone[b]

Polymerization process	Polymer stereoregularity	$[\Phi]_D^{25\,(c)}$ experimental	$[\Phi]_D^{25\,(c)}$ background	$[\Phi]_D^{25\,(c,\,d)}$ ($>$C$=$O) $n \rightarrow \pi^*$
Anionic[e]	low	$+11 \cdot 7$	$+28 \cdot 5$	$-16 \cdot 8$
Spontaneous (probably radical)	practically absent	$+15 \cdot 6$	$+21 \cdot 0$	$-5 \cdot 4$

(a) In CHCl₃ solution.
(b) Polymerized monomer optical purity 95%.
(c) Referred to one monomeric unit.
(d) Calculated assuming λ_0 background = 190 mμ and K background = $8 \cdot 88 \cdot 10^{-6}$ for the sample of low stereoregularity, and $6 \cdot 55 \cdot 10^{-6}$ for the atactic sample.
(e) Initiator LiAlH₄.

Table 7. Relationship between optical rotation and stereoregularity in poly-[(S)-2-methylbutyl]-vinyl-ether[a]

Fraction[b]	m.p. (°C)	I.R.[d] crystallinity index	$[\Phi]_D^{25\,(e,\,f)}$	$\lambda_{max}^{(h)}$ (mμ)	$[\Phi]_{\lambda_{max}}^{27}$ (e,g)
Acetone sol.	<25	0·48	$+5 \cdot 5$	258	$+21 \cdot 1$
Acetone ins., diethyl ether sol.	115–120[c]	0·53	$+5 \cdot 9$	244	$+31 \cdot 1$
Diethyl ether ins., benzene sol.	135–140[c]	0·87	$+6 \cdot 5$	222	$+63 \cdot 8$

(a) Polymerized monomer optical purity > 99%.
(b) Obtained by boiling solvent extraction.
(c) Determined by i.r. spectroscopy.
(d) D_B 827 cm⁻¹/D_B 771 cm⁻¹.
(e) Referred to one monomeric unit.
(f) In toluene solution.
(g) In n-heptane solution.
(h) Wavelength corresponding to the maximum of the o.r.d. curve.

occasional because in polyvinyl ethers $[\Phi]_D$, as shown by o.r.d. both of polymers and models, is given by the contributions of opposite sign of at least two chromophoric systems. The above contributions vary probably to a different extent with stereoregularity.

Investigating complexes both of polymers[22] and models[16] with Lewis acids it has been possible to attribute the negative contribution to the $n \rightarrow \sigma^*$ transition[23] of the ethereal oxygen both in III $(n=0)$ and III $(n=1)$. If we admit that the positive contribution is chiefly given by the chromophoric systems connected with the hydrocarbon back bone, we must conclude that this last contribution is more influenced by the stereoregularity than the contribution arising from the $n \rightarrow \sigma^*$ transition of the ethereal oxygen. For the above reason in the case of III $(n=1)$ where the o.r.d. curve shows a maximum arising from the superimposition of the contributions of opposite sign to the rotation, the wavelength of the maximum is displaced toward shorter wavelengths by increasing the stereoregularity. In this case the wavelength of the maximum can be taken as an indication of the relative stereoregularity[24] of the different fractions (*Table 7*).

From the above discussion we can conclude that, in general, relationships exist between rotatory power and stereoregularity in vinyl polymers and hence relationships between conformational equilibria and stereoregularity; however the existence of the above relationships can be proposed only after a thorough investigation of o.r.d. in the largest possible wavelength range, confirming the origin of the maxima by c.d. measurements, and not on the basis of $[\Phi]$ measured only at a few wavelengths.

4. RELATIONSHIP BETWEEN $[\Phi]$ AND CONFORMATION

Interesting indications on the relationships between $[\Phi]_\lambda$- and polymer chain conformation have been obtained in the case of poly-α-olefins on the basis of conformational analysis, semi-empirical calculations of optical activity per monomeric unit[6], and statistical mechanical calculation of the macromolecular conformation[25].

Conformational analysis, carried out according to well-established methods used in low-molecular-weight compounds[2], has shown that in the case of isotactic polymers IV $(n=0)$ and IV $(n=1)$, in which large differences in $[\Phi]$ have been observed between polymers and low-molecular-weight models, only two conformations having highly positive and one having highly negative optical rotation are allowed when the asymmetric carbon atoms of the lateral chains has (S) absolute configuration (*Table 8*). In an ideal isotactic polymer the allowed conformations of the monomeric units can give rise only to a left-handed or to a right-handed helical conformation of the principal chain. Despite the small energy difference, mainly of entropic origin, calculated for monomeric units included in left-handed and right-handed helical conformation of principal chain sections, the comparison between $[\Phi]$ calculated by a semi-empirical method[2], which gives excellent results in the case of low-molecular-weight paraffins, and experimental value shows that helical conformation of the thermodynamically-favoured screw sense largely prevails—at least in the case of IV $(n=0)$ and IV $(n=1)$ (*Table 9*)[25]. Statistical mechanical calculations give a consistent explanation of the

Table 8. Conformational analysis of the monomeric unit of isotactic optically-active poly-α-olefines

Polymers		Number of conformations(a)				$\Delta E°$ (a) cal/mole	Absolute configuration of the asymmetric carbon atom of the lateral chain	More-favoured helical conformation of the main chain
		Total staggered	allowed	n_F (b)	n_U (c)			
Poly-(S)-3-methyl-1-pentene	(IV, n=0)	81	3	2	1	400	S	left
Poly-(S)-4-methyl-1-hexene	(IV, n=1)	243	3	2	1	400	S	left
Poly-(S)-5-methyl-1-heptene	(IV, n=2)	829	11	6	5	130	S	left
Poly-(S)-6-methyl-1-octene	(IV, n=3)	2487	21	11	10	60	S	left

(a) For the monomeric unit in an isotactic enchainment.
(b) Number of conformations allowed to the monomeric unit included in the more-favoured helical conformations.
(c) Number of conformations allowed to the monomeric unit included in the less-favoured helical conformation.
(d) $\Delta E° = RT \ln (n_F/n_U)$, at 300°K.

Table 9. Semi-empirical calculation of optical activity referred to one monomeric unit of poly-α-olefines and model compounds

Polymers		$[\Phi]_D$ calc.			$[\Phi]_D^{25}$ exp. (d) (e)	Model compounds (f)	$[\Phi]_D$ calc. (c)	$[\Phi]_D^{25}$ exp.
		(a)	(b)	(c)				
Poly-(S)-3-methyl-1-pentene	(IV, n=0)	+180	−240	+40	+161	(S)-2,3-dimethyl-pentane	−15·0	−11·4
Poly-(S)-4-methyl-1-hexene	(IV, n=1)	+240	−300	+60	+288	(S)-2,4-dimethyl-hexane	+20·0	+21·3
Poly-(S)-5-methyl-1-heptene	(IV, n=2)	+228	−225	+22	+68·1	(S)-2,5-dimethyl-heptane	+10·0	+11·7
Poly-(S)-6-methyl-1-octene	(IV, n=3)	+240	−192	+34	+20·4	(S)-2,6-dimethyl-octane	+14·3	+14·4

(a) Average among the values calculated for the allowed conformations inserted in a left-handed helical sequence.
(b) Average among the values calculated for the allowed conformations inserted in a right-handed helical sequence.
(c) Average among all the allowed conformations according to Brewster [J. Amer. Chem. Soc. 81, 5475 (1959)].
(d) Referred to one monomeric unit, in aromatic hydrocarbon solution.
(e) For the monomer optical purity see Table 2.
(f) For references on hydrocarbons see P. Pino, Adv. Polymer Sci. 4, 393 (1965).

above facts and in the case of completely isotactic poly-(S)-4-methyl-1-hexene [IV $(n=1)$] yield the following very detailed picture: the macro-molecules are formed by relatively-long left-handed helical sections, the average length of which corresponds at 300°K to about 24 monomeric units, alternated with short right-handed helical sections the average length of which corresponds to 2–3 monomeric units[3] (Luisi, Pino 1968). The "con-formational reversals", connecting sections spiralled in opposite screw senses, continuously flow along the main chain because of the low potential barriers existing between different conformations. As a consequence the spiralled sections continuously change their length, only the average length of more favoured and less favoured helical sections remaining constant at a given temperature. The differences between the energy per monomeric unit included respectively in left-handed and right-handed helical conformation is 300–500 calories at 300°K, not very far from that calculated on the basis of purely entropic factors, admitting the same statistical weight for each allowed conformation. The average energy of the couple of monomeric units involved in the conformational reversals is 800–1100 calories per monomeric unit higher than that of the average between the energies of couples of monomeric units included respectively in the thermodynamically more favoured and less favoured helical sections[3] (Luisi, Pino 1968).

Shifting from a completely isotactic macromolecule to a macromolecule containing in the main chain a certain number of configurational inversions, a decrease in the prevalence of the thermodynamically most favoured screw sense is expected[3] (Luisi[3]). These theoretical aspects could give a plausible explanation of the experimentally-observed decrease of the [Φ] by decreasing stereoregularity.

The above model is in agreement with the optical rotation experimentally measured in the solid state[26] where the existence of helical conformation has been clearly demonstrated by x-ray analysis. Furthermore it has enabled us to foresee correctly the results of the following different experiments, some of which are still in progress.

According to the model, the high optical activity observed in the polymers IV $(n=0,1)$ is due to a particular position of the conformational equilibrium in which conformations having high rotation of the same sign largely prevail. The same phenomenon should occur in low-molecular-weight paraffins, in which only few conformations having high optical rotation of the same sign are allowed.

This situation can be foreseen by conformational analysis for the (3S; 5S)-2,2,3,5-tetramethylheptane for which only one conformation having [Φ]$_D$ $-180°$ is allowed and (3R: 5S)-2,2,3,5-tetramethylheptane for which only two conformations having respectively $+180°$ and $+60°$ are allowed[27]. As shown in *Table 10* the values found for the optically-pure compounds are respectively $-100°$ and $+140°$. These values are of the same order of magnitude of rotation found in IV $(n=0,1)$, corresponding to the rotation calculated for the allowed conformations of their monomeric units included in left-handed helical sections of the macromolecule.

The existence of helical conformations in solution seems to be confirmed also by the optical activity of copolymers of optically-active α-olefins with

Table 10. Molar optical rotation for some optically-active poly-α-olefins and low-molecular-weight model compounds

Structure of compound	Poly-(S)-3-methyl-1-pentene	Poly-(S)-4-methyl-1-hexene	(3S:5S)-2,2,3,5-tetra-methylheptane	(3R:5S)-2,2,3,5-tetra-methylheptane
Structure of compound	C_2H_5 a| H—|—CH_3 b| CH_2—|—CH_2〜 c| H	C_2H_5 a| H—|—CH_3 b| H—|—H c| CH_2—|—CH_2〜 H	C_2H_5 a| H—|—CH_3 b| H—|—H c| CH_3—|—$C(CH_3)_3$ H	C_2H_5 a| H—|—CH_3 b| H—|—H c| $(CH_3)_3C$—|—CH_3 H
Allowed conformations[a] and their molar rotation calculated according to Brewster[b]	a b [Φ]D T T −240[h] T T +120[i] G G' +240[i]	a b c [Φ]D T T G' −300[h] T T G' +180[i] G' T G' +300[i]	a b c [Φ]D T G' G −180	a b c [Φ]D T T T +180 G G' G' +60
Average molar rotation calculated according to Brewster[b]	+40	+60	−180	+120
[Φ]D²⁵ experimental	+161[c, d]	+288[c, e]	−100[f, g]	+140[f, g]

(a) T = +180, G = +60, G' = −60 assuming as zero of internal rotation angles the coplanar *cis* conformation of bonds.
(b) J. H. Brewster *J. Amer. Chem. Soc.* **81**, 5475 (1959).
(c) Referred to one monomeric unit.
(d) Polymerized monomer optical purity 91%.
(e) Polymerized monomer optical purity 93%.
(f) Neat.
(g) See reference 28.
(h) in a right-handed helix.
(i) in a left-handed helix.

monomers not containing asymmetric carbon atoms as 4-methyl-1-pentene and styrene.

(S)-4-methyl-1-hexene has been copolymerized with different amounts of 4-methyl-1-pentene and the acetone-insoluble, ethyl acetate-soluble fraction has been investigated[28].

The amount of 4-methyl-1-pentene units present in the fraction was determined by i.r. analysis. Optical activity of the fraction was compared with that of a mixture of acetone-insoluble, ethyl acetate-soluble poly-(S)-4-methyl-1-hexene and acetone-insoluble, ethyl acetate-soluble poly-4-methyl-1-pentene containing a corresponding percentage of (S)-4-methyl-1-hexene and 4-methyl-1-pentene monomeric units.

Table 11. Comparison between optical rotation in cyclohexane solution of some samples of copolymer[(a)] (S)-4-methyl-1-hexene[(b)]/4-methyl-1-pentene and mixtures of the two homopolymers[(d)] having the same composition

Composition of copolymer samples and homopolymer mixtures (%) of (S)-4-methyl-1-hexene-m.u.[(c)]	$[a]_D^{25}$ copolymer	$[a]_D^{25}$ homopolymers mixture	$[\Phi]_D^{25\,(e)}$	$[\Phi]_D^{25\,(f,\,g)}$
71·0	+239	+188	+316	+165
48·1	+210	+130	+396	+140
24·9	+147	+ 71	+515	+ 89

(a) Acetone ins., ethyl acetate sol. fraction.
(b) Polymerized monomer optical purity 93%.
(c) $\dfrac{\text{(S)-4-methyl-1-hexene m.u.}}{\text{(S)-4-methyl-1-hexene m.u. + 4-methyl-1-pentene m.u}} \cdot 100$ determined by i.r. spectroscopy, taking for poly-(S)-4-methyl-1-hexene D B 964 cm⁻¹ and for poly-4-methyl-1-pentene D B 918 cm⁻¹.
(d) Acetone ins., ethyl acetate sol. fractions.
(e) Referred to one monomeric unit of poly-(S)-4-methyl-1-hexene, calculated assuming that in the copolymer the optical rotation derives only from (S)-4-methyl-1-hexene monomeric unit.
(f) Referred to one monomeric unit of poly-4-methyl-1-pentene, calculated attributing to the (S)-4-methyl-1-hexene monomeric unit in the copolymer $[\Phi]_D^{25}$ + 249, corresponding to that of poly-(S)-4-methyl-1-hexene sample used for the homopolymers mixture.
(g) Value calculated by the Brewster method for the allowed conformation of 4-methyl-1-pentene monomeric unit inserted in a left-handed helical sequence $[\Phi]_D$ + 240.

Preliminary data reported in *Table 11* show that the specific rotation of the copolymers is much higher than that of the mixture of the two homopolymers. Attributing all the optical activity to the (S)-4-methyl-1-hexene monomeric units present in the copolymer, the value of $[\Phi]_D^{25}$ found for them is much higher than that ever found for the most isotactic poly-(S)-4-methyl-1-hexene[3] (Pino 1965) prepared up to now. Therefore the 4-methyl-1-pentene units must contribute to the optical activity of the copolymers.

Supposing that (S)-4-methyl-1-hexene monomeric units have the same optical activity in the copolymer and in the homopolymer having the same solubility behaviour of the copolymer, $[\Phi]_D$ values referred to one 4-methyl-1-pentene monomeric unit are found which decrease with increasing percent of 4-methyl-1-pentene in the copolymer. The above values are of the same sign and of the same order of magnitude, calculated by the Brewster method[2], for the allowed conformation of 4-methyl-1-pentene monomeric unit included in a left-handed helical section of poly-(S)-4-methyl-1-hexene.

Interesting results have been also obtained in the copolymerization of (R)-3,7-dimethyl-1-octene and styrene[29], comparing the o.r.d. and c.d.

Table 12. Comparison between optical rotatory power of some fractions of a copolymer (R)-3,7-dimethyl-1-octene[a]/styrene and a low-molecular-weight model compound: (3S: 9S)-3,9-dimethyl-6-phenyl-undecane[b]

	Copolymer[c]			Model compound
Fraction[d]	Copolymer composition, (%) styrene m.u. [e]	$[\alpha]_D^{25}$ [g]		$[\alpha]_D^{25}$ [f]
Acetone sol.	42·0	−27·1		
Acetone ins., diethyl ether sol.	17·9	−66·6		+14·8
Diethyl ether ins., cyclohexane sol.	3·5	−84·6		

(a) Polymerized monomer optical purity 75%.
(b) Prepared from (S)-1-chloro-3-methylpentane having $[\alpha]_D^{25}$ + 19·0 (neat), optical purity 95%.

(c) Monomer mixture composition = $\dfrac{\text{moles sytrene}}{\text{moles styrene + moles (R)-3,7-dimethyl-1-octene}} \cdot 100 = 4\cdot75.$

(d) Obtained by boiling-solvent extraction.

(e) $\dfrac{\text{styrene m.u.}}{\text{styrene m.u. + (R)-3,7-dimethyl-1-octene m.u.}} \cdot 100$ estimated by u.v. spectra taking for ε (polystyrene) 262 mμ

= 230, for ε[poly-(R)-3, 7-dimethyl-1-octene] 262 mμ = 0.
(f) In *n*-heptane.
(g) In cyclohexane.

of the copolymer with the o.r.d. and c.d. of (3S: 9S)-3,9-dimethyl-6-phenyl-undecane which, in our opinion, is a suitable low-molecular-weight model.

The acetone-insoluble, diethyl ether-soluble fraction of the copolymers (*Table 12*) which, on the basis of a very rough quantitative analysis based on the u.v. maximum at 262 mμ[30] contained about 14% by weight of styrene, was used for o.r.d. and c.d. measurements. A multiple Cotton effect appears in the region of 260 mμ which has been confirmed by c.d. measurement[29] (*Figure 4*).

The same multiple Cotton effect in the region of the forbidden $\pi \rightarrow \pi^{*}$[31] transition of the benzene chromophore has been found also in the low-molecular-weight model according to the fact that the phenyl group is

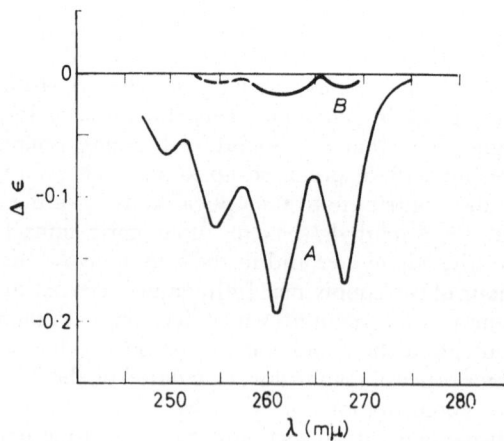

Figure 4. Circular dichroism curves: (A) (R)-3,7-dimethyl-1-octene-styrene copolymer containing 17·9% of styrene m.u. (B) (3R: 9R)-3,9-dimethyl-6-phenyl-undecane calculated on the basis of measurements carried out on its antipode.

placed in both cases in an asymmetric environment. However $\Delta\epsilon$ of c.d. maxima at 267 and 262 mμ of the polymers are at least 10 times higher than corresponding $\Delta\epsilon$ in the model. In our opinion the relatively large values of $\Delta\epsilon$ found for the polymer in the region of about 260 mμ can be suitably explained by assuming that the styrene monomeric units are inserted in helical sections of the macromolecule principal chain which is prevailingly spiralled in a single screw sense, so that the styrene monomeric units, for thermodynamic reasons, assume few conformations having high optical activity of the same sign as indicated by the relatively large Cotton effect. However the above explanation should be regarded as preliminary and more experimental data are needed for a better understanding of the above facts.

Final remarks

In the present paper we have considered some results obtained in the field of optically-active synthetic polymers: in general optical activity is highly dependent on structure in vinyl polymers when asymmetric carbon atoms are present in the lateral chains in α or β position with respect to the principal chain. In these cases a remarkable dependence of optical activity on stereoregularity has been found. The relationship between optical activity and conformations of the macromolecules has been clarified only in the case of poly-α-olefins. For these polymers a model has been proposed which is consistent with all the experimental facts so far obtained and suggests further experiments which should give us a deeper understanding of the behaviour of this type of polymers in solution. The presence of helical conformation which has been ascertained in solution for poly-α-olefins, has not been proved up to now for the other high polymers investigated; a deeper knowledge of the conformational analysis of oxygenated compounds and suitable semi-empirical calculations of optical activity for the same compounds should favour further progress in this field.

In conclusion the results achieved up to now show that optical activity, o.r.d. and c.d. are very powerful tools for the investigation of macromolecular conformation in solution and further interesting progress may be expected in the investigation of both optically-active addition and condensation polymers.

Acknowledgements

The authors express their gratitude to Prof. L. Lardicci, Dr. F. Ciardelli, Dr. O. Pieroni and Dr. C. Carlini for their kind permission to report in this paper some not yet published data. Their advice during the preparation of this paper is also very gratefully appreciated.

Particular acknowledgement is due to the staff of the Soc. Jouan of Paris for their assistance in performing some c.d. measurement by a Roussel-Jouan Dichrograph CD 185.

Notes added in proof

1. *Refers to page 471*

In a recent paper [*Makromol Chem.* **105**, 18 (1967] K. J. Liu, J. S. Lignowski and R. Ullman have investigated polymethacrylates of optically active alcohols and attribute the features of the u.v. spectra of the isotactic poly-methylmethacrylate[11] to the presence of terminal phenyl groups in the macromolecules. However, A. M. Liquori does not agree with the above explanation, as the polymers used by him did not contain low molecular weight fractions.

2. *Refers to page 476*

Ultraviolet spectra of poly-(S)-lactic acid and its low molecular weight models have been recently investigated by M. Goodman and M. D'Alagni [*Polymer Letters* **5**, 515 (1967)] and by R. C. Schulz and A. Guthmann [*Polymer Letters* **5**, 1099 (1967)].

References

[1] M. V. Volkenstein. *Configurational Statistics of Polymer Chains*, Interscience, New York, 1963, p. 139.

[2] J. H. Brewster. *J. Amer. Chem. Soc.* **81**, 5475 (1959).

[3] P. Pino. *Adv. Polymer Sci.* **4**, 393 (1965);
P. L. Luisi, *Polymer Letters* (1968), in the press.
P. L. Luisi and R. Rino. *J. Chem. Phys.* (1968), in the press.

[4] R. C. Schulz and E. Kaiser. *Adv. Polymer Sci.* **4**, 236 (1965);
M. Goodman, A. Abe, and Y.L. Fan, in *Macromolecular Reviews* **1**, 1 (1967).

[5] I. Tinoco. *Adv. Chem. Phys.* **4**, 113 (1962);
R. A. Harris. *J. Chem. Phys.* **43**, 959 (1965);
R. Ullman. *J. Polymer Sci.* C, **12**, 317 (1966).

[6] P. Pino, F. Ciardelli, G. P. Lorenzi, and G. Montagnoli. *Makromol. Chem.* **61**, 207 (1963).

[7] See for instance G. Holzwarth and P. Doty. *J. Amer. Chem. Soc.* **87**, 218 (1965).

[8] W. Moffitt. *J. Chem. Phys.* **25**, 467 (1956);
W. Moffitt, D. D. Fitts, and J. G. Kirkwood. *Proc. Nat. Acad. Sci.* **43**, 723 (1957);
E. R. Blout, in *Optical Rotatory Dispersion* by C. Djerassi, McGraw-Hill, 1960.

[9] F. Ciardelli, E. Benedetti, and O. Pieroni. *Makromol. Chem.* **103**, 1 (1957).

[10] W. M. Pasika and R. Brandon. *Polymer* **6**, 503 (1965).

[11] M. D'Alagni, P. De Santis, A. M. Liquori, and M. Savino. *Polymer Letters* **2**, 925 (1964).

[12] See for instance W. D. Closson and P. Haug. *J. Am. Chem. Soc.* **86**, 2384 (1964).

[13] P. Pino, G. P. Lorenzi, and L. Lardicci. *Chim. e l'Ind.* **42**, 712 (1960);
P. Pino, G. P. Lorenzi, E. Chiellini, and P. Salvadori. *Atti Accad. Nazl. Lincei, Rend.* [8], **39**, 196 (1965).

[14] O. Pieroni, F. Ciardelli, C. Botteghi, L. Lardicci, P. Salvadori, and P. Pino. *I.U.P.A.C. Symposium on Macromolecular Chemistry*, Bruxelles, 1967, Preprint 8/23.

[15] P. Maroni. *Ann. Chim.* **2**, 757 (1957).

[16] P. Salvadori, L. Lardicci, G. Consiglio, and P. Pino. *Tetrahedron Letters* 5343 (1966).

[17] E. I. Klabunovsky, M. I. Shvartsman, and Yu. I. Petrov. *Vysokomol. Soed.* **6**, 1579 (1964);
Izvest. Akad. Nauk SSSR **223** (1966).

[18] E. Chiellini, M. Osgan, P. Pino, and P. Salvadori, in preparation.

[19] R. C. Schulz and J. Schwaab. *Makromol. Chem.* **87**, 90 (1965).

[20] O. Pieroni, F. Ciardelli, P. Salvadori, and P. Pino, in preparation.

[21] L. Lardicci, P. Salvadori, C. Botteghi, and P. Pino. *Chem. Commun.* 381 (1968).

[22] P. Pino, G. P. Lorenzi, and E. Chiellini. *Symposium on Macromolecular Chemistry*, Prague, 1965. Preprint, 455.

[23] S. F. Mason. *Quart. Reviews* **15**, 287 (1961).

[24] P. Pino, P. Salvadori, and E. Chiellini, in preparation.

[25] P. Pino and P. L. Luisi. *J. Chim. phys.* **65**, 130 (1968).

[26] P. Pino, G. P. Lorenzi, and O. Bonsignori. *Chim. e l'Ind.* **48**, 760 (1966).

[27] S. Pucci, M. Aglietto, P. L. Luisi, and P. Pino. *J. Amer. Chem. Soc.* **89**, 2787 (1967).

[28] F. Ciardelli, C. Carlini, E. Benedetti, and P. Pino, in preparation.

[29] P. Pino, C. Carlini, E. Chiellini, F. Ciardelli, and P. Salvadori. *Chim. e l'Ind.* **50**, 257 (1968).

[30] G. Loux and G. Weill. *J. Chim. phys.* **61**, 484 (1964).

[31] L. Verbit. *J. Amer. Chem. Soc.* **88**, 5340 (1966).

ADVANCES AND TRENDS IN THE CHEMISTRY OF POLYMERS WITH A CONJUGATED SYSTEM

A. A. BERLIN

Institute of Chemical Physics, The Academy of Sciences of the U.S.S.R., Moscow, U.S.S.R.

INTRODUCTION

The chemistry of polymers with a conjugated system (PCS) has developed since 1957–58, when systematic work was begun on the synthesis and investigation of various types of polyconjugated systems with delocalized π-electrons in the main chain of the macromolecules[1,2].

Since then the chemistry of PCS has attracted the attention not only of investigators working on polymer chemistry, but of scientists studying various branches of theoretical chemistry and physics, particularly the physics of semiconductors. Such a wide interest in the chemistry of polymers with a conjugated system is connected primarily with the characteristic magnetic and semiconductor properties of these polymers and their high thermal and radiation stability[2-7].

The great interest in the electrophysical properties of PCS has resulted in a rather one-sided research development in this field of polymer science. In fact, most of the work has been connected with studies of the semi-conductor properties of PCS without proper attention having been paid to their individual structures or to their chemical reactivity. Structure–reactivity relationships in conjugated systems play an important role, as a study of this relationship reveals the peculiarity of the chemical transformations in the systems with wide conjugated chains and enables one to find new routes for the synthesis and modification of polymers[2-4].

THE FORMATION AND STRUCTURE OF POLYCONJUGATED SYSTEMS

By increasing the number of delocalized π-electrons in the molecule of a compound with a system of conjugated bonds the energy levels become closer and the energy of the electron transition from the highest occupied to the lowest unoccupied level (ΔE) is reduced. This reduces the energy of singlet–triplet (E_{st}) transition and consequently causes an increase in the reactivity of a polyconjugated system, especially its reactivity towards radical and electrophilic reagents. In *Figure 1* are presented the experimental curves, showing the change of ΔE as a function of the number of π-electrons (N_π) in molecules of individual compounds with a conjugated system. As can be seen, the most rapid decrease of ΔE is observed for the first few members of a homologous series. Later, the curves $\Delta E = f(N_\pi)$ reach an asymptotic value at rather high energy values (poly-*p*-phenylenes,) or slowly approximate to small values of ΔE (polyenes, polyallenes).

i

Such regularity can be realised, because the growth of the conjugated chain is connected with the decrease of the length of the alpha carbon–carbon bonds and this causes an increase of the inter-repulsion of electron shells C—H or other bonds of neighbouring atoms. Thus at definite values

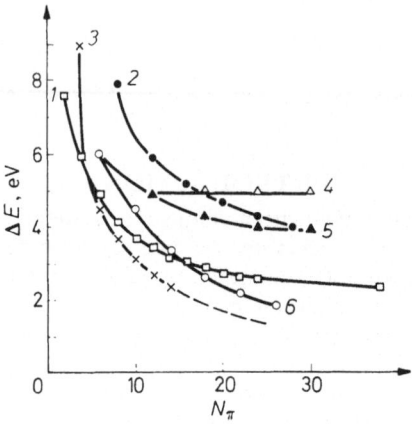

Figure 1. Energy shift (ΔE) *vs.* number of conjugated π-electrons (N_π) for: 1. polyenes, 2. polyines, 3. cumulenes, 4. *m*-polyphenylenes, 5. *p*-polyphenylenes and 6. polyacenes.

of N_π it is better to have a non-coplanar disposition of the next link with regard to the previous group of conjugated links (n_c), which will henceforth be called "the conjugated block". The energy disadvantage of long conjugated blocks follows from thermodynamic considerations too. From what has been said above it follows that a rapid decrease of enthalpy (ΔH) is possible at comparatively low n_c during the growth of the chain. But at the same time the entropy of the system decreases with the growth in length of the polymer macromolecule with the conjugated system (PCS). So under fixed conditions the most probable values of n_c will be realized and the common length of the polymer chain may exceed the value of the conjugated block.

We have confirmed these conclusions by investigating n.m.r. absorption spectra and luminescence spectra of polyphenylenes, polyazophenylenes and polyphenylacetylenes, obtained by thermal (temp. 150°) and catalytic polymerization (temp. 40–70°, R_3Al $TiCl_3$).

As a result of this work, it was shown that a conjugated system of several links is obtained at moderate temperatures and that the dimension of the macromolecules considerably exceeded the value of the average conjugation block.

A steep fall in absorption (in the region of 250–280 nm) as well as the nature of luminescence spectra in the region 400–700 nm indicate the presence in PCS of some distribution, which is a function of the length of the conjugated blocks (i.e. n_c).

It is curious that the average length of the conjugated blocks for a given type of PCS depends on the conditions of synthesis of a polymer. For example,

ii

polyphenylacetylene, obtained by thermal polymerization, consists of blocks with $n_c = 3–4$, whereas the same compound obtained by a catalytic process with the catalyst $(C_2H_5)_3Al.TiCl_3$ includes blocks with $n_c = 4–7$. Moreover both polymers differed from one another in their physical and chemical properties[8,9].

It is essential that during the thermal treatment of PCS (polyphenyl-acetylene, polydiphenylbutadiene at temperatures 300–400°) or in the high-temperature polymerization of arylacetylenes (tolane, at 300–400°; phenylacetylene, at 300°) the degradation and the formation of PCS take place of a length of the conjugated block of about the same value as the degree of polymerization ($n_t \simeq n_p$). Such soluble block polymers have a considerably high thermal stability, electro-conductivity and paramagnetism[3–5,14].

The presence of conjugated blocks in PCS suggests that it is characteristic of the presence of a new type of structural isomerism (apparently, stereo-isomerism) in PCS, caused by the difference in dimension, location and alternation of conjugated blocks.

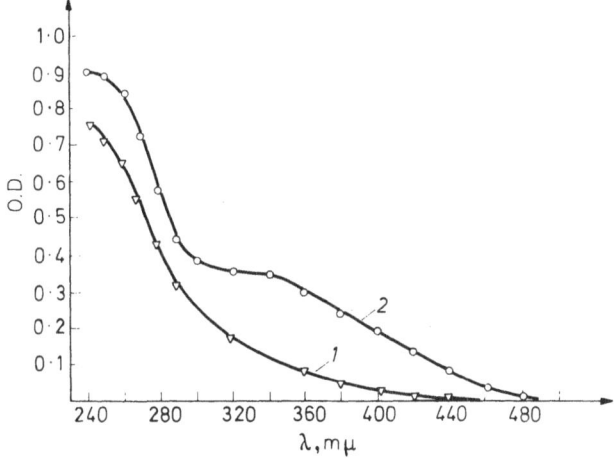

Figure 2. Absorption spectra of polyphenylacetylenes obtained by thermal (*1*) and catalytic (*2*) polymerization (in CHCl$_3$ solution).

The comparison of properties of two fractions of nearly equal molecular weight ($\bar{M}_n = 1050$ and 1200), and with the same composition and infrared spectra, but various average values for conjugated blocks of polyphenylenes, obtained by thermal and catalytic polymerization respectively illustrates "the structural isomerism of polyconjugation" (SIP) (*Figures 2* and *3*). As we can see these polymers differ strongly in their absorption spectra and their reactivity to α,α-diphenyl-β-picrylhydrazyl (DPPH). Apparently, in the chemistry and physics of polyconjugated systems it is necessary to take into consideration the effect of the isomerism on the structure caused by the presence of conjugated blocks.

iii

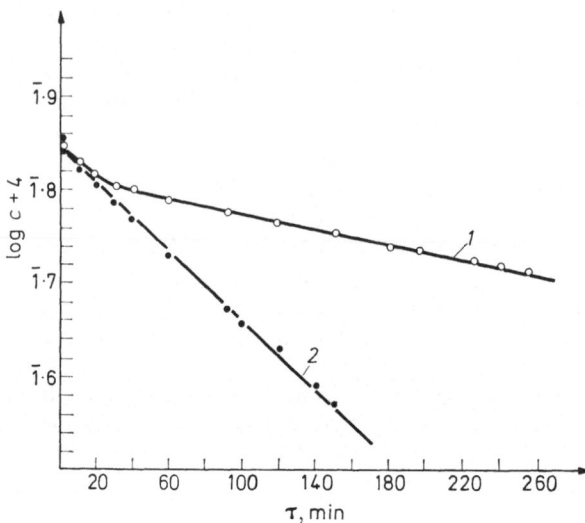

Figure 3. Comparison of reactivity of polyphenylacetylenes obtained by thermal (1) and catalytic (2) polymerization.[The logarithm of concentration of α,α-diphenyl-β-picrylhydrazyl (log c) against the time of reaction, τ (min)]

Figure 4 illustrates the decrease of the ionization potential (I) and the increase of the electron affinity (A) of individual compounds with the energy shift, i.e. the numbers of π-electrons. If we now consider the chain growth during the synthesis of PCS, it will become obvious that in the process of the formation of polyconjugated macromolecules (independent of the mechanism of polymerization) conditions favourable for the formation of charge transfer complexes (CTC) are created since this process is accompanied by the reduction of $I - E$. Two conclusions follow from this fact (i) the inactivation of the active centre, leading to the polymerization in consequence of inter- and intramolecular delocalization of π-electrons; (ii) the increase of the probability of S–T transition and the ease of the excitation of polyconjugated molecules during synthesis and further energy influences.

In fact, the use of numerous polymerization reactions for the synthesis of PCS has not allowed high-molecular products to be obtained[2-7]. Regardless of the chemical mechanism and conditions of carrying out the process, oligomer compounds, with $\bar{M}_n \rightleftharpoons 8000{-}10000$, were formed in all cases.

The investigation of the thermal initiated and radiation polymerization kinetics of arylacetylene and other derivatives of acetylene, not containing labile atoms in the α-positions indicated the monomolecular character of the breaking and the independence of $\bar{M}_n$ on the temperature, the amount of initiatior and dose power[14]. Some increase of $\bar{M}_n$ (5000–8000) is obtained in anion-coordination catalysis[13] and in carrying out the process with anion catalyst in an electron-donor solution[15]. The polymers thus formed contain stable paramagnetic centres, characteristic of PCS[2-7].

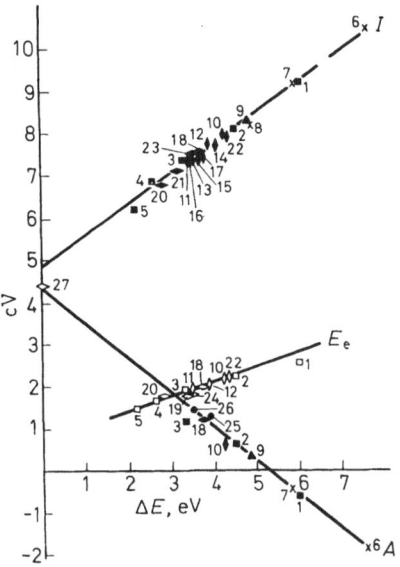

Figure 4. The dependence of ionization potential (I), electron affinity (A) and activation energy (E) of electro-conductivity on energy Shift (E_e)

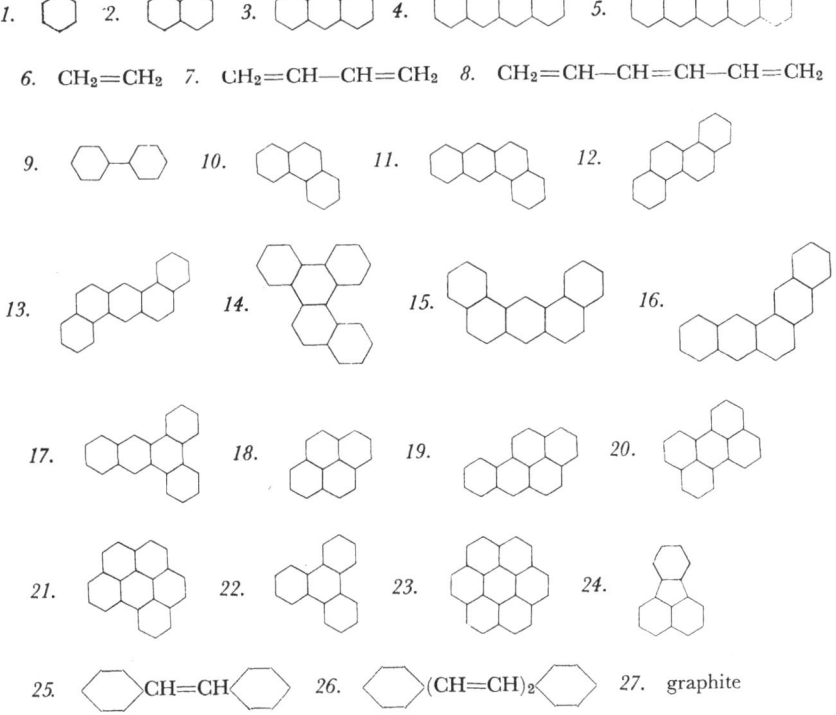

1. *2.* 3. 4. 5.

6. $CH_2=CH_2$ 7. $CH_2=CH—CH=CH_2$ 8. $CH_2=CH—CH=CH—CH=CH_2$

9.   10.   11.   12.

13.   14.   15.   16.

17.   18.   19.   20.

21.   22.   23.   24.

25. $CH=CH$ 26. $(CH=CH)_2$ 27. graphite

V

On the other hand it was found that the derivatives of acetylene, containing hydrogen in an α-position with respect to the triple bond, in catalytic polymerization with Ziegler–Natta's catalyst, form colourless or lightly-coloured high-molecular polymers with $\bar{M} = 60\,000\text{–}80\,000$ without paramagnetic centres, possessing high elastic properties and which are typically dielectrics[2, 5, 16]. It was determined experimentally that in the latter case the polymerization is accompanied by an allylic rearrangement and, as a result, the continuity of the conjugation chain is broken[14].

$$CH{\equiv}C \qquad \ldots -CH_2-C-CH_2-C-CH{=}C- \ldots$$

$$\begin{array}{cccc} | & \| & \| & | \\ CH_2 \longrightarrow & CH & CH & CH_2 \\ | & | & \| & | \\ R & R & R & R \end{array}$$

Thus the presence of conjugated blocks in polymers of the first type and their absence in the second type of polymers are decisive factors in the growth of the chain, the formation of paramagnetic centres and the electrophysical properties.

The mechanism of chain breaking in the polymerization of acetylene derivatives, which form PCS, has not been fully determined so far. But there is some basis for assuming that a disappearance of the active centre is connected with the transfer of the chain to the monomer, which, for polyvinylene, is an electron acceptor.

The conditions for polyconjugated chain growth, mentioned above, and the deactivation of the leading centre apply also to the polycondensation reaction. For example, in the synthesis of polyazines, conditions are created when an electron-acceptor aldehyde group is inactivated[17], apparently, as a consequence of the complex formation, in which the transfer of charge is realized by p-electrons of nitrogen, situated in the polyazine chain.

It should be added that the formation of PCS during synthesis leads, in many cases, to the formation of products, insoluble in the reaction mixture, which may hinder further progress of the reaction. Such examples were observed in the oxidative dehydro-polycondensation of some diacetylenes[7], during the synthesis of polymethine and the polymethinephenylenes by Wittig's reaction[18], in the production of polyazophenylenes and other polymers by the diazosynthesis[19], in the polymerization of acetylene[20], in obtaining polyphenylenes by Kovacic's method[21], and in others[2-7].

From what has been said above it became necessary to elaborate the synthesis of soluble PCS and develop methods to transform polyconjugated oligomers into chain and space-cross-linked polymer substances, possessing desirable physical and chemical properties.

At present the following ways of modification of oligomeric polyconjugated systems are under investigation. (i) The method of pseudo-radical polymerization or block co-polymerization of polyconjugated systems with electron-acceptor monomers, activated by paramagnetic centres. (ii) Formation of polyconjugated oligomers with functional groups (HO, NH_2, COOH, and others), having the ability to undergo polycondensation reactions or migration block co-polymerization. (iii) Methods based on oxidation–reduction transformation of polyconjugated systems, containing electron exchange groups ("cube polymers").

The first way may be illustrated by the co-polymerization of polynuclear aromatic compounds (anthracene) polyphenylene, polyazophenylene, poly-phenylacetylene etc., with di- and tetra-functional monomers, containing electron-acceptor double and triple bonds (*Figure 5*)[3–5,22–24].

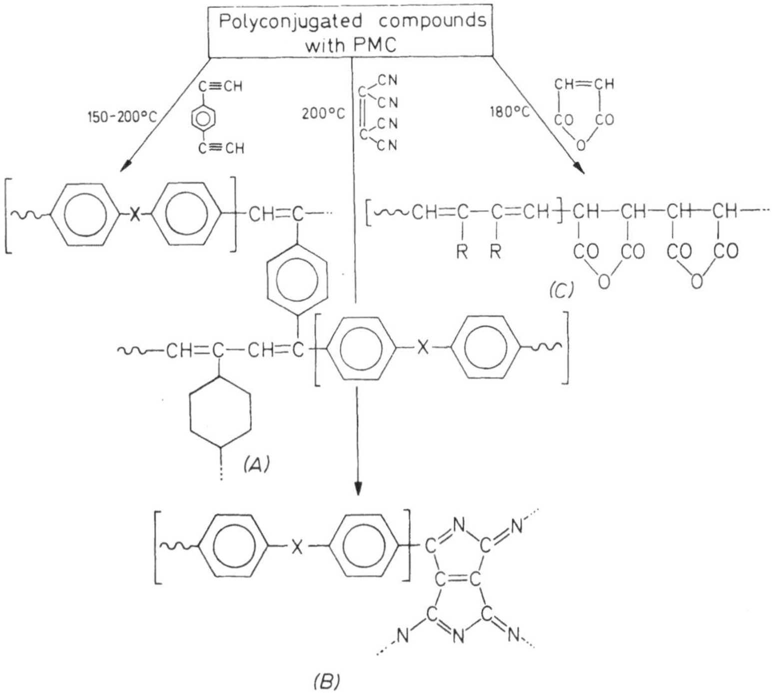

Figure 5. Schematic representation of pseudo-radical polyreaction of polyconjugated systems.

As we can see from the schemes cited, block co-polymerization of polyconju-gated systems, activated by paramagnetic centres allows soluble and fusible oligomers to be turned into rather thermostable (up to 400–500 °C), cross-linked or chain polymers with a branched system of conjugation (*A,B*) or to block copolymers with vinyl monomers (*C*). In the latter case, polymers are obtained which combine the properties of polyconjugated and saturated blocks. The mechanism of this process, which is specific for polyconjugated systems, has not been investigated. There is some basis for supposing that it is connected with the formation of donor–acceptor complexes or adducts.

The second way of transforming polymers with a conjugated system is shown schematically in *Figure 6*.

In this way one can obtain polymer substances with a cross-linked struc-ture by the interaction of carboxy-containing polyconjugated oligomers, for example, with oxides or salts of polyvalent metals (*A*) or by the condensa-tion polyhydroxyphenylenes[3, 25].

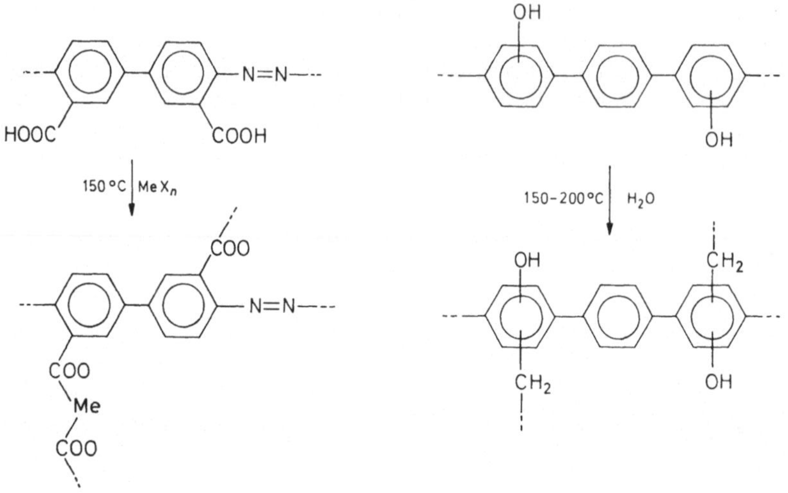

Figure 6. Some types of transfer of polyconjugated polymers, containing functional groups.

According to Marvel's data polyhydroxyphenylenes, cured by formaldehyde, lose only 30 per cent by weight at 900°C.

By using polyconjugated oligomers containing end functional groups it may be possible to obtain various types of chain block co-polymers.

The recently-projected third way of transforming polyconjugated oligomers opens up very interesting prospects (*Figure 7*).

In this case, a water-soluble leucoform of 'vat dye-like' polymer may be used for the following oxidizing dehydration in air; alternatively an oxidizer may be employed. In this process insoluble polymers are formed with a conjugated system possessing a high stability, and electron-exchangeable, catalytic and semiconductive properties. To this group of polymers are attributed polymeric indigo (recently synthesized[26]), poly-5,5'-bis-isatyl (thiophene)-indophenine[27], polyquinonedioxine[28] and, apparently various polyarylene-quinones[3–5, 29]. These processes for obtaining thermostable polymer materials with a conjugated system may be as important as the two-stage cyclopolycondensation and pyrolysis of saturated high-molecular compounds, realized on the industrial scale.

PARAMAGNETISM AND THE EFFECT OF LOCAL ACTIVATION

The other consequence of the formation of charge transfer complexes (CTC) in the synthesis of polyconjugated systems is the comparative ease of their transition in the excited state.

This enables a local unpairing of π-bonds in the fraction with the longest conjugation blocks and the coplanarity of neighbouring links is broken. This causes a degeneration of the singlet–triplet state[24] and the formation of stable unexcited double radicals. In such systems unpaired electrons are delocalized along the conjugation block at the expense of intermolecular

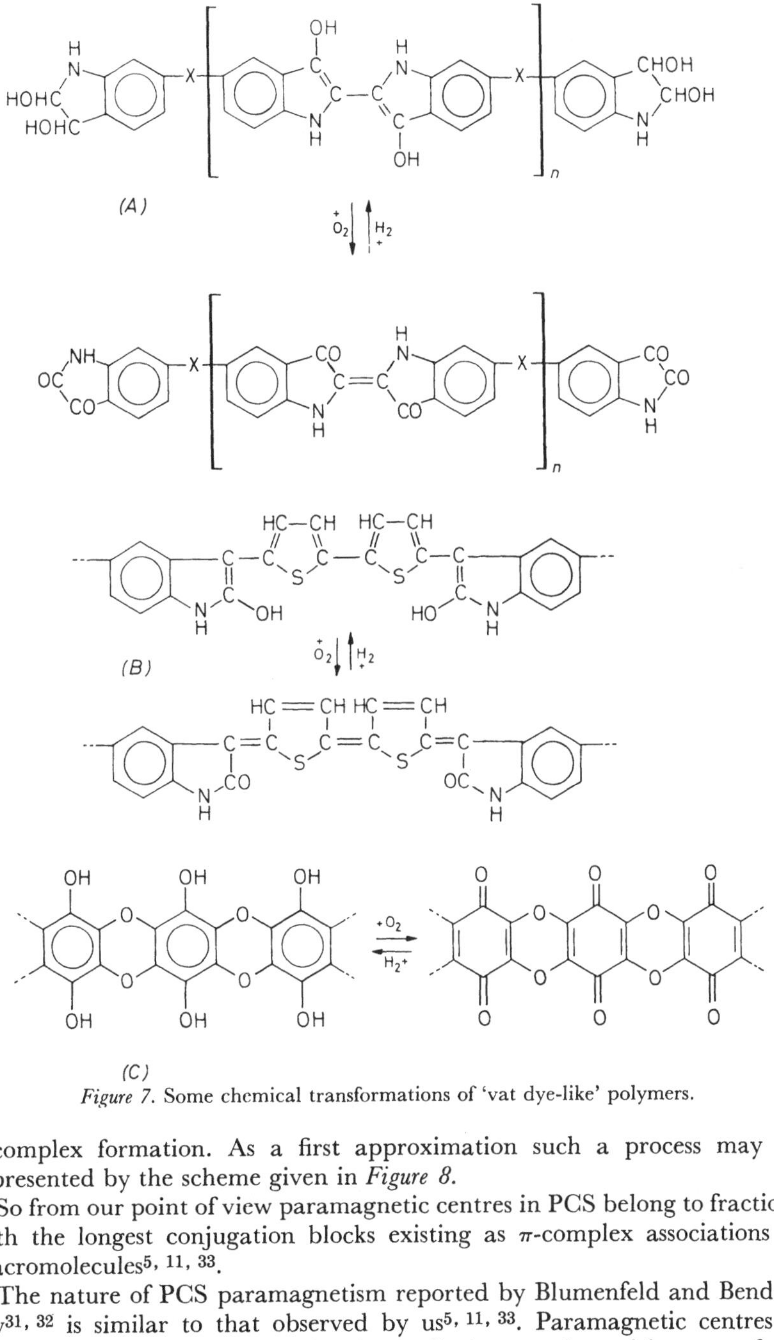

(A)

(B)

(C)

Figure 7. Some chemical transformations of 'vat dye-like' polymers.

π-complex formation. As a first approximation such a process may be represented by the scheme given in *Figure 8*.

So from our point of view paramagnetic centres in PCS belong to fractions with the longest conjugation blocks existing as π-complex associations of macromolecules[5, 11, 33].

The nature of PCS paramagnetism reported by Blumenfeld and Bendersky[31, 32] is similar to that observed by us[5, 11, 33]. Paramagnetic centres of polyconjugated systems as stable ion-radicals were formed because of the electron transfer between macromolecules with a long sequence of conjugated bonds.

ix

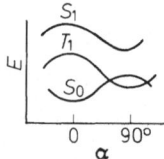

Figure 8. The scheme of probable formation of paramagnetic centres of polyconjugated polymers, where *C* signifies the break of coplanarity, *E* is the energy of the polyconjugated system, α is the rotation angle around the C—C bond. S_0 the singlet level, S_1 the excited singlet level, and T_1 the triplet level.

We cannot discuss here the numerous data, confirming the correctness of the proposed mechanism of formation of the PCS paramagnetic fraction. We will only state that the presence of stable paramagnetic centres is the inalienable property of substances with a branched conjugation system. By an appropriate energy influence (thermolysis, photolysis, radiolysis) PCS can increase the dimensions of conjugation blocks, and consequently, raise the concentration of paramagnetic centres (PMC)[3-12].

In some cases of course, the formation of the so-called "trapped" radicals occurs side by side with PMC. But they can be distinguished from PMC by a detailed analysis of e.p.r. spectra or be removed by numerous reprecipitations and by thermal treatment of the substrate.

Investigations of the PMC influence on the chemical and physical properties of compounds with π-bonds led to the discovery of the phenomenon called 'the local activation effect'[3,4,25]. In general, this effect is demonstrated by such facts as the influence of PMC on polyconjugated systems, which form π-complexes with diamagnetic molecules, and by their reactivity, capacity for polyaddition, and by physical and chemical properties. During these processes PMC do not undergo visible changes. We have observed the local activation effect in processes of rather different types. Thus, for instance, it was indicated that polymers with a system of π-conjugation (polyphenylene, polyazophenylene, polyphenylacetylene), polynuclear, aromatic hydrocarbons (anthracene and its homoloques, etc.) and also monomers with

x

electrophilic multiple bonds (p-diethylbenzene, tetracyanoethylene, maleic anhydride, etc., are capable of polymerization and copolymerization.

Besides this fact it was found that the autocatalytic character of onium polymerization of γ-chloropyridine and similar compounds was connected with the accumulation of paramagnetic polymer formed in the process of polyreaction[34]. The analogous activating action of PMC was probably observed in the polymerization of pyridine, proceeding with decyclization[35] and in catalysis by polymer semiconductors[52].

The activating influence of PMC is to some extent displayed in the low temperature pyrolysis of aromatic hydrocarbons[36,37]. It was found that the introduction of one per cent of paramagnetic polyanthracene ($M_n = 1000-1200$, containing 5×10^{17} spin/g) into anthracene, reduces the induction period of pyrolysis (at 450°), greatly increases the speed of the process, increases the production of gaseous products by more than 30 times and changes their composition, and also accelerates the accumulation of para-magnetic polymer in the solid products obtained by heat treatment (*Figure 9*).

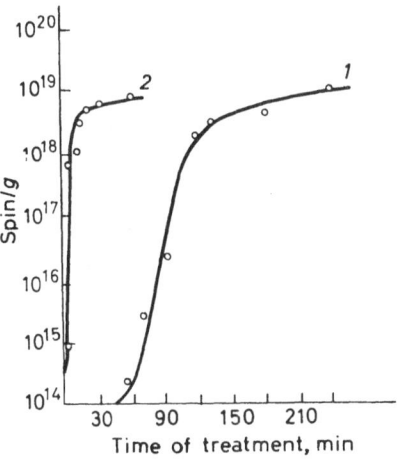

Figure 9. Accumulation of paramagnetic centres in the course of thermal treatment of anthracene at 450°C. The concentration of paramagnetic centres is plotted as a function of the time of thermal treatment for: *1* anthracene, and *2* anthracene with 0·1% paramagnetic polyanthracene (PMC = 5×10^{17} spin/g).

The increase in reactivity of compounds with a π-conjugated system, containing PMC, with reference to stable radicals, was investigated in reactions of anthracene, polyphenylene, polyphenylacetylene and polyazophenylene with α,α-diphenyl-β-picryl-hydrazyl (DPPH)[38,39].

With a PMC content, corresponding to $10^{16}-10^{17}$ (spin/g) both anthracene and π-conjugated polymers react quickly with DPPH in benzene or tetrachloromethane solution at 40–70° with the conversion of DPPH to diphenylpicrylhydrazine (DPPH-H). However, a narrow e.p.r. singlet, characteristic for PMP remained unchanged.

Investigation of the kinetics of this reaction showed that it proceeds with rather small activation energy ($E_{eff} = 9$–11 kcal/mole). Taking into account the heat of complex formation of DPPH the solvent activation energy does not exceed several kcal/mole.

The difference in reactivity of the catalytic and thermal forms of poly-phenylacetylene is apparently connected with the difference in the dimension and distribution of conjugation blocks (see above). Since both polymers react with DPPH with approximately the same activation energy it may be assumed that structural factors, in general, influence the pre-exponential factor of Arrhenius.

In the presence of PMC, substances with a conjugated system actively react with peroxides and hydroperoxide compounds, so that their mono-molecular decay is accelerated and the bonding radicals are converted into π-complexes[44].

The capacity of PCS to interact with free radicals may be the means of estimating the reactivity of their homologous fractions.

A kinetic investigation has been made of the interaction of homologous fractions of polyphenylacetylene with DPPH and peroxy radicals, formed in the reaction of cumene with oxygen[41].

Investigation of the e.p.r. absorption spectra and luminescence of poly-meric fractions showed the increase of free spin concentration in one gram of a sample with increasing average molecular weight and with a rather small change in the average dimension of the conjugated block (for example $\bar{M}_n = 2400$, n_c 4–6; $\bar{M}_n = 7600$; $n_c = 4$–7). The investigation of the relative fraction reactivity at constant temperature (60°) and initial con-centration of the polymer and DPPH ($2 \cdot 05$–10^{-4} mole/litre and $0 \cdot 9 \times 10^{-4}$ mole/litre) leads to the conclusion that the reaction velocity increases with the rise of $\bar{M}_n$ and the PMC content (*Figure 10*).

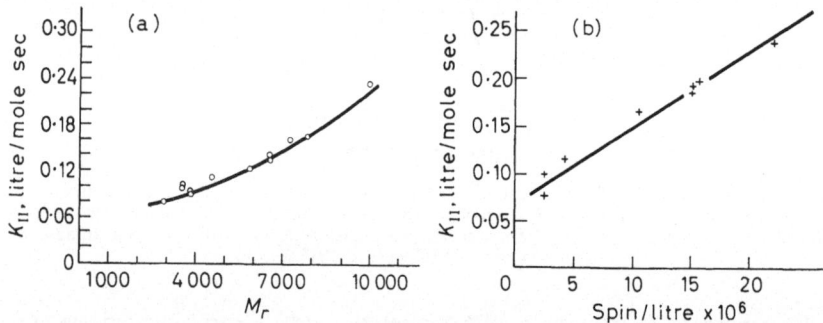

Figure 10a. Rate constants of diphenylpicrylhydrazyl interaction with homologous fractions of polyphenylacetylene, obtained by catalytic polymerization with $(C_2H_6)_3Al$. $TiCl_3$ ($M_n = 5000$). The rate constant (litre/mole.sec) is plotted as a function of the PMC fraction.

Figure 10b. Rate constants of the interaction of homologous fractions of polyphenyl-acetylene as a function of the concentration of PMC in fractions.

An analogous result was also obtained for the reaction of polyphenyl-acetylene with peroxy radicals[41]. Apparently the difference in reactivity of homologous fractions is due mainly to the effect of local activation and to a lesser extent to the variation in the dimensions of conjugated blocks. Such a conclusion follows from the fact that the differences in the dimensions of conjugated blocks n_c for different fractions are insignificant, while the PMC content roughly rises with the increase of $\bar{M}_n$[41].

The increase of radical reactivity of π-conjugated systems in the presence of PMC is demonstrated to some extent in their behaviour as inhibitors of thermoxidative destructions of a number of low-molecular weight compounds (paraffin hydrocarbons, siloxane liquid, esters, plasticizers) and high-molecular weight substances (polyvinyl chloride, polymer olefins, rubbers, polysiloxanes, polyamides, polyacrylates, polyethylene-terephthalate, esters of celluslose, etc.)[42,43].

Thus it was established that the strong inhibition effect of naphthacene and pentacene in the thermal oxidation of paraffins (at 160–200°C) is caused by the presence of paramagnetic fractions in these hydrocarbons.

An increase of polymer concentration beyond a definite limit leads to a decrease of its inhibition activity in the case of a prolonged heat-treatment of anthracene. The same extreme regularity was observed with the introduction of soluble paramagnetic fractions of polyanthracene into anthracene without heat treatment (*Figure 11*).

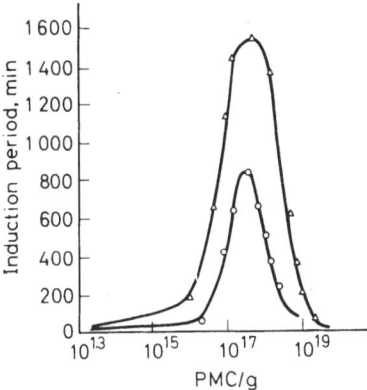

Figure 11. The dependence of the induction period on the inhibitor content of PMC during the oxidation of kerosine at 160°C. △, Unfractionated thermolysed anthracene. ○, The complex of pure anthracene with a polyanthracene paramagnetic fraction. The concentration of inhibitor is 1%.

On the other hand, an increase of anthracene concentration for a fixed paramagnetic fraction leads to an increase of induction period of the thermal oxidation (*Figure 12*). Thus it is evident that the effect under consideration is displayed in cases where there is an activator (paramagnetic fraction), a substrate (anthracene, polyphenylene, etc.) and, in the case of a chemical process, a reagent (radicals, electro- and nucleophilic substances). It should be noted that during the inhibited oxidation of paraffin hydrocarbons as well as during the activation of the PMC interaction of π-conjugated systems with stable radicals (see above), the e.p.r. signals intensity of the PMC present in the reacting medium remains unchanged. In addition, the unreacted paramagnetic fraction at the end of the reaction maintains the capacity to activate the inhibitor in its repeated addition to substrate.

All this evidence points to a catalytic role of PMC in cases displaying the effect of local activation which have been investigated. It was established experimentally that the effect of local activation is specific and resembles fermentation catalysis in this respect. This effect is observed more effectively

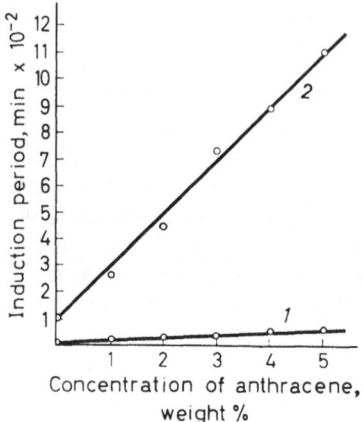

Figure 12. Change of the induction period in the thermal oxidation of kerosine as a function of the progressive increase of anthracene at the constant quantity (0·1%) of polyanthracene activator containing PMC = 3–5 × 10^{17} spin/g.

in those cases where there is a close conformity between the structure of the 'paramagnetic activator' and the substance being activated. For instance it was found that a paramagnetic fraction isolated from anthracene exposed to heat treatment, activates anthracene more effectively than the inhibitor of thermal oxidation but its influence on the reactivity of polyphenylene, polyazophenylene and polyphenylacetylene is fairly small. This result was obtained by the introduction of polymers with a conjugated system into anthracene. It is curious to note that the paramagnetic products of thermolysis of polyphenylacetylene, containing blocks of condensed aromatics do not activate polyphenylacetylene but are effective activators of anthracene.

The polarity of the reaction medium has considerable influence on the effect of local activation and it has been established that the efficiency of the stabilizing action of polyconjugated systems containing PMC, greatly increases in the presence of polar polymers (phenol-formaldehyde resin, polyesters, polyamides) and polar low-molecular weight compounds (ester plasticizers) during the inhibition of thermal-oxidative destruction. The increase in efficiency of an inhibitor was also observed in the case of the oxidation of non-polar substances (e.g. kerosine) in polar medium (nitroanisole), which has no influence on the kinetics of thermal-oxidation.

With respect to the addition of PMC the greatest effect is obtained, all other things being equal, in those cases where the activator is combined with the substrate in solution with subsequent removal of the solvent. Of course, in those cases, when PMC is formed during the synthesis (polymers with π conjugated system, thermolyzed or irradiated substances with delocalized π-electrons), the necessity for such an operation is removed.

On the basis of absorption and luminescence spectra, it may be assumed that π-complexes are formed by combination of PMC with diamagnetic molecules, related in structure[44, 45].

The fact that an inhibitor is not destroyed during the induction period, is specific of the inhibiting action of PCS. At the end of this period, in the presence of inhibitor PCS peroxides accumulate. Moreover their maximum concentration greatly exceeds that of the peroxides which accumulate with the oxidation in the absence of inhibitor[43]. In other words, PCS may be strong inhibitors and accumulators of peroxides at the same time; some increase in efficiency is also characteristic for PCS in the stabilizing action with the addition of substances which can react with peroxides to give non-radical products of decay (phosphites, sulphides) (*Figure 13*).

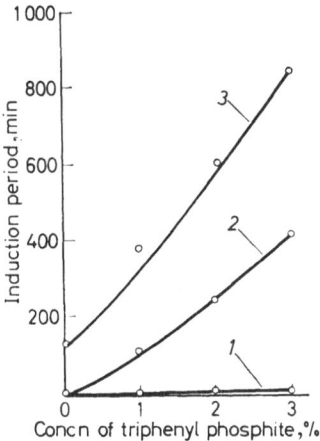

Figure 13. Dependence of the induction period on the concentration of triphenyl phosphite in the oxidation of kerosin (t = 180°): (*1*) without addition, (*2*) with 2% of anthracene and (*3*) with the addition of 2% of anthracene complex with paramagnetic polyanthracene (the concentration of PMC = 5×10^{17} spin/g).

From all the experimental data, the scheme shown in *Figure 14* is proposed as a first approximation to explain the inhibiting action of PCS[32,36,37].

Apparently, at the first stage of the inhibitor action the formation of charge transfer complexes (CTC) takes place between the more high-molecular PMP and the diamagnetic molecules of inhibitor (anthracene) surrounding them, and is accompanied by the polarization of the inhibitor molecules. The possibility of $S-T$ transition in molecules with a conjugation system is increased by the influence of PMP.

In the excited biradical state, it interacts with the radical RO_2., which "sticks" to the inhibitor forming a stable complex, with perhaps electron transfer and the formation of an ionic pair. This process is accompanied by the release of PMP, which forms a complex with a new molecule of anthracene and so on. The equilibrium is displaced to the left by cooling the system (in a stage of induction period), thereby releasing the original anthracene and causing the free radicals $RO_2^{\cdot}$ to recombine. This would seem to explain the constant quantity of anthracene determined experimentally over a period.

When all or most of the anthracene is used up in the formation of complexes with radicals $RO_2^{\cdot}$, interaction of these complexes with newly formed radicals $RO_2^{\cdot}$ occurs, leading to the formation of anthraquinone and, consequently, to the anthracene being used up. The possibility of the

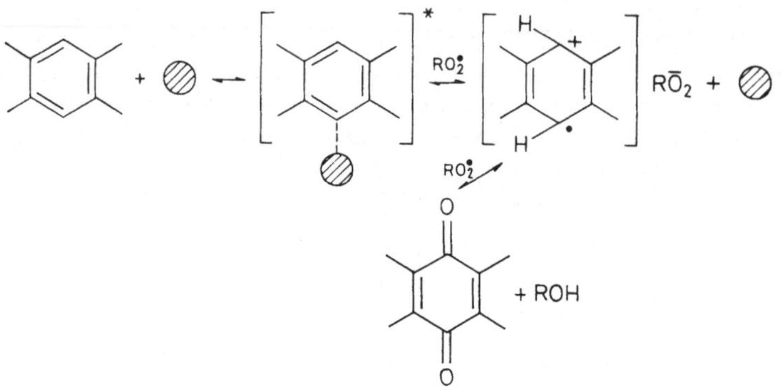

Figure 14. Probable mechanism of the inhibiting action of compounds with conjugated system activated by PMC. ⊘ Polyconjugated system with PMC.

apparent transition of molecules into the singlet state, in the formation of complexes with PMC by thermal influence, was confirmed by us by an investigation of the *cis–trans* isomerization of dimethylmaleate activated with paramagnetic PCS[46]. It is known that this reaction proceeds according to the triplet mechanism. As a result, it was shown that this transformation is also activated by the PMC, which causes decrease of the activation energy of the process (see *Figure 15*).

In the investigation of PCS as a stabilizer against thermo- and photo-oxidation of industrial polymers, a relative increase of the efficiency of their inhibiting action with temperature was noted[42, 43]. However, it has become possible to compensate the dependence of efficiency on temperature by a change of PMC concentration. Such specificity of PCS directly follows from the theoretical considerations stated above and we reach the conclusion that PCS may be effective stabilizers against the high-temperature degradations of photo- and radiation-oxidation industrial polymers. Towards phenol resins, polyamides, polyesters, polyurethanes, polysiloxanes, cured oligomeric ester acrylate, polyvinyl chlorides, etc., paramagnetic PCS was found to have an efficient inhibiting action at 200–300° against oxidation and irradiation of industrial polymers[2, 3, 15, 36].

The line of development in this field led to the investigation of the possibility of formation of polyconjugated fractions and PMC in some industrial polymers on thermal treatment or on irradiation and measurement of their influence on the destructive processes of polymer substances. The tendency of high-polymers and, especially, polymers with aromatic nuclei to change colour (up to black) when undergoing chemical transformation, leading to the formation of π-conjugated blocks served as a basis for these investigations.

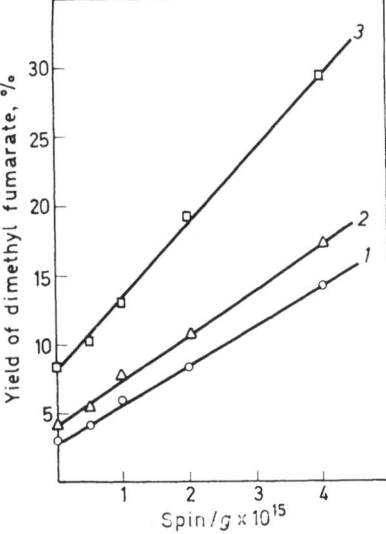

Figure 15. The initiation of *cis–trans* isomerization of dimethyl maleate by polyphenylacetylene (containing PMC $= 10^{17}$ spin/g). The dimethyl fumarate yield (%) is plotted as a function of the proportion of polyphenylacetylene, expressed in terms of spin concentration (g. of dimethyl maleate). The curves *1, 2,* and *3* correspond to 225°, 240°, and 255°C.

As a result of this work it was shown that conjugated blocks containing unpaired spins, and which are formed by thermolysis of phenol-formaldehyde resins[51] and by photolysis of polyurethanes and aromatic polyamides[53] are the effective stabilizers of their own polymers or others (*Figures 16* and *17*). This may also apply to such polymers as poly-pyromellitimides, poly-benzimidazoles, aromatic esters, aromatic polyamides, etc. Thus, for

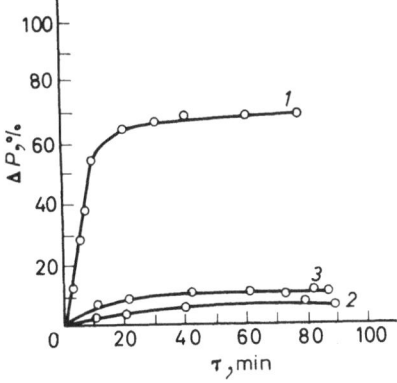

Figure 16. The influence of the addition of polymers, containing PMC, on the kinetics of thermo-oxidation of phenol-formaldehyde resin (at 300°C and a pressure of 760 mm Hg) for: (*1*) resin without additions, (*2*) resin with the addition of 5% of polymer with conjugated system containing $5 \cdot 10^{17}$ spin/g, and, (*3*) resin with the addition of thermolysed resin, containing 2×10^{18} spin/g.

example, it was determined that the formation of coloured products and PMC (10^{17}–10^{18} spin/g) in the polycondensation of the anhydride of 1,2,4,5-naphthalenetetracarboxylic acid with benzidine is observed at the stage of the formation of acid polyamides. The subsequent cyclization with increase of

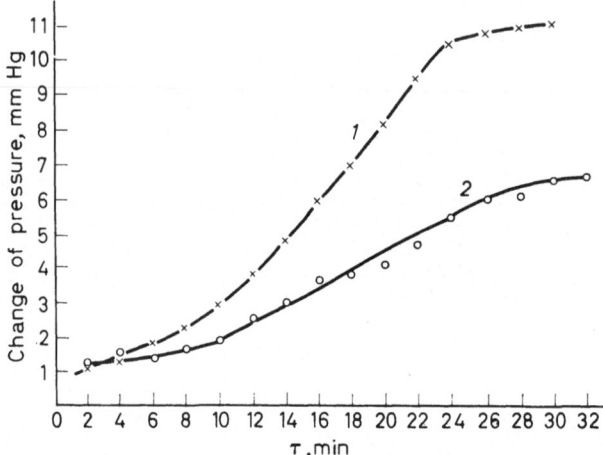

Figure 17. Influence of the paramagnetic fraction of photolysed polyurethane based on toluylene di*iso*cyanate (containing PMC $= 2 \times 10^{16}$ spin/g) on the photo-oxidative destruction of polyurethane [at 300 mm Hg and 25°C, with addition of 1% acetone-soluble fraction]. The change of pressure (mm Hg) is plotted as a function of the time of photo-oxidation (min).

temperature (200–250°) leads to the accumulation of PMC and a more pronounced bathochromic shift in the absorption spectra.

Apparently in the formation of aromatic polyimides and polybenzimidazoles, the amide-isoamide tautomerism, causing the formation of conjugated blocks and PCS, plays an essential role (*Figure 18*). In the light of the

Figure 18. Possible role of isoamide structures in the formation of aromatic polyimides and polyaroyliden-bis-benzimidazole.

Properties	Polymer (see Figure 18)			
	I	II	III	IV
PMC content spin/g	1.8×10^{18}	2.7×10^{18}	1.0×10^{18}	1.1×10^{18}
ΔH	12·5	4·0	11·0	5·0
E eV/mol	0·31	0·47	—	
σ_0 cm^{-1}, g^{-1}	1.6×10^{-2}	2.5×10^{-1}	—	7.0×10^{-2}
$- \log \sigma_{150}$	$- 5.5$	$- 5.7$	—	$- 5.5$
Temperature of thermodegradation in air	—	530	—	380

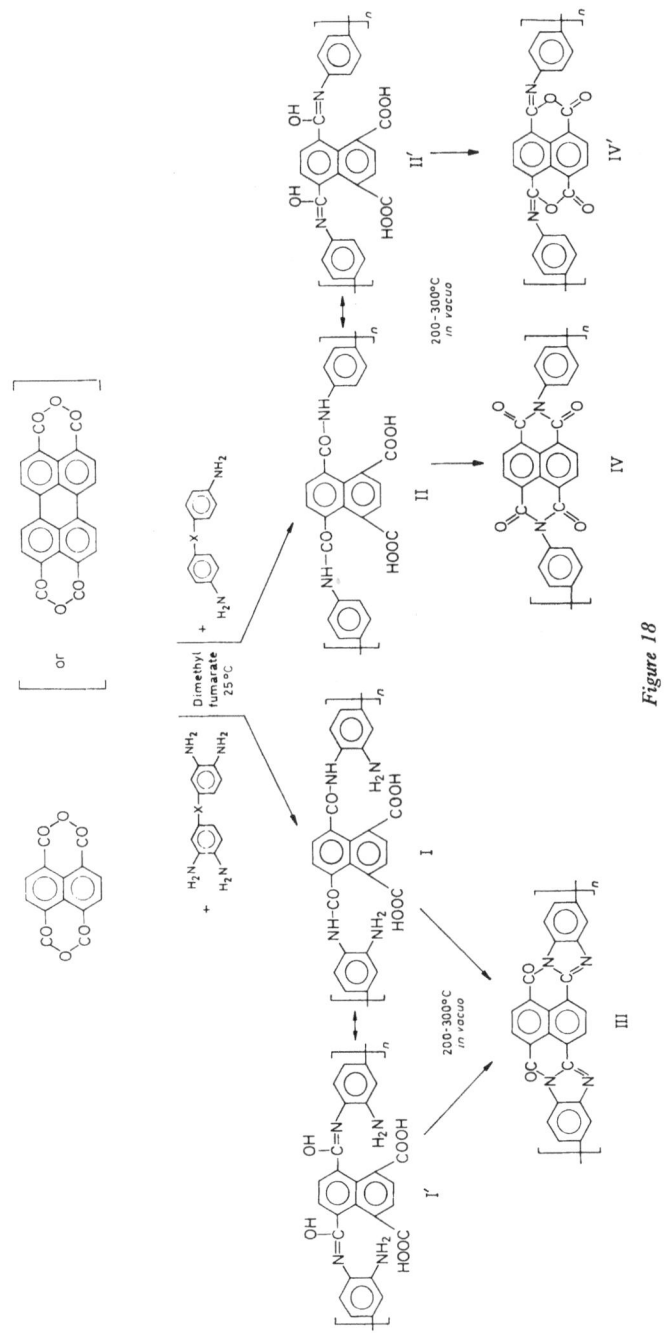

Figure 18

foregoing it may be supposed that a high thermal stability of such polymers is to an essential degree conditioned by the presence of conjugated blocks, and PMC, which play the role of stabilizers against thermo- and photo-oxidative polymer degradation.

The display of the effect of local activation is not limited to the field of chemical processes. Indeed the influence of the addition of PMC on some chemical and physical properties of PCS was demonstrated in our laboratory. Thus with the introduction into anthracene of various quantities of polymer PMC obtained by its thermolysis, the intensity of auto-fluorescence, the time of current drop, and the temperature of crystallization are changed to a similar degree. These facts testify that the same centres are responsible for these effects and in particular for the unirradiated exiton's decay and the capture of free carriers.

Investigation of absorption and luminescence spectra showed that paramagnetic polyanthracene formed stable complexes with anthracene and its homologues. Molecules of anthracene, contained in the complex, apparently, undergo a perturbation (complexes, containing 10^{16}–10^{17} spin/g, i.e. one spin among 10^3–10^5 diamagnetic molecules); the alcoholic solution is characterized by a higher lifetime in the excited state. The subsequent increase of concentration leads to the quenching of fluorescence. It is notable that some increase of the inhibiting capacity of anthracene in the thermal oxidation of low-molecular and high-molecular weight substances is observed when the concentration of PMC is about 10^{16}–10^{18} spin/g. On investigation of the electrical properties of complexes of poly-anthracene's paramagnetic fractions with anthracene it was found that the conductivity σ and the activation energy of conductivity E_{el} depended to a large extent on the concentration of PMC[47].

The observed extreme dependence of optical, electrical and also chemical properties of PCS on the concentration of PMC appear to be due to the same cause. There is reason to suppose that the increase of the concentration of PMC increases the polarization, decreases the ionization potential and increases the probability of S—T transition.

The discovery of the influence of PMC on the physical properties of polyconjugated systems stimulated work on the possibility of achieving the directional regulation of sub-molecular structures and physico-mechanical properties of some industrial high polymers.

Investigations with phenol–formaldehyde resins, aromatic polyesters and some other high polymers revealed that the introduction of a small quantity of PCS roughly changes the sub-molecular structure and increases the physico-mechanical properties of polymer materials (plastics, fibres). Some of the data show that there is an improvement in mechanical properties of amorphous aromatic polyesters (polyarylates) with the addition of PCS up to 1–1·5 per cent. Usually the introduction of more than 3–5 per cent of PCS does not lead to an improvement of properties[48].

Electron-microscopic investigations of the charcoal replicas shaded by palladium carried out on a section of polyarylates etched by oxygen, indicated that the introduction of 0·1–1 per cent of PCS led to the transition from globule to a dense molecular form. The dimensions of such forms grow with the amount of PCS introduced up to a definite limit (1–2 per cent)

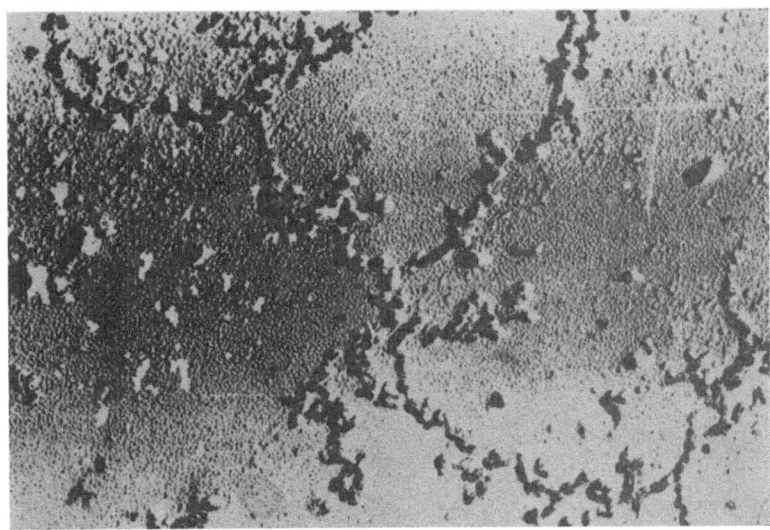

Figure 19. Structure of aromatic polyester before the introduction of a polymer with a conjugated system (× 18500).

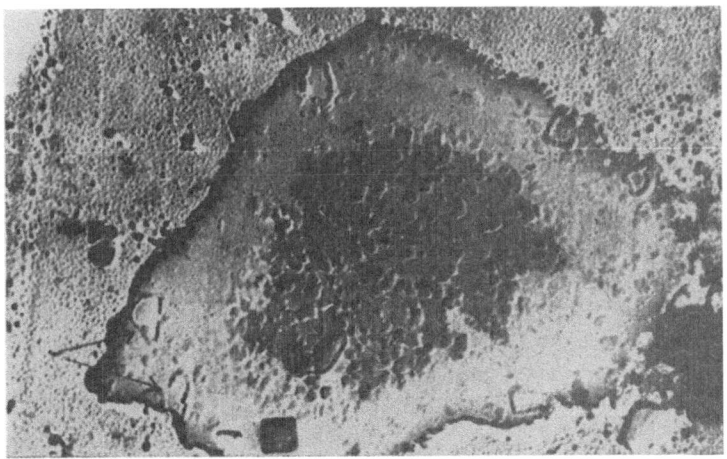

Figure 20. Structure of aromatic polyester after the addition 0·1–1% of polymer with conjugated system (content of PMC = 1–5·10¹⁷ spin/g) × 18900.

and reach 5–10μ in some cases. A more careful examination of these structural forms elucidated that they have a greater density in the centre than at the periphery and that they contain heterogeneous particles (apparently PCS) (*Figures 19* and *20*).

The structures described above remain intact in the long ageing period

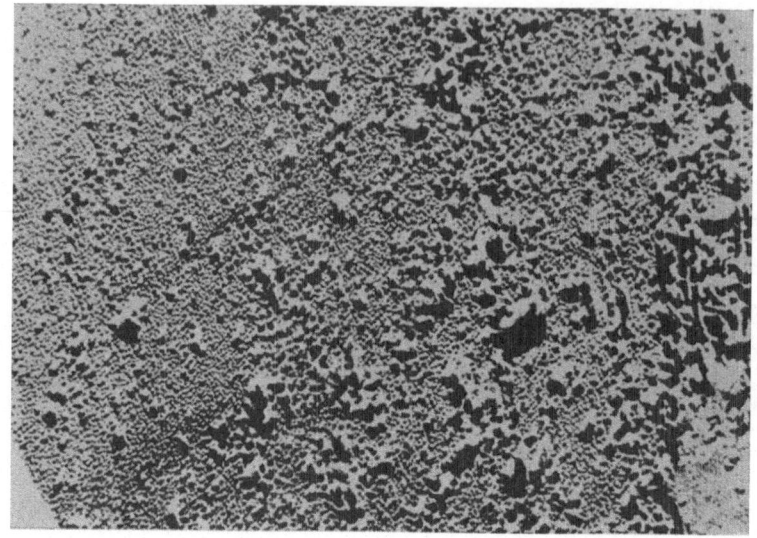

Figure 21. Change of structure of aromatic polyester, containing PMC after photooxidative ageing.

of polyarylate but break up with a decrease in physico-mechanical properties (*Figure 21*). Similar results were obtained with phenol–formaldehyde resins, obtained by curing resol or novolak–hexamethylene-tetramine mixtures in the presence of polyphenylacetylene and polyphenylene ($X = 210^{18}$ spin/g, $M_n = 3000$) pyrolyzed at $400°$ ($X = 210^{17}$ spin/g, $\bar{M}_n = 1400$), and without pyrolysis ($X = 310^{16}$ spin/g, $\bar{M}_n = 1000$)[51].

It was found that the introduction of PCS up to 0·5–1 per cent increased the mechanical properties of cured resins and that this effect represents a maximum in the function of the proportion and number of PMC (*Figure 22*). The investigation of replica from etched sections allowed the appearance of some regularity and compact packing of structural elements to be detected (*Figures 23* and *24*).

The influence of paramagnetic PCS on the sub-molecular structure of amorphous, linear and cross-linked polymers may now be considered proved. They show the same strong influence on crystalline high polymers, influencing their dimensions and crystalline form.

The phenomenon described is of obvious practical importance when one considers that the addition of PCS affects both the physico-mechanical properties and the stability of high-polymers with regard to thermo- and photo-oxidation and radiation influences (see above).

The mechanism of the influence of PCS on the structure and properties of valence-saturated polymers has not been elucidated so far. Since this e ffect was observed first in amorphous high polymers, it cannot be connected with the influence of PCS on the crystallization and dimensions of spherolites as was observed with the addition of polar substances to polymer olefins and other crystalline polymers[49, 50].

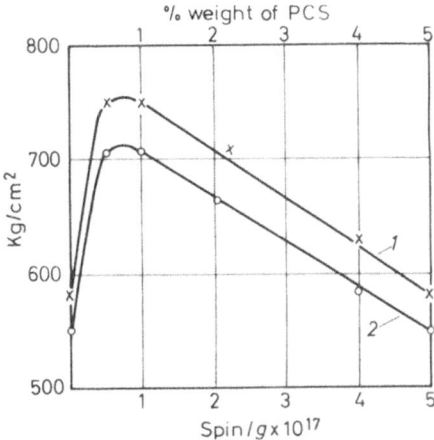

Figure 22. Influence of the addition of thermolysed polyphenylacetylene (PMC content 10^{19} spin/g) on the bending strength of phenol–formaldehyde resin. In the lower diagram the bonding strength in kg/cm² is plotted as a function of the concentration of PMC in spin/g and in the upper diagram against the proportion of the addition in weight %.

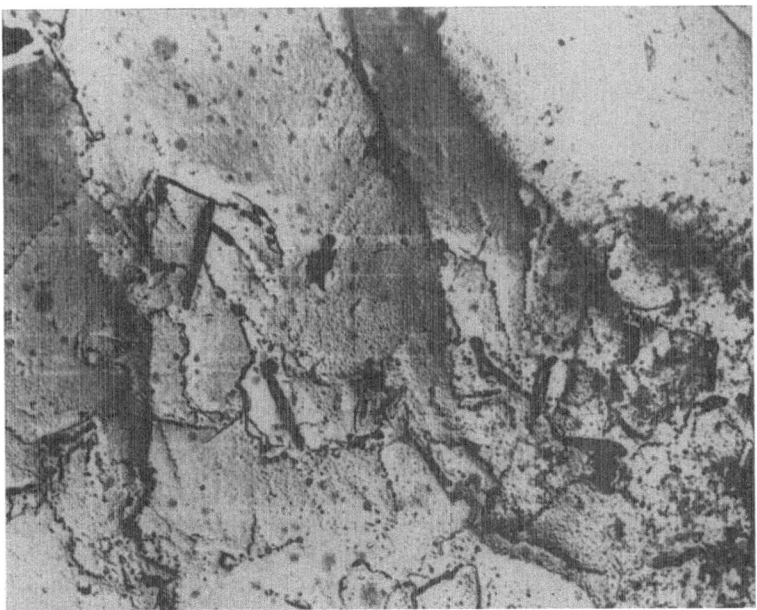

Figure 23. Structure of phenol-formaldehyde resin before introduction of polymer with a conjugated system ($\times$ 14600)

So the phenomenon of local activation includes various processes caused by the influence of stable PMC on electron excitation and the configuration of electron shells, with formation of macromolecular complexes.

The mechanism of the effect of local activation is still uninvestigated.

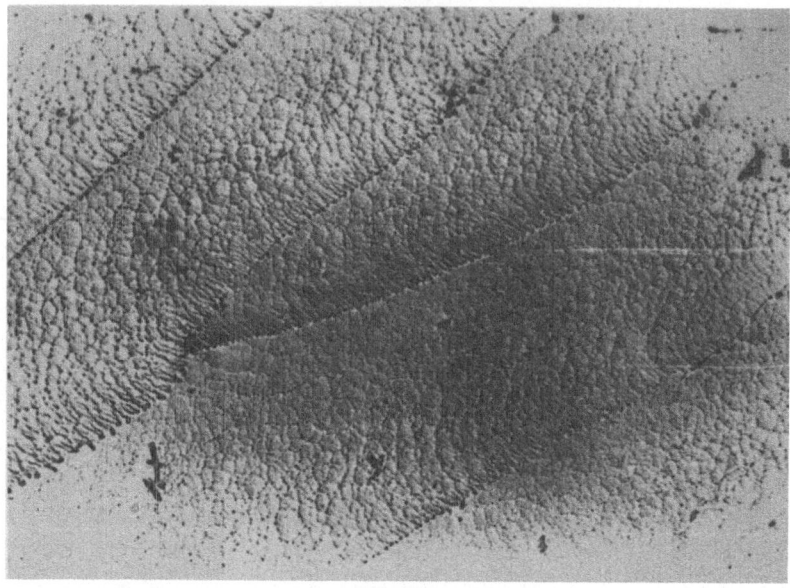

Figure 24. Structure of phenol-formaldehyde resin after the introduction of 0.1–1% polymer with conjugated system.

In our opinion it is not connected with the influence of a heterogeneous magnetic field of unpaired electrons of PMC on the value of the spin-orbitral bond, but is more probably connected with the formation of π-complexes between the activating substance and PMC.

In such π-complexes the role of the electron donor apparently belongs to the paramagnetic polymer so long as there are unpaired spins present. In such complexes reversible interaction between π-electrons of the substrate and unpaired electrons of PMC is possible.

In the case of matrix elements between various states, it can be proved that the reversible interaction of the substance with a conjugated system (a substrate) and PMC leads to undisappeared matrix element between excited states of the complex. These elements correspond to transition $S_0 - S_2$ and $S_0 - T_1$ and, therefore their transition becomes a permitted one. Because of the population increase of the triplet level in the complex PMC-substrate it is more profitable to visualize this process by the triplet excited state as a collision of a complex with reagent in the transition state.

Summarizing, we come to the conclusion that the effect of local activation includes a wide range of phenomena, connected with the influence of unpaired spins of paramagnetic particles of ion-radical or radical nature on the electron shell, in the formation of complexes with diamagnetic particles. This influence causes a change of reactivity, structure and physical properties of the compounds.

This effect may be caused not only by organic PMC, but also by biradicals (for example, O_2), forming complexes with the substrate or by paramagnetic ions. It should be emphasized that the local activation effect has great importance not only in the chemistry of polymers, but also in the solution of some real problems of molecular biology and, perhaps medicine.

References

[1] A. A. Berlin and V. P. Parini. *Izv. Vish. Uch. zaved., Ser. Khim. i khim, tekhn.* **N4**, 122 (1948).

[2] N. N. Semenov. *Report of the VIII Mendelejev Congress, Moscow*, Izd AN SSSR, 1959.

[3] A. A. Berlin. *Khim i Techn. Polymerov* **N 7–8**, 139 (1960); 56, 621 (1961).

[4] A. A. Berlin. *Khim. Prom.* **N 5**, 375; **N 6**, 444 (1960).

[5] A. A. Berlin. *Khim. Prom.* **N 12**, 23 (1962).

[6] *Organic Semiconductors*, V. Topchiev ed., Izd. AN SSSR, 1963.

[7] A. M. Sladkov and U. P. Kudrjavchev. *Usp. Khim.* 5,509 (1963).

[8] G. Pohl and I. A. Bornmann. *Khim i Tekhn. Polymerov* **N 11**, 81 (1961).

[9] A. A. Dulov, *Usp. Khim.* **235**, 1853 (1966).

[10] *Semiconductive Polymers with Conjugated Bonds* J. M. Paushkin ed., CNI.ITI Nevtekhim, Moscow, 1966.

[11] A. A. Berlin. *Izv. Akad. Nauk SSSR, Otd. Khim. Nauk* **1**, 59 (1965).

[12] A. A. Berlin, I. A. Drapkın, M. I. Cherkashin, P. P. Kisilicha, M. G. Chauser and A. N. Chegir. *Vysokomolekul. Soedin* in the press.

[13] V. F. Gachovsky, P. P. Kisilicha, M. E. Cherkashin and A. A. Berlin. *Vysokomolekul. Soedin*, in the press.

[14] A. A. Berlin, M. I. Cherkashin, V. G. Aseev and M. I. Scherbakova. *Vysokomolekul. Soedin.* **6**, 1773 (1964).

[15] R. I. Jahimovich, E. A. Shilov and G. F. Dvorko. *Dokl. Akad. Nauk USSR* **166**, 388 (1966).

[16] A. A. Berlin, M. I. Cherkashin, E. J. Vainshtein and V. Sh. Mashkovski. *Vysokomolekul. Soedin.* **5**, N 9, 1354 (1963).

[17] A. V. Topchiev, V. V. Korshak, B. E. Davidov and B. A. Krenchel. *Dokl. Akad. Nauk SSSR* **147**, 645 (1962).

[18] S. M. Makin, G. A. Lapichkji and A. M. Koldunov. *Jurnal Vses. Khim. Obsch. im. Mendeleeva VIII* 6, 708 (1963).

[19] A. A. Berlin, B. I. Liogonki and V. P. Parini. *Vysokomolekul. Soedin.* **3**, 1491 (1961); **4**, 662 (1962).

[20] M. L. Green, M. New'en and J. Wilkinson. *Chem. & Ind.* 1136 (1960).

[21] P. Kovacic and A. Kyriakis. *Tetrahedron Letters* 467; (1962) see also *J. Polymer Sci.* **A-2**, 1193 (1964).

[22] A. A. Berlin, V. A. Vansiachkji and B. I. Liogonki. *Dokl. Akad. Nauk. SSSR*, **144**, N 6, 1316 (1962); see also *Izv. Akad. Nauk. SSSR Otd. Khim. Nauk* 1654 (1963).

[23] A. A. Berlin and N. G. Matveeva. *Dokl. Akad. Nauk SSSR* **167**, 51 (1966); *Vysokomolekul. Soedin.* **8**, 736 (1966).

[24] B. G. Zadonchev, I. M. Cherkashin and A. A. Berlin. *Izv. Akad. Nauk SSSR Otl. Khim. Nauk.* 2065 (1967).

[25] P. Cassidy, C. Marvell and S. Ray. *J. Polymer Sci.* **A-3**, 1553 (1965); *Khim i Tekhn. Polymerov* **N 11**, 3 (1965).

[26] A. A. Berlin, B. I. Liogonki and A. N. Zelenetski. *Izv. Akad. Nauk SSSR* 225 (1967).

[27] J. Schopov. *Polymer Letters* **4**, 1023 (1966).

[28] A. A. Berlin, B. I. Liogonki, A. A. Gurov and E. F. Razvadovski. Vysokomolekul. Soedin. A IX; N3, 533 (1967).

[29] A. A. Berlin, B. I. Liogonki and A. B. Ragimov. *Izv. Akad. Nauk SSSR, Otd. Khim. Nauk* 1883 (1962); 4593 (1964); see also *Vysokomolekul. Soedin.* **7**, 652 (1965); **8**, 540 (1966).

[30] *Steric Effects in Organic Chemistry* M. S. Newman ed., Izd. Inostrliter. Moscow, 1960.

[31] V. A. Benderskij and L. A. Blumenfeld. *Dokl. Akad. Nauk SSSR* **144**, 813 (1962).

[32] L. A. Blumenfeld and V. A. Benderskij. *Zh. Strukt. Khim.* **3**, 405 (1963); **3**, 415 (1963); **7**, 5 (1965).

[33] A. A. Berlin. *International Symposium on Macromolecular Chemistry* (Prague) Preprint p. 281, 1965.

[34] A. A. Berlin, E. F. Razvodovskij and G. V. Korolev. *Vysokomolekul. Soedin.* **6**, 1838 (1964); see also A. A. Berlin and E. F. Razvadovskij. *J. Polymer Sci.* Part C N 16 p. 369 (1967).

[35] V. P. Kovaleva, D. A. Topchiev, V. A. Kabanov and V. A. Kargin. *Izv. Akad. Nauk SSSR. Otd. Khim. Nauk* 387 (1963).

[36] A. A. Berlin, V. A. Grigorovskijaia, V. P. Parini and H. Gafurov. *Dokl. Akad. Nauk SSSR* **156**, 1371 (1964).

[37] A. A. Berlin, V. A. Grigorovskaja and V. E. Skurat. *Vysokomolekul. Soedin.* **8**, 1976 (1966).

[38] A. A. Berlin and V. A. Vonsiatskij. *Dokl. Akad. Nauk SSSR* **156**, N 3, 627 (1964).

[39] V. A. Vonsiatskij, G. E. Koliaev and A. A. Berlin. *Izv. Akad. Nauk SSSR Otd. Khim Nauk* 304 (1964).

[40] A. A. Berlin, S. I. Bass and V. V. Jarkin. *Izv. Akad. Nauk SSSR Otd. Khim Nauk* 1352 (1966).

[41] A. A. Berlin, S. I. Bass, N. R. Belova, E. G. Saschko and N. Ivanov. *Vysokomolekul. Soedin.* in the press.

[42] S. I. Bass and A. A. Berlin. *International Symposium on Macromolecular Chemistry* (Prague) Preprint p. 205, 1965.

[43] A. A. Berlin and S. I. Bass. *Ageing and Stabilization of polymers* A. S. Kuzminskij ed., Izd. 'Khimija' M. 1966.

[44] H. M. Gafurov, V. F. Mulikov, V. F. Gachkovskij, V. P. Parini, A. A. Berlin and L. A. Blumenfeld *Zh. Strukt. Khim.* **6**, 649 (1965).

[45] V. F. Gachkovskij, H. M. Gafurov, L. A. Blumenfeld and A. A. Berlin. *Zh. Strukt. Khim,* in the press.

[46] A. A. Berlin, V. P. Parini and K. Almambetov. *Dokl. Akad. Nauk SSSR* **166**, N 3 595 (1966).

[47] A. A. Berlin, H. M. Gafurov, G. S. Golubkov and V. A Talikov. *Izv. Akad. Nauk SSSR, Otd. Khim. Nauk*, in the press.

[48] A. A. Berlin, I. P. Pavlova and L. A. Radivilova *Mekhanika Polymerov*, in the press.

[49] V. A. Kargin, T. I. Sogolova, N. Ja. Rapoport-Molodscheva. *Dokl. Akad. Nauk SSSR*, **156**, 6, 1406 (1964).

[50] V. A. Kargin, T. I. Sogolova and N. I. Kubanova. *Dokl. Akad. Nauk SSSR* **162**, 5, 1092 (1965).

[51] A. A. Berlin, R. M. Aseeva and K. Almanbetov. *Plaste u. Kautschuk.* **15**, N2, 91 (1968).

[52] J. Jellard, M. Nechtschern, M. Soutif and P. Tranyard. *Bull. Soc. chim. Fr.* 2209 (1963).

[53] A. A. Berlin, V. K. Beliakof, L. V. Nevskij and O. G. Tazakanov. *Plaste u. Kautschuk.* **15**, N2, 91 (1968).